Federico Montesino Pouzols, Diego R. Lopez, and Angel Barriga Barros

Mining and Control of Network Traffic by Computational Intelligence

T0135249

Studies in Computational Intelligence, Volume 342

Editor-in-Chief

Prof. Janusz Kacprzyk
Systems Research Institute
Polish Academy of Sciences
ul. Newelska 6
01-447 Warsaw
Poland
E-mail: kacprzyk@ibspan.waw.pl

Further volumes of this series can be found on our homepage: springer.com

Vol. 319. Takayuki Ito, Minjie Zhang, Valentin Robu, Shaheen Fatima, Tokuro Matsuo, and Hirofumi Yamaki (Eds.)
Innovations in Agent-Based Complex Automated Negotiations, 2010
ISBN 978-3-642-15611-3

Vol. 321. Dimitri Plemenos and Georgios Miaoulis (Eds.)
Intelligent Computer Graphics 2010
ISBN 978-3-642-15689-2

Vol. 322. Bruno Baruque and Emilio Corchado (Eds.)
Fusion Methods for Unsupervised Learning Ensembles, 2010
ISBN 978-3-642-16204-6

Vol. 323. Yingxu Wang, Du Zhang, and Witold Kinsner (Eds.)
Advances in Cognitive Informatics, 2010
ISBN 978-3-642-16082-0

Vol. 324. Alessandro Soro, Vargiu Eloisa, Giuliano Armano, and Gavino Paddeu (Eds.)
Information Retrieval and Mining in Distributed Environments, 2010
ISBN 978-3-642-16088-2

Vol. 325. Quan Bai and Naoki Fukuta (Eds.)
Advances in Practical Multi-Agent Systems, 2010
ISBN 978-3-642-16097-4

Vol. 326. Sheryl Brahnam and Lakhmi C. Jain (Eds.)
Advanced Computational Intelligence Paradigms in Healthcare 5, 2010
ISBN 978-3-642-16094-3

Vol. 327. Slawomir Wiak and Ewa Napieralska-Juszczak (Eds.)
Computational Methods for the Innovative Design of Electrical Devices, 2010
ISBN 978-3-642-16224-4

Vol. 328. Raoul Huys and Viktor K. Jirsa (Eds.)
Nonlinear Dynamics in Human Behavior, 2010
ISBN 978-3-642-16261-9

Vol. 329. Santi Caballé, Fatos Xhafa, and Ajith Abraham (Eds.)
Intelligent Networking, Collaborative Systems and Applications, 2010
ISBN 978-3-642-16792-8

Vol. 330. Steffen Rendle
Context-Aware Ranking with Factorization Models, 2010
ISBN 978-3-642-16897-0

Vol. 331. Athena Vakali and Lakhmi C. Jain (Eds.)
New Directions in Web Data Management 1, 2011
ISBN 978-3-642-17550-3

Vol. 332. Jianguo Zhang, Ling Shao, Lei Zhang, and Graeme A. Jones (Eds.)
Intelligent Video Event Analysis and Understanding, 2011
ISBN 978-3-642-17553-4

Vol. 333. Fedja Hadzic, Henry Tan, and Tharam S. Dillon
Mining of Data with Complex Structures, 2011
ISBN 978-3-642-17556-5

Vol. 334. Álvaro Herrero and Emilio Corchado (Eds.)
Mobile Hybrid Intrusion Detection, 2011
ISBN 978-3-642-18298-3

Vol. 335. Radomir S. Stankovic and Radomir S. Stankovic
From Boolean Logic to Switching Circuits and Automata, 2011
ISBN 978-3-642-11681-0

Vol. 336. Paolo Remagnino, Dorothy N. Monekosso, and Lakhmi C. Jain (Eds.)
Innovations in Defence Support Systems – 3, 2011
ISBN 978-3-642-18277-8

Vol. 337. Sheryl Brahnam and Lakhmi C. Jain (Eds.)
Advanced Computational Intelligence Paradigms in Healthcare 6, 2011
ISBN 978-3-642-17823-8

Vol. 338. Lakhmi C. Jain, Eugene V. Aidman, and Canicious Abeynayake (Eds.)
Innovations in Defence Support Systems – 2, 2011
ISBN 978-3-642-17763-7

Vol. 339. Halina Kwasnicka, Lakhmi C. Jain (Eds.)
Innovations in Intelligent Image Analysis, 2010
ISBN 978-3-642-17933-4

Vol. 340. Heinrich Hussmann, Gerrit Meixner, and Detlef Zuehlke (Eds.)
Model-Driven Development of Advanced User Interfaces, 2011
ISBN 978-3-642-14561-2

Vol. 341. Stéphane Doncieux, Nicolas Bredèche, and Jean-Baptiste Mouret (Eds.)
New Horizons in Evolutionary Robotics, 2011
ISBN 978-3-642-18271-6

Vol. 342. Federico Montesino Pouzols, Diego R. Lopez, and Angel Barriga Barros
Mining and Control of Network Traffic by Computational Intelligence, 2011
ISBN 978-3-642-18083-5

Federico Montesino Pouzols, Diego R. Lopez,
and Angel Barriga Barros

Mining and Control of Network Traffic by Computational Intelligence

 Springer

Dr. Federico Montesino Pouzols
Dept. of Information and Computer Science
Aalto University
P.O. Box 15400
FI-00076 Aalto
Finland
E-mail: fedemp@cis.hut.fi
http://www.cis.hut.fi/ fedemp/

Prof. Angel Barriga Barros
Instituto de Microelectrónica de Sevilla
c. Americo Vespucio s/n
41092 Sevilla
Spain
E-mail: barriga@us.es
http://www2.imse-cnm.csic.es/ barriga/

Dr. Diego R. Lopez
RedIRIS, Red.es, Edif. Bronce
Pza. Manuel Gomez Moreno s/n, Planta 2.
E-28020 Madrid
Spain
E-mail: diego.lopez@rediris.es
http://www.rediris.es

ISBN 978-3-642-42399-4   ISBN 978-3-642-18084-2 (eBook)

DOI 10.1007/978-3-642-18084-2

Studies in Computational Intelligence   ISSN 1860-949X

© 2011 Springer-Verlag Berlin Heidelberg

Typeset & Cover Design: Scientific Publishing Services Pvt. Ltd., Chennai, India.

Printed on acid-free paper

9 8 7 6 5 4 3 2 1

springer.com

Preface

As other complex systems in social and natural sciences as well as in engineering, the Internet is difficult to understand from a technical point of view. The structure and behavior of packet switched networks is hard to model in a way comparable to many natural and artificial systems. Nonetheless, the Internet is an outstanding and challenging case due to its incredibly fast development and the inherent lack of measurement and monitoring mechanisms in its core conception. In short, packet switched networks defy analytical modeling.

It is generally accepted that Internet research needs better models. A great deal of development in network measurement systems and infrastructures have enabled many advances throughout the last decade in understanding how the basic mechanisms of the Internet work and interact. In particular, a number of works in Internet measurement have led to the first results in what some authors call Internet Science, i.e., an experimental science that studies laws and patterns in Internet structure. However, many mechanisms are still not well understood. As a consequence, users experience performance degradations and networks cannot be used to their full potential. For instance, it is a common experience to see real-time applications perform poorly unless (or even if) the network is largely overprovisioned.

This monograph deals with applications of computational intelligence methods, with an emphasis on fuzzy techniques, to a number of current issues in measurement, analysis and control of traffic in packet switched networks. The general approach followed here is to address concrete problems in the areas of data mining and control of network traffic by means of specific fuzzy logic based techniques. The set of problems has been chosen on the basis of their practical interest in current networking systems as well as our aim at providing a unified approach to network traffic analysis and control. Of course, not all open issues are addressed here but the set of methods we propose and apply provides a fairly comprehensive approach to current open problems. This set of methods is in addition open to countless extensions to address current and future related problems.

Data mining and control problems are addressed. In the first class we include two issues: predictive modeling of traffic load as well as summarization and inductive analysis of traffic flow measurements. In the second class we include other two issues: active queue management schemes for Internet routers as well as window based end-to-end rate and congestion control. While some theoretical developments are described, we favor extensive evaluation of models using real-world data by simulation and experiments.

The field of computational intelligence embraces a varied number of computational techniques such as neural networks, fuzzy systems, evolutionary systems, probabilistic reasoning and also computational swarm intelligence, artificial immune systems, fractals and chaos theory and wavelet analysis. Some if not all of the areas covered by the term computational intelligence are also often referred to as soft computing. As opposed to operations research, also known as hard computing, soft computing techniques require no strict conditions on the problems and do not provide guarantees for success. This is a shortcoming that is compensated in practice by the robustness of soft computing methods, a widely accepted fact.

Fuzzy inference systems (FIS for short, also commonly referred to as fuzzy rule-based systems or FRBS) play a central role in this monograph. FIS are used for tasks such as performance evaluation, prediction and control. However, in addition to fuzzy inference based techniques we apply other computational intelligence methods and complementary techniques including nonparametric statistical methods, OWA operators, association rules mining algorithms, fuzzy calculus, nearest neighbor methods, support vector machines and neural networks.

Fuzzy logic is a precise logic of imprecision, based on the concept of fuzzy set. Fuzzy logic integrates numerical and symbolic processing into a common scheme. This way, it allows for the inclusion of human expert knowledge into mathematical models, i.e., it provides a mathematical framework into which we can translate the solutions that a human expert expresses linguistically.

FIS are rule-based modeling systems. Fuzzy inference mechanisms have been shown to be an effective way to address problems that are subject to uncertainty and inaccuracy For modeling and control, one major reason to use fuzzy systems is that fuzzy rules can be expressed in a linguistic manner and are thus comprehensible for humans. This is what makes it possible to use a priori knowledge. In addition, fuzzy inference based models can be interpreted and thus evaluated by experts. Many methods to generate different kinds of fuzzy inference models with an interpretability-accuracy trade-off have been proposed.

An additional key feature of fuzzy inference systems is that they are universal approximators. Also, so-called neuro-fuzzy systems combine FIS with the learning capabilities of artificial neural networks (ANNs), often using the same learning algorithms that were initially developed for ANNs. Neuro-fuzzy systems offer the computational power of nonlinear computational intelligence techniques and can also provide a natural language approach to solving a number of current issues around the analysis and control of network traffic. On the one hand, the rule based structure of FIS allows for the incorporation of domain expert knowledge. On the other hand, the ability to learn allows neuro-fuzzy systems to be used on problems where no a priori or expert knowledge based rule-based solutions seem feasible or one is primarily interested in inducing an interpretable model from data. In addition, efficient hardware implementations can be developed in an structured and systematic manner.

This monograph is organized as follows. In chapter 1 we introduce and provide concise descriptions of the core building blocks of Internet Science and other related networking aspects that will be used throughout the next chapters. Chapter 2 describes a methodology for for building predictive time series models combining statistical techniques and neuro-fuzzy techniques.

Data mining of network traffic is the topic of chapters 3 and 4 where we focus on two related issues: traffic load prediction and analysis of traffic flows measurements.

In chapter 3 we investigate first the predictability of network traffic at different time scales, following a quantitative approach based on statistical techniques for nonparametric residual variance estimation. With an extensive experimental background of a wide set of diverse and publicly available network traffic traces, it is shown that, in some cases, it is possible to predict network traffic with a satisfactory accuracy for a wide range of time scales. Then, the methodology described in chapter 2 is applied to diverse network traffic traces. The methodology is compared against least squares support vector machines (LS-SVM), Ordered Weighted Averaging Aggregation Operators (OWA)-induced nearest neighbors and optimally pruned extreme learning machines (OP-ELM). These methods are applied to an extensive set of time series derived from publicly available traffic traces. The methodology proposed is shown to provide advantages in terms of accuracy and interpretability. Further, it has been implemented in a tool integrated into the Xfuzzy development environment.

In chapter 4 a method and a tool for extracting concise linguistic summaries about network statistics at the flow level are described. In addition, a procedure for mining extended linguistic summaries from network flow collections is developed and the results for a number of publicly available traces are discussed. The theory of linguistic summaries has been extended for traffic statistics summarization and new tools for linguistic analysis of traffic traces at the flow level have been developed.

Chapter 5 deals with control of network traffic in routers, by means of active queue management schemes, as well as on an end-to-end basis, by means of window based techniques. First it is proposed an scheme for implementing end-to-end traffic control mechanisms through fuzzy inference systems. A comparative evaluation of simulation and implementation results from the fuzzy rate controler as compared to that of traditional controlers is performed for a wide set of realistic scenarios. Then, fuzzy inference systems for traffic control in routers are designed. A particular proposal has been evaluated in realistic scenarios and is shown to be robust. The proposal is compared against the random early detection (RED) scheme. It is experimentally shown that fuzzy systems can provide better performance and better adaptation to different requirements with mechanisms that are easy to modify using linguistic knowledge.

Finally, chapter addresses 6 the practical implementation of some of the fuzzy inference systems proposed in previous chapters. Both architectural and operational constraints are considered. The chapter focuses on an open FPGA-based hardware platform for the implementation of efficient fuzzy inference systems for solving networking analysis and control problems. A feasibility study is conducted in order to show that the techniques developed can be deployed in current and future network scenarios with satisfactory performance. The major contribution is the development of a platform and a companion development methodology that does not only fulfill operational requirements but also addresses the scalability and flexibility challenges posed by current routing architectures. In addition, evidence for the feasibility of real implementations is provided.

In conclusion, this monograph describes computational intelligence based methods and tools for addressing a number of current issues around network traffic measurement, modeling and control. Besides developing methods, special attention is paid to a number of practical aspects that have a determining impact on the adoption of novel methods and mechanisms for traffic analysis and control.

Espoo, Finland and Sevilla, Spain   Federico Montesino Pouzols
September 2010   Diego R. Lopez
                                                                   Angel Barriga Barros

Acknowledgements

The first author is supported by a Marie Curie Intra-European Fellowship for Career Development (grant agreement PIEF-GA-2009-237450) within the European Community's Seventh Framework Programme (FP7/20072013). Most of this work was done while the first author was with the Microelectronics Institute of Seville, IMSE-CNM, CSIC. This work was supported in part by the European Community under the MOBY-DIC Project FP7-IST-248858 (www.mobydic-project.eu). The research presented here has been supported in part by a PhD studentship from the Andalusian regional Government, project TEC2008-04920, from the Spanish Ministry of Education and Science, as well as project P08-TIC-03674 from the Andalusian regional Government.

This monograph is based in part upon the Ph.D. dissertation of the first author, directed by the second and third authors, and completed in 2009 at the Department of Electronics and Electromagnetism of the University of Seville and the Microelectronics Institute of Seville, CSIC. We would like to thank all the colleagues that made this work possible. In particular, we would like to acknowledge the members of the thesis jury, Professors Jose Luis Huertas, Iluminada Baturone and Plamen Angelov, and Drs. Amaury Lendasse and Santiago Sanchez-Solano. Their comments and encouraging suggestions helped improve this monograph and motivated new research directions.

The extensive and computationally expensive analysis of network measurements performed in this monograph would not have been possible without the facilities and support from the e-Science infrastructure managed by the Centro Informático Científico de Andalucía (https://eciencia.cica.es/). A special thanks should go to Ana Silva for her support.

We would like to acknowledge a number of institutions and individuals that have made this research possible by providing measurement infrastructures and repositories of network traces. In particular, our work has benefited from the use of measurement data collected on the Abilene network as part of the Abilene Observatory Project (http://abilene.internet2.edu/observatory/). We acknowledge the MAWI Working Group from the Wide Integrated Distributed Environment (WIDE) project (http://tracer.csl.sony.co.jp/mawi/) for kindly providing their traffic traces.

We also used data sets from the Internet Traffic Archive (http://ita.ee.lbl.gov/), an initiative by the Lawrence Berkeley National Laboratory and the ACM Special Interest Group on Data Communications (SIGCOMM), as well as the Community Resource for Archiving Wireless Data (CRAWDAD) at Dartmouth (http://crawdad.cs.dartmouth.edu). We are also indebted to the Cooperative Association for Internet Data Analysis (CAIDA, http://www.caida.org), for providing a number of data collections. This work uses the following traces from CAIDA:

- The CAIDA OC48 Traces Dataset - August 2002, January 2003 and April 2003, Colleen Shannon, Emile Aben, kc claffy, Dan Andersen, Nevil Brownlee http://www.caida.org/data/passive/.
- The CAIDA Anonymized 2007 and 2008 Internet Traces - January 2007 and April 2008, Colleen Shannon, Emile Aben, kc claffy, Dan Andersen, http://www.caida.org/data/passive/passive_2007_dataset.xml.

Support for CAIDA's OC48 and Internet Traces is provided by the National Science Foundation, the US Department of Homeland Security, DARPA, Digital Envoy, and CAIDA Members.

Contents

1   **Internet Science** 1
   1.1 Modeling the Internet 1
   1.2 Measurement Systems and Infrastructures 4
       1.2.1 Active Systems 4
       1.2.2 Passive Systems 6
       1.2.3 Publicly Available Measurements 6
   1.3 Network Traffic 7
       1.3.1 Traffic Models 8
       1.3.2 Transport Layer Models. TCP 11
       1.3.3 Models of Applications and Services 12
       1.3.4 Network Simulation 12
       1.3.5 Performance Metrics 14
       1.3.6 Congestion 15
   1.4 Traffic Control 16
       1.4.1 End-To-End Traffic Control 19
       1.4.2 Traffic Control in Routers 20
   1.5 Time Series Models for Network Traffic 26
       1.5.1 Short-Memory Stochastic Models 28
       1.5.2 Long-Memory Stochastic Models 31
       1.5.3 Mean Square Error Predictors 34
       1.5.4 OWA-Induced Nearest Neighbor Models 36
       1.5.5 Least Squares Support Vector Machines 36
       1.5.6 Extreme Learning Machine 38
       1.5.7 Prediction Performance Metrics 38
   1.6 Conclusions 41
   References 41

2   Modeling Time Series by Means of Fuzzy Inference Systems 53
   2.1   Predictive Models for Time Series 53
   2.2   Nonparametric Residual Variance Estimation: Delta Test 55
   2.3   Methodology Framework for Time Series Prediction with
       Fuzzy Inference Systems 55
       2.3.1   Variable Selection 57
       2.3.2   System Identification and Tuning 59
       2.3.3   Complexity Selection 60
   2.4   Case Study and Validation: ESTSP'07 Competition
       Dataset 61
   2.5   Experimental Results 67
       2.5.1   Poland Electricity Benchmark 67
       2.5.2   Sunspot Numbers 71
       2.5.3   Aggregated Incoming Traffic in the Internet2 Backbone
            Network 73
       2.5.4   Santa Fe Time Series Competition: Laser Dataset 73
       2.5.5   Mackey-Glass Series 78
       2.5.6   NN3 Competition 80
       2.5.7   Discussion 80
   2.6   Conclusions 83
   References 83

3   Predictive Models of Network Traffic Load 87
   3.1   Models for Network Traffic Load 87
   3.2   Analysis of Traffic Traces 89
   3.3   Series of the Internet Traffic Archive 93
       3.3.1   LBL Traces 93
       3.3.2   Bellcore Traces 94
       3.3.3   DEC Traces 99
   3.4   Application to Recent Traffic Time Series 99
       3.4.1   Backbone Traffic 99
       3.4.2   Exchange and Peering Traffic 111
       3.4.3   Intercontinental Traffic 116
       3.4.4   Access Point Traffic 120
       3.4.5   Wireless Traffic 130
   3.5   Discussion 130
   3.6   Conclusions 142
   References 143

4   Summarization and Analysis of Network Traffic Flow Records 147
   4.1   Network Traffic Measurement Systems 147
   4.2   Flow Measurement and Statistics: NetFlow and IPFIX 149
   4.3   Linguistic Summaries 152

4.4   Definition of Linguistic Summaries of Network Flow
      Collections 154
      4.4.1   Defining Linguistic Labels from a Priori
              Knowledge 156
      4.4.2   Automatic Definition of Linguistic Labels by
              Unsupervised Learning 158
      4.4.3   Quantifiers 159
4.5   Summarization of NetFlow Collections 159
      4.5.1   On-Line Summarization of NetFlow Collections 159
      4.5.2   Data Mining Summaries of NetFlow Collections 167
      4.5.3   Experimental Results 168
      4.5.4   Predefined Set of Summaries 170
      4.5.5   Identifying Attribute Labels by Clustering 174
      4.5.6   Mining Association Rules for Extracting Linguistic
              Summaries.. 183
      4.5.7   Discussion 183
4.6   Conclusions 185
References 186

5 **Inference Systems for Network Traffic Control** 191
5.1   Network Traffic Control 191
5.2   Simulation Scenarios 192
5.3   Fuzzy End-To-End Rate Control for Internet Transport
      Protocols 200
      5.3.1   Related Work 202
      5.3.2   End-To-End Window Based Rate Control and a Fuzzy
              Generalization.................................... 203
      5.3.3   Design of a Fuzzy End-To-End Window Based Rate
              Controler 205
      5.3.4   Development Methodology and Tool Chain 213
      5.3.5   Simulation Results 214
      5.3.6   Implementation Results 219
      5.3.7   Discussion 222
5.4   Active Queue Management by Means of Fuzzy Inference
      Systems 226
      5.4.1   Approach and Related Work 226
      5.4.2   Development Methodology and Tool Chain 229
      5.4.3   Fuzzy Internet Traffic Control of Aggregate Traffic 230
      5.4.4   Fuzzy Controler of Best-Effort Aggregate Traffic 231
      5.4.5   Simulation Results 233
      5.4.6   Implementation Results 250
      5.4.7   Discussion 255
5.5   Conclusions 256
References 256

6   Open FPGA-Based Development Platform for Fuzzy Inference
     Systems 263
     6.1   Fuzzy Inference Systems for High-Performance Networks 263
     6.2   Routing Architectures 264
           6.2.1   High-End Routing Hardware 269
           6.2.2   Expected Evolution 272
           6.2.3   Architectures and Platforms for Research 273
     6.3   Inference Rate of Software Implementations 274
     6.4   Hardware Implementation of Fuzzy Inference Systems 275
     6.5   Development Platform for Fuzzy Inference Systems with
           Applications to Networking 277
           6.5.1   Development Methodology and Design Flow 282
           6.5.2   Application to Internet Traffic Analysis and
                   Control 285
     6.6   Computational Intelligence Based Processing Subsystems in
           Routing Architectures 296
     6.7   Conclusions 298
     References 299

Index 305

Acronyms

ACK	Acknowledgment
AF	Assured Forwarding
AQM	Active Queue Management
LS-SVM	Least Squares Support Vector Machines
AR	Autoregression, Autoregressive model
ARX	Autoregressive model with eXogenous inputs
ARMA	Autoregression with Moving Average
ARIMA	Autoregression with Integrated Moving Average
ASIC	Application Specific Integrated Circuit
ATM	Asynchronous Transfer Mode
BGP	Border Gateway Protocol
BTC	Bulk Transfer Capacity
CAIDA	Cooperative Association for Internet Data Analysis
CBQ	Class Based Queuing
CBR	Constant Bit Rate
CoS	Class of Service
DCCP	Datagram Congestion Control Protocol
DNS	Domain Name System
DS	Differentiated Services
EF	Expedited Forwarding
ELM	Extreme Learning Machine
ECN	Explicit Congestion Notification
FCFS	First-Come First-Served
FIFO	First-In First-Out
FIM	Fuzzy Inference Module
FPGA	Field Programmable Gate Array
FPI	Fuzzy Proportional Integral
FTP	File Transfer Protocol
HTTP	HyperText Transfer Protocol
IETF	Internet Engineering Task Force
IOB	Input/Output Block

IP	Internet Protocol
IPPM	IP Performance Metrics
IRTF	Internet Research Task Force
ISP	Internet Service Provider
ITU-T	International Telecommunication Union, Telecommunication Standardization Sector
IXP	Internet eXchange Processor
LRD	Long-Range Dependence
LUT	Look-Up Table
MAC	Medium Access Control
MF	Membership Function
MPLS	Multi Protocol Label Switching
NARX	Nonlinear autoregressive model with eXogenous inputs
NCL	Network Classification Language
NP	Network Processor
NPU	Network Processing Unit
NTP	Network Time Protocol
OPB	On-Chip Peripheral Bus
OSPF	Open Shortest-Path First
OWA	Ordered Weighted Average
PI	Proportional Integral
QoS	Quality of Service
RED	Random Early Detection
RFC	Request For Comments
RIO	RED In/Out
RSVP	Resource ReSerVation Protocol
RTP	Real-Time Streaming Protocol
RTT	Round-Trip Time
SACK	Selective Acknowledgment
SAPE	Symmetric Absolute Percentage Error
SMAPE	Symmetric Mean Absolute Percentage Error
SCTP	Stream Control Transmission Protocol
SLA	Service Level Agreement
SoC	System-on-a-Chip
SoPC	System-on-Programmable-Chip
SVM	Support Vector Machines
TCAM	Ternary Content-Addressable Memory
TM	Traffic Management
ToS	Type of Service
TCP	Transport Control Protocol
UDP	User Datagram Protocol
VBR	Variable Bit Rate
VoIP	Voice Over IP
VOQ	Virtual Output Queuing

Chapter 1
Internet Science

Abstract. The structure and behavior of packet switched networks is difficult to model in a way comparable to many natural and artificial systems. Nonetheless, the Internet is an outstanding and challenging case because of its incredibly fast development, unparalleled heterogeneity and the inherent lack of measurement and monitoring mechanisms in its core conception. In short, packet switched networks defy analytical modeling. This chapter is intended to introduce and provide concise descriptions of some of the building blocks of what some authors call Internet Science [21, 104], i.e., the study of laws and patterns in Internet structure. Additional related aspects that will be used throughout the next chapters are discussed as well. We will briefly define and describe the most relevant concepts about Internet performance and measurement that will be used throughout the next chapters. However, we will not get into details about all the networking concepts this monograph deals with. We refer to [37] for a good overall and in-depth analysis of traffic measurement and performance analysis. There are also a number of research papers that provide good insight into more specific topics. Among these, we highlight [21], where some key mathematical concepts in Internet traffic analysis are discussed. It is also out of the scope of this monograph to analyze in detail the mathematical aspects of most of the concepts this monograph deals with, and in particular those related to traffic control. For this, we refer the interested reader to [153] and [15]. Some of the most relevant and seminal research papers in this area can also be consulted [134, 132, 129, 171, 71].

1.1 Modeling the Internet

Analyzing and modeling traffic in packet switched computer networks can turn into a daunting task due to the virtually unlimited amount of data. There are both spatial and temporal issues. Considering the spatial dimension, the amount of end nodes, routers and switches can be of the order of several thousands even in local area networks [22]. Regarding the temporal dimension, the volume of data is huge even in medium-sized low-speed subnetworks for todays standards: a traffic trace taken

F.M. Pouzols et al.: Mining & Control of Network Traffic by Computational Intelligence, pp. 1–51.
springerlink.com

during a week on a gateway of an university in 1995 added up to 89 GB of data corresponding to 439 millions of packets [24].

The complexity of modeling the Internet of today and the foreseeable future can be understood considering the sustained exponential increase of traffic and nodes observed throughout the years [65] as well as the fast evolution of network protocols and applications. Currently, capturing packet header traces in fast links for a few minutes or hours may produce of the order of hundreds of GBs or even several TBs of data [38].

The recent development of high performance hardware for IP packet capture up to 10 Gb/s [47] has made it possible to record traffic traces in backbone nodes of current high-speed networks. However, it is not feasible to use such a huge volume of information for research and operation tasks. Filtering and preprocessing methods are required. Often, data volumes have to be reduced by 12 orders of magnitude, from 10^{12} bytes down to a report of 10 lines of text [48]. It is also common to reduce huge volumes of traffic measurement data down to a set of a few graphs and tables [145].

The difficulties in this field are clear if we consider the analysis and modeling of wide area networks and the Internet in particular. In addition, there is a lack of measurement and monitoring mechanisms in the Internet architecture [164], which has been defined in a rather unstructured manner through an aggregation of protocols, technologies and applications developed independently. This architecture, that has been called a cooperative anarchy [123], defies measurement and characterization. As Willinger and Paxson point out, *"it is difficult to think of any other area in the sciences where the available data provide such detailed information about so many different facets of behavior"* [170].

In this sense, technologies based on the Simple Network Management Protocol (SNMP) and the concept of network flow have seen a great deal of development and deployment during the last years [37]. Still, many efforts are required to enable macroscopic analysis of the Internet.

During the last decade, some areas, such as switching techniques and topology design, have seen fast development. However, systems and infrastructures for traffic measurement are still in early stages of development and scarcely deployed. The fast evolution and great diversity of the Internet together with the long periods of time required to analyze measurement data have a drastic consequence: experiments and studies based on traffic measurements are already obsolete when finished and specially when published [32]. Thus, it is hardly feasible to implement measurement and analysis systems that can be used to support other infrastructures.

A number of works in Internet measurement [124, 32] have led to the first results in what some authors call Internet Science [21]: an experimental science that studies laws and patterns in Internet structure [104]. Traditional statistical inference techniques often used to analyze networks are limited. Instead, Internet research require inference methods for searching for law-like relationships across large collections of high-volume data sets that generalize to a wide range of conditions [170]. That is, scientific inference is required in order to unveil *traffic invariants*. This requires building intuition and physical understanding rather than using conventional black-box descriptions and data fitting techniques.

At first sight Internet Engineering might seem a more precise term for this area of research since the current Internet is the result of applying diverse engineering disciplines. However, issues and questions currently posed require an approach more close to that of the experimental sciences. This area involves theories as well as techniques and infrastructures for measurement, analysis and modeling.

Broadly speaking, three main aspects in Internet measurement, analysis and modeling have to be addressed in order to construct models of the Internet as a whole:

1. Traffic.
2. Topology.
3. Effect of protocols on traffic and topology.

In particular, Internet traffic modeling comprises macroscopic characterization as well as multi-scale modeling. Throughout the last years, many developments have shed some light on traffic dynamics. As a result, long-range dependencies, self-similarity and power-laws and wavelets have been established as common modeling tools. These aspects will be overviewed in the next sections. Often, traffic and topology are analyzed as orthogonal aspects. For instance, the obvious effect of routing protocols on traffic dynamics and congestion episodes is not well understood. In fact, the last research efforts towards an in-depth analysis of this interactions, the so-called traffic-sensitive routing, were abandoned several years ago. The adaptive routing protocols designed were found to be highly unstable [167].

Analysis and data mining of topology related measurements are commonly performed off-line and require cooperation from operators. operators, etc.). The objective of these studies is to identify invariants that help understand how topologies evolve. For instance, at the application level, it has been found that two randomly chosen documents on the web are on average 19 clicks away from each other [4]. Research on the overall topology of the Internet has been successful in revealing and validating the so-called jellyfish model: the network is compact, i.e, 99% of pairs of nodes are within 6 hops, there exists a highly connected center, there exists a loose hierarchy, and one-degree nodes are scattered everywhere. In summary, the network has the tendency to be one large connected component. Power laws appear in other settings, such as WWW pages and peer-to-peer networks. In short, the topology of Internet is described by power-laws, its growth is slowing down (following a sigmoid curve), it is compact, becomes denser with time, and looks like a jellyfish [49, 101].

Major advances in Internet modeling include the identification of self-similarity and long-range dependencies in traffic as well the use of power-laws to describe the global topology of the Internet. But many issues are still open: spatio-temporal correlations, interest and group behavior, anomaly detection, etc. From the data mining viewpoint, there are many modeling challenges, including massive multidimensional data, time-space correlations, and case dependent phenomena.

1.2   Measurement Systems and Infrastructures

Network performance depends on and can be measured in terms of a number of parameters such as capacity, available bandwidth, delay, jitter, packet loss and packet disorder. These and other network parameters are related in a complex manner and to a varying extent. Measuring the network is crucial to understanding the Internet behavior and designing control mechanisms for improving performance.

Unfortunately, the original Internet architecture has little or no support for measurement. End hosts and their applications, however, have a limited capability in accessing and acquiring information about the network behavior. To them, end-to-end measurement of the network behavior is usually the only available information.

A number of factors have led to a surge in research of Internet measurement systems and infrastructures during the last years. The outcomes of these research activities have a positive impact in two areas. First, experimental support is provided for a better understanding of network traffic dynamics. Second, the availability of measurement infrastructures enables the development of measurement based traffic control and quality of service mechanisms.

In particular, nodes and protocols in the current Internet provide very little support for performance measurement. In addition, a number of new applications would greatly benefit from dynamic adaptation mechanisms based on network measurement. Also, improved methods and tools for network performance monitoring and troubleshooting are sought.

In fact, besides the development of novel techniques and tools within current architectures, firm proposals have been made [164] towards introducing modifications in network layer protocols as well as switching and routing equipment so that better support for measurement tasks is available in basic infrastructures.

In order to study the dynamics of Internet traffic both on-line and off-line techniques are required. These techniques and the infrastructures that support them are usually based on counting interesting events such as sessions, connections, arrivals of packets or cells to a node for a given period of time.

Current measurement systems [37, 124, 131] can be classified into two main types: active and passive. The former are of a distributed nature and are usually accessible to end users and applications. The latter are centralized and often restricted to network operators and engineers. The current challenges in this area are to increase the maturity of these systems, to deploy measurement infrastructures and to enable generalized macroscopic analysis of the Internet.

1.2.1   Active Systems

Active measurement systems work by sending probe traffic from an end node in order to measure parameters such as round-trip time and packet loss percentage [118, 124, 136]. Active measurement tools inject probe packets into the network and analyze the response. Following a particular network model, some characteristics are estimated, such as propagation delay and a number of metrics related to bandwidth.

Active measurement tools can not only provide network operators with useful information on network characteristics and performance, but also can enable end users (and user applications) to perform independent network auditing, load balancing, and server selection tasks, among many others, without requiring access to network elements or administrative resources.

The research community is developing a set of metrics and techniques for active bandwidth measurement, including concise reporting to users [146]. Many of them [136] are well understood and can provide accurate estimates under certain conditions.

Some institutions are currently undertaking initiatives to deploy test platforms for active and passive bandwidth estimation as well as other related techniques. Also, some partial measurement and evaluation studies of bandwidth estimation tools have been published [147, 116, 86, 158].

The models underlying active systems often rely on a large number of parameters difficult to model in an independent manner. As a consequence, these systems suffer from errors and accuracy limitations in measurements and estimations, especially regarding timing accuracy in general purpose platforms [95, 2].

The network model chosen for designing an active measurement tool has a determining impact on the applicability and performance of the tool. Thus, research on active measurement tools [95, 160, 5], and specially of those that estimate bandwidth related metrics by probing the network [86, 46], has been very active during the last years. This area has made important contributions to the understanding of network traffic dynamics, particularly in the case of the behavior of aggregated flows in router queues.

The first attempt at using bandwidth estimates for application adaptation purposes reported in the literature can be tracked back to 1996, when BPROBE/CPROBE were introduced as tools for server selection tasks. Soon after appeared pathchar, introduced in 1997 as a per-hop network capacity estimation tool.

For about a decade, a number of bandwidth estimation methods and tools have been developed. These tools show a wide spectrum of requirements and characteristics, such as accuracy and intrusiveness. Underlying models, metrics definitions, terminologies as well as measurement and processing methodologies also differ.

A number of techniques for estimating bandwidth capacity and available capacity have been developed: variable packet size (VPS), packet pairs, packet trains, packet tailgating, ALBP (Asymmetric Link Bandwidth Probing), self-loading streams, to name a few. Implementations of these techniques can be found in a number of tools [86, 46, 116]. The performance of each technique usually provides insights on how the network reacts to a certain traffic pattern. Note that some tools also estimate parameters related to bandwidth, such as the ADR (asymptotic dispersion rate). The tool thrulay [146] further elaborates on the same idea and combines application level measurement of available bandwidth capacity and round-trip time.

1.2.2 Passive Systems

Passive measurement systems are based on recording data at a network node, i.e., no probe packets are sent. While passive systems do not require cooperation or coordination among end nodes, the quality and relevance of data decisively depends on the location of the measurement point. Thus, cooperation between network operators [118, 32] is a prerequisite of passive measurement infrastructures.

Passive systems are a field for the application of analysis and interpretation techniques for large volumes of data where measurements are often missing and inaccurate. These systems run in network nodes and particularly in routers gathering data usually through sampling procedures applied to traffic as traverses the network in real-time. These measurements are usually transfered to collection points following standards such as SNMP and NetFlow. The NetFlow technology is further discussed in chapter 4 where a novel method for summarizing network flow collections is described.

Passive systems enable global analysis of subnetworks at the infrastructure level. They make it possible to detect the emergence and growth of new applications, protocols and related traffic patterns. Some of the main current areas of research in traffic analysis based on passive measurement systems can be listed as follows:

- Analysis of the interactions between macroscopic traffic dynamic and routing algorithms. In particular, the analysis of routing tables in the BGP protocol [138, 139, 161] is key for understanding traffic flows between service providers and autonomous systems.
- Analysis of the distribution of traffic over the address space (both IPv4 and IPv6). This is a requirement for building maps of the address space assigned to institutions and service providers as well as the set of addresses that can be globally accessed.
- Analysis of the dynamic characteristics linked to protocols, applications and technologies. This area becomes more and more important as different novel services are deployed on the Internet.
- Development of tools and hardware support for traffic measurement and analysis [47, 43, 81].
- Privacy and security related procedures and techniques, including anonymization of network traces.

1.2.3 Publicly Available Measurements

Traces are one of the main outcomes of measurement infrastructures. The use of common traces recorded by both active and passive measurement infrastructures are key reproducible research and comparison of results in general. Traces may comprise data about topology, traffic, specific applications and a variety of heterogeneous measurements.

In this sense, the recent availability traffic traces of high-speed networks, specially at OC48 and OC192 speeds, requires a great deal of effort and cooperation among different agents. Cooperative measurement projects and infrastructures also allows for wide scale analysis of networks.

A remarkable initiative in this context is the Day in the Life of the Internet series of events held in 2007 and 2008, that gathered together institutions from several continents in order to record continuous traffic traces in a coordinated manner for a considerable large period of time, spanning more than 50 hours in some cases.

In this monograph we will use a wide set of publicly available network traffic traces obtained through passive monitoring. These traces are usually made of a sequence of packet headers (possibly including part or all the payload as well). Some other traces only provide a restricted set of data about each received packet, in particular the arrival time and size, as well as some other specially relevant data such as TCP flags. In chapters 3 and 4 we will analyze traffic traces from two perspectives. First, time series models for traffic load as derived from these traces are designed. Then, a method for summarizing flow collections derived from these traces is described.

Some traces have an historical relevance such as the Bellcore traces and the traces taken at the Lawrence Berkeley National Laboratory. The first were the empirical basis for finding self-similarity and long-range dependence in Ethernet traffic [69, 106] whereas the second were instrumental in showing that the Poisson model fails to capture the general behavior of traffic in wide area networks [134]. It is interesting to note that the limitations of the Poisson model in the communications field, though often overlooked and usually not dealt with in the literature, were well-known by practitioners since more than 2 decades before.

1.3 Network Traffic

The problem of modeling Internet traffic is both interesting in its own right and useful for a variety of applications, including congestion control and protocol design. It is out of the scope of this monograph to review all the proposed descriptive and predictive approaches to modeling Internet traffic. For an in-depth and exhaustive overview we refer the interested reader to a general book on traffic measurement [37] as well as a number of research papers on the topic [71, 36, 140, 141, 128, 41, 109]. In this section, we overview some of the most relevant, often antagonistic, models for network traffic with the focus on those models that can shed some light on the modeling of network traffic from a time series modeling point of view.

Network traffic can be analyzed either from the perspective of the network and transport layers and the impact of generic metrics on the performance perceived by users [118], or from application specific viewpoints, such as Web traffic [120], peer-to-peer traffic [119] and multimedia traffic [121]. Here we will discuss the most important issues in modeling network traffic, network performance metrics and the concept of congestion in a general manner.

1.3.1   Traffic Models

Data obtained by measurement systems are usually processed using statistical tools in order to obtain as much information as possible [162]. This way, in the case of a video or audio application network flow, packets can be distributed over time following an exponential, subexponential or light-tailed distribution [132, 134]. This process leads to the extraction of empirically derived analytic models of traffic [129] and helps identifying invariants.

The natural step after network measurements are gathered is to analyze them and run simulations [65]. Network measurement enables analysis of data as well as realistic simulation of networks. By identifying and reproducing invariants in network traffic in simulation scenarios a better understanding on how these invariants impact traffic dynamics can be obtained.

Describing traffic properties for supporting analysis and simulation tasks requires simple models that capture different levels of abstraction and time scales. That is, different levels of detail in simulation systems, represented by application sessions, connections, transfers, packets, etc. In an analogous manner, simulations can be run with different levels of detail, ranging from analytical models to more detailed behavioral simulation at the session and packet levels.

Let us now overview some of the traffic models that have been applied to and developed for packet switched networks. Teletraffic theory originally embraced all the mathematics applied to the design, control and management of the public switched telephone network (PSTN). Techniques belonging to the fields of queuing theory, statistical inference, performance analysis, mathematical modeling and optimization were used to lay out teletraffic theory. The natural step with the advent of the Internet was to extend this theory in order to include data networks. This way, Internet engineering (emcompassing the design, control, operation and management of the global Internet) would become part of teletraffic theory. However, Internet practitioners have emphasized engineering and experimental deployment rather than rigorous mathematical modeling and application of theories. In fact some in the Internet community would say that the Internet works because "it ignored mathematics -in particular, teletraffic theory-" [170].

Teletraffic theory has been remarkably successful in the case of the PSTN. Conventional PSTN is however a highly static environment where the notion of limited variability is well-defined and ever-present. Typical users, generic behavior and averages are proper descriptions of the overall system performance. In addition the most widely used models are specially practical from an engineering viewpoint. These models are parsimonious and additionally the few required parameters can be easily estimated in practice.

These factor led to the belief that a universal law in voice networks established the Poisson nature of call arrivals for aggregated traffic. According to this assumption, call arrivals are mutually independent and the interarrival times are exponentially distributed. Poison models are the first model widely applied to communications traffic.

The application of Poison models dates back to the early telephone networks and the pioneering works by Erlang and others. In general, a Poisson process is characterized as a renewal process with interarrival times A_n exponentially distributed with rate parameter λ. If $X = (X_t : t \geq 1)$ is the number of arrivals in successive, non-overlapping time intervals of length $\Delta t > 0$, then X is the increment process of a Poisson process with parameter λ if and only if the random variables X_t are i.i.d. with:

$$P[X_t = n] = e^{(-\lambda t)} (\lambda t)/n!$$

In this formulation, Poisson process are described as a counting process where the number of arrivals in different intervals is statistically independent.

The so-called Poisson law has been widely accepted for several decades. The same applies to the following laws: call durations follow an approximately exponential distribution, there is a high predictability in growth rates, network control and operation are fully centralized (so information about the global state of the network is available), and services are strictly monitored and regulated. However, the high stability of telephone networks was compromised by the advent of fax in the 1980s. This was due to the fundamentally different statistical properties of fax transmissions. With the popularization of TCP/IP networks and the WWW, teletraffic theory was no longer able to cope with data transmissions in a satisfactory manner.

Still, the first formal models proposed for Internet traffic were based on traditional teletraffic theory [134]. However, in the Internet, the engineering reality overcomes traditional teletraffic analytical modeling. Since self-similarity and long-range dependencies were first formally identified in data traffic [106] a number of studies have shown extensive evidence of the failure of Poisson models in the Internet. Poisson models have thus been rejected for characterizing packet arrival processes in the Internet [128, 134] at different levels of aggregation (ranging from local area networks to backbones).

The relevant mathematics for the PSTN deals with limited variability in both time and space, i.e., traffic processes are either independent or have exponentially decaying temporal correlations, and the distributions of traffic related properties have exponentially decaying tails.

In contrast, the mathematics relevant to packet switched networks has to deal with extreme variability. In many cases, very bursty at many different time scales (or fractal-like) behavior can be identified in network traffic load over a wide range of time scales from milliseconds to tens of seconds and beyond, i.e., traffic is self-similar [128].

More formally, a discrete-time, covariance-stationary, zero-mean stochastic process $X = (X_t : t \geq 1)$ is *exactly self-similar* or fractal with scaling (Hurst) parameter $H \in [0.5, 1)$ if, for all levels of aggregation $m \geq 1$,

$$X^{(m)} = m^{H-1} X,$$

where the equality should be understood in the sense of finite-dimensional distributions. The aggregated processes $X^{(}m)$ are defined as follows:

$$X^{(m)} = m^{-1}(X_{(m-1)k+1} + \ldots + X_{km}), \ k \geq 1.$$

For this kind of process, it is easy to show that the following relationship holds:

$$var(X^{(m)}) = km^{2H-2}.$$

That is, there is a relationship between a quantity Q of the underlying process, traffic load, and the resolution m that follows:

$$Q(\tau) \approx k_f \tau^{f(D)},$$

where $f(\cdot)$ is a simple function of D, and D is a fractal dimension. Thus, such processes are fractal.

In addition, the resulting linear log-log plot representation of $var(X^{(m)})$ versus m is the so-called variance-time plot, which is one of the methods commonly applied to identify the Hurst parameter of traffic time series.

Many evidences suggest that traffic in packet switched networks is self-similar and fractal in nature. A plausible explanation is that self-similarity is a consequence of the power-law distribution of different types of traffic workload, such as flow durations, web transfers, file sizes and even the way users interact with networked applications [128, 36, 37].

The heavy-tailed property exhibited by the distribution of flow sizes and durations is an invariant for an aggregate property of flows. It does not provide any information on the packet-level behavior of traffic sources. However, direct links between connection sizes and durations with infinite variance and fractal scaling in aggregate network traffic have been mathematically proven. Thus, this invariant has been key in finding a physical explanation of the observed fractal nature of aggregate traffic. A heavy-tailed distribution is defined as follows:

$$P[X > x] \propto x^{-\alpha},$$

as $x \to \infty$, and $0 < \alpha < 2$. The fact that this kind of distribution governs different traffic workloads can be explained in a generic manner by Zipf's law [128, 36].

Poisson models cannot cope with high variability at the packet level. However, there is evidence that these models are satisfactory for human interactions with networked applications [36]. That is, the times at which users start interactions with applications conform to a memoryless process with an arrival rate that can be satisfactorily approximated as constant over time intervals of many minutes or perhaps an hour [170]. In addition, some works have shown the usefulness of time-varying Poisson models for small time scales in networks with a high level of traffic aggregation [97, 180, 23]. The argument that network traffic tends to Poisson as the level of aggregation increases is disputed though, as only a few limited studies support it.

In this context, it is currently widely recognized that better theoretical models with more extensive experimental basis are required [10, 64] in order to enable full understanding of the dynamics of Internet traffic.

1.3.2   Transport Layer Models. TCP

Modeling the dynamics of transport layer flows, and TCP flows in particular, is a central problem in Internet traffic research. Applications of predictive performance models range from peer-to-peer and content distribution networks (CDN) to grid computing. Most traffic in the current Internet, in terms of flows, packets and octets, is due to TCP connections [114]. Models for TCP dynamics have been developed following either of two approaches known as model based and equation based [162, 169].

Modeling TCP performance has also deep implications in transport protocol design. Preventing congestion collapse in the Internet and guaranteeing fairness at least in a TCP-compatible manner are two key aspects that should be addressed when developing new standard transport protocols [58]. In a similar way, TCP models have a significant impact on the design of active queue management and mechanisms for differentiated quality of service provisioning. Additional implications of TCP models include the definition of a meaningful set of evaluation scenarios and conditions for transport protocols [8, 7].

Some simple equation based models [169] point out the dramatic effect of packet loss on the performance of TCP. These models establish the relationship between the transfer rate of a TCP flow, T, and the packet loss rate, p, as follows:

$$T \propto \frac{1}{\sqrt{p}}.$$

Further elaborating on the same simple model, a basic formulation of the expected average TCP transfer rate can be established as follows [78, 112]:

$$E[T(s, t_{RTT}, p)] = \frac{s}{t_{RTT} \sqrt{\frac{2Dp}{3}}},$$

where s is the maximum segment size, t_{RTT} is the round trip time, p is the packet loss rate, and D denotes the number of data units (TCP segments) acknowledged for each ACK packet, The t_{RTT} of a TCP connection between a sender and a receiver is defined as the time elapsed between the instant a packet is sent by the source to the instant the corresponding ACK from the receiver is received by the source.

However, obtaining equations for modeling and predicting the stationary behavior of TCP in a general manner is a complex problem. A number of solutions have been proposed. To date, the most complete model that has been extensively evaluated through experimentation [75, 168] defines the following equation for the TCP-compatible transfer rate:

$$E[T(s, t_{RTT}, p, t_{RTO})] =$$

$$\min \left(\frac{sW_m}{t_{RTT}}, \frac{s}{t_{RTT}\sqrt{\frac{2Dp}{3}} + t_{RTO}\min\left(1, 3\sqrt{\frac{3Dp}{8}}\right)p(1+32p^2)} \right),$$

where W_m is the maximum size of the TCP congestion window and t_{RTO} is the packet retransmission timeout of the TCP protocol under the particular conditions given by the parameters of the equation. Note the application of this model requires the sender to know the parameters of the equation. Thus, it is necessary that TCP receivers provide the required information. As a particular case, where there is no packet loss, the expected rate is given by the ratio $\frac{W_m s}{t_{RTT}}$.

The model above can be extended to multicast networks [168], which requires the definition of a variety of feedback mechanisms so that senders are informed about network conditions at the reception points.

Nonetheless, accurate modeling of TCP is an increasingly complex problem due to the many variants proposed throughout the years [156, 133, 92] and the intricate evolution of the standard variants [98, 93, 20, 155]. Therefore, there are many open issues in the design of TCP variants that can cope with technological and architectural changes in the Internet [10]. Some TCP variants recently proposed will be overviewed in chapter 5, where a new approach to end-to-end congestion control is described based on fuzzy logic.

1.3.3   Models of Applications and Services

During the last years, the diversity of network conditions and traffic patterns that can be found in the Internet has been progressively increasing [10, 64]. Thus, the development of schemes for generating flexible aggregate flows and topologies is key for modeling applications and services.

Characterizing the dynamics of specific types of traffic linked to particular services and applications is key for providing proper definitions of the quality of service requirements of current and foreseeable network applications. In addition, the definition of traffic models for different applications is crucial not only for characterization purposes but also to enable the development of realistic simulation and emulation environments.

In particular, extensive studies have addressed traffic patterns for widespread applications, such as web [120], bulk transfers by FTP and similar protocols [26], peer-to-peer applications [119], and voice and video applications [121, 143].

1.3.4   Network Simulation

Simulation of network scenarios can help overcome the limitations of measurement and experimentation. In particular, simulation models make it possible to explore new protocols, environments and architectures. By simulation is also possible to explore complex scenarios that would otherwise be difficult or impossible to analyze.

Nonetheless, there does not exist a complete suite of simulation scenarios that can be deemed as sufficient to demonstrate that a new protocol or mechanism will perform properly in the future evolving Internet. Instead, simulations are limited to exploring specific aspects of new proposals or the behavior of the Internet, as well as advancing the understanding of traffic dynamics. The role of network simulation is thus to explore scenarios in order to build understanding of dynamics, to illustrate a point, or to explore for unexpected behavior [65]. Simulations however can be misleading when used for producing quantitative performance comparisons.

In particular, network simulation is the most proper method for addressing many of the open issues in traffic dynamics, specially the complex interactions between topologies and traffic, as well as the central role of adaptive congestion control.

Simulating the behavior of the global Internet or a significant part of it is an immense challenge. This is due to its heterogeneity and fast evolution. Experience shows that techniques that were studied using partial models were not implemented eventually because of doubts about their limitations [64]. Thus, the variety of scenarios and conditions taken into consideration for the simulation and evaluation of new systems is a key factor for their eventual acceptance.

A sound network model for simulation comprises all the aspects that can have an impact on a simulation or experiment. These include the topology, traffic generation patterns, behavior of protocols at every layers of the protocol stack, queue control mechanisms, among many other possible factors. In general, it is useful to lay out simulations in such a way that *invariants* can be identified by exploring the simulation parameter space [65].

However, many research works rely on simulations with assumptions that are not experimentally proven. These include long-lived and large flows, simple topologies with often only one congested link, small range of round-trip times for the simulated flows, most traffic flowing in a single direction through the congested link and negligible amount of reverse traffic.

Instead, the use of a number of well known invariants can help designing realistic simulation scenarios. These invariants include diurnal patterns of activity, self-similarity in packet arrival processes, Poisson session arrivals, log-normal connection sizes, heavy-tail distributions and topological invariants of the global Internet derived from the Earth's geography and the distribution of human population [65].

A large amount of techniques and methodologies for network simulation have been proposed and applied throughout the years and further research is being carried out. These techniques and methodologies include discrete event, web-based and agent-based simulation schemes, Petri nets, fluid-flow based simulation, specific languages for simulation and overlay networks among many others. In particular, the use of advanced simulation tools, such as ns-2 [88], SSF Net [154] and OM-NET++ [163], and emulators, such as Netbed/Emulab [68], Planetlab [135], NIST Net [25], iproute2 [80] and dummynet [142]), to name only a few, is key for addressing the aforementioned problems [65]. In chapters 5 and 6 we will describe how we have used some simulation and emulation environments in order to test new traffic control mechanisms.

1.3.5 Performance Metrics

In order to assess the performance and reliability of networks, a set of parameters are usually measured or indirectly estimated from measurements. When these parameters are unambiguously specified, whether qualitatively or quantitatively, they are identified as performance metrics.

Even though the definition of these parameters can be unambiguous, there many not be clear procedures for their effective measurement. This way, measuring some of the most common network performance metrics, such as connectivity, delay, loss pattern and reordering pattern, poses different practical issues. Moreover, a certain parameter that describes network performance, such as packet reordering, may have several associated metrics. Also, the definition of metrics usually depends on the network model under which they are interpreted.

Defining metrics that provide quantitative and unbiased information about network parameters is required in order to develop tools for network quality, performance and reliability evaluation. Currently, the IP Performance Metrics (IPPM) group of the IETF is working together with the T1A1.3, SG 12 and SG 13 groups of the ITU-T towards laying out and standardizing quantitative metrics on data delivery by transport protocols. The objective is to obtain metrics that provide quantitative information about performance avoiding any ambiguity. The metrics, considered for both end-to-end paths and subnetworks can be listed as follows:

- Connectivity.
- One-way delay and loss rate.
- Round-trip delay and loss rate.
- Delay variation.
- Loss pattern.
- Packet reordering.
- Bulk transfer rate.
- Capacity and bandwidth of links.

The IPPM working group has defined through a series of RFC documents a large number of richly parameterized metrics in order to address the many possible objectives of network measurement procedures. Often, the ultimate purpose is to report a concise set of metrics describing a network's state to an end user. Elaborating on this idea, the Internet Draft on reporting metrics to users [146] defines a small set of metrics that are robust, easy to understand, orthogonal, relevant, and easy to compute.

The standardization process for this metrics considers not only their formal definition but also documentation and measurement procedures. There is however the need for establishing procedures for measuring individual metrics and interpreting their values as relevant properties for different classes of service, such as bulk transfer, periodic and multimedia flows.

Nonetheless, this standardization effort embraces only low level metrics, i.e., those that characterize the network regardless of transport protocols and applications. That is, the definition of metrics for characterizing different traffic patterns (such as VoIP applications) are beyond the scope of these working groups and are thus left within the general domain of network modeling.

Moreover, some parameters, in particular the available link capacity [46, 45], have not been considered due to the lack of mature and accepted definitions, models and measurement procedures. Additionally, high level parameters such as inter-flow fairness, congestion control and resource sharing metrics are equally outside of the scope of this standardization effort. Some of these aspects are being addressed by other IETF groups in a more general manner [67, 7].

1.3.6 Congestion

Congestion control has been identified as a critic function for the growth and evolution of the Internet [63, 10]. In the past, some global congestion collapse episodes have been experienced in the Internet [58, 19]. In order to avoid congestion collapse and provide proper management of different kinds of traffic, congestion control mechanisms have to be implemented.

The following sentence may be a good summary of the general notion of congestion: "We have seen that as a system gets congested, the service delay in the system increases." Here, the service delay can be considered at a number of levels: application level responsiveness, server response time, etc. For instance, the amount of congestion in terms of packet loss may be low whereas that low loss at the network layer is the reason behind a high degree of congestion as perceived by users at the application layer. In packet switched networks, the performance degradation is dramatical beyond a certain congestion point, as depicted in figure 1.1.

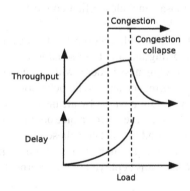

Fig. 1.1 Throughput and delay increase as the load increases up to the congestion point

This effect is particularly severe for reliable transport services, such as TCP transfers, because of the progressive increase of packet retransmissions [59, 92]. This can lead to a state where the percentage of packets that arrive at their destination decreases drastically as the traffic generated by end nodes increases. Parameters such as the packet loss rate in an end-to-end connection or a link as well as the transmission delay can indicate congestion. However there is a lack of formal definitions of congestion and congestion collapse. There is no quantitative estimation of the concrete parameters nor the scale at which a degraded network condition can be considered a congestion episode. In fact, the perception of congestion by end users depend, among other factors, on the traffic patterns and quality of service requirements of the applications in use.

A very general definition of the concept of congestion can be stated as follows: a decrease in utility, from the perspective of a given traffic source, due to increased load [153]. The problem can be looked at from both the source and network perspective. The convenience and practical need for end-to-end congestion control mechanisms has been extensively documented [59] and procedures for the evaluation of congestion control mechanisms [67] have been recently layed out.

The congestion control schemes currently deployed in the Internet as well as the proposed alternatives follow either of the two following approachers:

- Congestion control distributed among the end nodes, implemented by the transport protocols used by the applications.
- Centralized congestion control, implemented in routers as queue control mechanisms. However, these mechanisms have to work in a cooperative manner.

In the current Internet the dominant congestion control scheme is implemented as a transport layer mechanism, particularly in the TCP congestion control scheme. In order to address the limitations and drawbacks of this scheme, a large number of TCP variants have been proposed. Complementary active queue management schemes have been proposed as well. In particular, the RED mechanisms has been advocated for some years [61, 59, 19] though little deployment has happened so far. Other complementary proposals include explicit congestion notification [137]) and novel architectures [14, 89, 172].

However, there is lack of plausible theories, simulation procedures and experimental evidence for supporting any of these schemes, whether deployed or not, with enough efficiency and robustness under a wide range of conditions. Some authors [31] note the poor performance and cyclic behavior of TCP/IP systems. This drawbacks have been found in some works by means of simulation [111] and theoretical analyses [105]. However, these performance degradations happen only under very specific and unrealistic conditions. Thus, these effects are rarely seen in real networks. Anyway, the lack of experimentation and the ad-hoc nature of some of the congestion control mechanisms deployed in the Internet is generally accepted.

1.4   Traffic Control

Traffic control involves different tasks, such as control of flow, congestion and admission as well as quality of service (QoS) provisioning. In the current Internet, TCP implements end-to-end flow and congestion control. Admission control and QoS mechanisms are rarely found and only in very specific cases.

Congestion control has long been considered an important research problem in computer networks. Different types of congestion control algorithms have been defined for packet switching networks. In the taxonomy by Yang and Reddy [178], the standard TCP congestion control falls within the class of closed loop with implicit feedback schemes, whereas drop-tail and the most accepted active queue management schemes belong to the class of open loop with destination control schemes.

A comprehensive set of metrics for evaluating congestion control algorithms in the Internet has been defined as well [67]. These include throughput, delay, loss, response time, minimizing oscillations, fairness, convergence robustness for challenging environments, robustness to misbehaving users and to failures, deployability as well as metrics for specific types of transport and user-centric metrics. The relations among these parameters are complex and in general all of them can effect both end-to-end and router based traffic control mechanisms. For instance, the distribution of round-trip time can dramatically affect not only the data rates achievable by TCP flows sharing a link but also the utilization of network links.

In addition, congestion control and quality of service provisioning are two tightly related functions. It is possible and common to implement congestion control without taking into consideration QoS mechanisms. However, the design and deployment of architectures and mechanisms for QoS has a twofold justification:

1. From the viewpoint of the operation and requirement of the current Internet, the cooperative and distributed traffic control implemented in TCP requires complementary control schemes in the network layer [59, 63].
2. From an abstract viewpoint, considering the distribution of services and functions among the different network layers, tasks such as QoS provisioning and admission control correspond to the network layer.

In general, the tranport layer of networks that implement QoS provisioning allows for the applications to specify the required or wished quality. These specifications can be satisfied to a varying degree while keeping a balance between the many parameters that might be in conflict. No comprehensive set of transport level QoS parameters has been widely accepted. Also, there is no concrete definition of the way QoS specifications have to be processed and enforced under different network conditions.

A proposal of QoS parameters has been made in the X.214 recommendation of the ITU-T about the definition of transport services [91]. This is the most exhaustive list of QoS parameters among the different standardization efforts carried out to date. Thus, it can be considered as a reference. The parameters included inthis standard are listed as follows:

- Connection establishment delay.
- Connection establishment failure probability.
- Transfer rate.
- Transit delay of data.
- Residual error rate (including wrong, loss and duplicated data).
- Probability of failure of a data transfer.

- Connection release delay.
- Probability failure of connection release.
- Protection (regarding integrity and confidentiality).
- Priority.
- Resilience (against spontaneous connection termination due to internal problems of congestion).

Negotiation mechanisms are considered for all the parameters above during the transport connection establishment phase. Objective, acceptable and minimum values can be specified for the whole set of parameters.

However, QoS support is far from being complete and deployed in the real world. For instance, ATM networks only supported two QoS parameters: propagation delay and transfer rate. It is only since a few years that the required technolgoies are available in routing equipment. As an alternative, there have been proposals of adaptive bandwidth control [149, 150]. This schemes adjust the bandwidth reservation at the packet level time scale in order to guarantee QoS requirements.

The functions we have dealt with so far are directly related to transport protocols. That is, the higher network layer (above the network layer and the set of underlying routers, bridges and links as shown in figure 1.2), works on an end-to-end basis and provides applications with services that abstract the technologies, design and operation of the underlying network.

The basic function of the transport layer is to provide a communication service between processes, abstracting the underlying network. In fact, relying exclusively on TCP for implementing congestion control in the Internet is a disputed scheme. It is accepted that hybrid systems should be implemented where the network layer performs some congestion control functions [19].

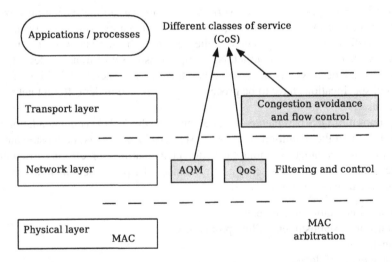

Fig. 1.2 Flow and congestion control and QoS at different network layers

1.4.1 End-To-End Traffic Control

We will focus on TCP congestion control mechanisms. It should be noted however that tens of alternative protocols have been proposed throughout the last years. In particular, from the viewpoint of traffic control TCP is nowadays a family of protocols rather than a particular protocol. Several modifications to the TCP traffic control mechanisms have been proposed throughout the years. In fact, versions of TCP currently deployed have little to do with its early versions or even versions widely used 10 years ago as far as traffic control is concerned. It is plausible to anticipate that TCP will go on evolving as the dominant transport protocol in the Internet for the next years while keeping an standard programming interface and header format.

Both flow control and congestion control are thus implemented in TCP. At the transport layer, the distinction between these two functions blurs [90]. Flow control between end nodes includes all the mechanisms by which the sender node limits the transfer rate in order not to overload the receiver and the network. Those mechanisms implemented with the aim of preventing global network overload are then referred to as congestion control mechanisms [178].

Most transport protocols designed during the last years, and TCP in particular, implement flow and congestion control in an intertwined manner. This way, the same mechanisms may implement flow and congestion control. This is a possible scheme for performing congestion control. In other architectures, congestion control is implemented at the network layer separated from flow control at the transport layer.

Congestion control at the transport layer is more complex than flow control at the link and network layers. This is due to the variability of the round-trip delay, packet reordering and other problems specific to end-to-end paths. Transport layer congestion control mechanisms are usually implemented based on sequence numbers and transmission windows. These two elements are equally used for implementing error control algorithms. The coupling between error control and congestion control is however a limitation in high-speed networks [52] that has motivated a number of recent proposals of modifications to TCP. In general, congestion control mechanisms at the transport layer can be classified into two kinds of techniques:

- *Sliding window.* This technique is based on the definition of a data window of whether static or variable maximum size that limits the amount of data in flight. Each time data are transmitted, the sender reduces the window size proportionally. When the maximum window size is static, the current size is increased when acknowledgment packets are received from the receiver. When the maximum window size is variable (known as credit schemes), the window is adjusted through decision procedures performed by the receiver. In this systems, the receiver informs the sender of the allowed window size.
- *Rate control.* This technique is based in the use of timers at the sender. Two basic variants are distinguished. In the first variant, the timers define the interval the sender has to wait between data bursts. In the second variant, the timers define the interval the sender has to wait between data units. The second option usually

provides better performance for flow and congestion control at the expense of the difficulty in implementing highly accurate timers.

The use of flow control techniques in order to implement congestion control at the transport level has a twofold objective:

- *Optimize* the use of resources, specially router input and output links. When no support for admission control is provided at the network layer, it is necessary to implement implicit admission control functions by identifying overload conditions and bottlenecks. To this end, current network conditions have to be inferred from a number of parameters such as the round-trip delay. The possible approaches to this end by mechanisms implemented in routers are analyzed in section 1.4.2.
- *Fairness* in resource sharing among end-to-end data flows and thus users. In order to accomplish this objective, transport layer is limited to guaranteeing that transport flows behave in a cooperative manner. It should be noted that so called misbehaving flows (or flows that do not conform to TCP congestion control principles) cannot be properly controled unless fairness techniques are implemented in routers. This is further discussed in section 1.4.2.

In order to fulfill these two objectives, TCP congestion control mechanisms follow two design principles:

- *Additive increase*. Initially, TCP connections use a bandwidth value lower than the available bandwidth. This value is progressively incremented in an additive manner until overload is detected. Since additions are performed each time acknowledgment packets from the receiver arrive at the sender, this scheme results in an increase exponential with time.
- *Multiplicative decrease*. In standard TCP, packet loss is taken as a sign of congestion. When packet loss is detected, the transfer rate is decreased exponentially (commonly by a factor of 2) and the increase process is initiated again.

This scheme guarantees cooperation among competing TCP end-to-end flows. The additive increase-multiplicative decrease (AIMD) scheme is further discussed in chapter 5, where a generalization based on fuzzy logic is described.

1.4.2   Traffic Control in Routers

Traffic control in routers are required in order to perform functions that are becoming more and more important as the Internet evolves. However, these functions have seen little implementation to date [63, 58], including protection against flows with no congestion control, misbehaving flows, large traffic bursts, denial-of-service attacks, incentives to flows performing congestion control, and service differentiation.

Router flow and congestion control schemes are based on queue management techniques. These techniques can be classified into two groups: active queue management (AQM) and class based queuing (CBQ).

CBQ systems can in principle be applied in a more general manner, both for congestion and admission control. However, CBQ schemes suffer from scalability issues. This is essentially because CBQ schemes are based on per-flow classification of packets. Thus, they are only generally applicable in edge or access routers [57].

The most simple queue management scheme is the First Come, First Served (FCFS) queue. It is the most deployed scheme in the current Internet where it is usually implemented by fast FIFO queues. This scheme is known in general as tail-drop. This scheme has two major drawbacks:

- As the network approaches an overload condition, queues are quickly filled and signals of congestion, i.e. packet loss, are only evident to end-nodes when queues are already full and packets are being dropped. Due to the bursty nature of network traffic this phenomena can be specially frequent, recurrent and severe.
- Synchronization among end-to-end data flows with different sources and destinations occur in certain network scenarios [19, 60, 62]. In these cases, bandwidth is shared unevenly in tail-drop queues.

The second problem can be addressed by alternative procedures for discarding packets, such as random selection. However, the first problem poses some conflicts among different parameters of traffic dynamics, including link utilization and round-trip delay due to queuing. Thus, active queue management techniques are sought that can help overcome the two aforementioned drawbacks [19].

A number of approaches to the problem of active queue management have been proposed. Some proposed algorithms [61, 84, 165, 9, 103] have been shown to provide significant performance improvements in terms of utilization and end-to-end delay variability. However, instability and oscillations can occur in some cases depending on configuration parameters. In addition, these algorithms suffer from performance degradation for some regions of the wide space of operating conditions of an AQM scheme [107].

In particular, Random Early Detection (RED) [61, 59] was the first firm proposed algorithm for AQM in the global Internet. It is also the most accepted algorithm and the common choice of router vendors. Deployment of RED in the real world is still very limited though. RED establishes a preventive strategy against congestion conditions, dropping packets before buffers are full so that the end-nodes respond to the packet loss events before queue are overloaded and wider congestion starts to occur. This way, the end-to-end delay is reduced as well, and less packets should be dropped because of buffer overload.

The RED algorithm has some issues though. [113, 53, 54]. These issues can translate into network instability and resource and performance degradation. Moreover, it has been shown that proper adjustment of the parameters of RED for a wide range of applications is a complex problem because of the dependence of the RED threshold value on the number of connections traversing a RED router. Because of this, several variants of RED have been designed [125, 166, 35, 82, 66], being Adaptive RED [66] the most popular among them.

With the aim of designing AQM algorithms that can be adapted to a wider range of network conditions, nonlinear and adaptive systems design techniques have been applied [108, 83, 56]. Alternative schemes have been proposed following a control theoretic approach as a way of overcoming the limitations of the heuristic approach taken by RED. Among them we cite algorithms based on PI controlers [84], new techniques such as random marking with exponential distribution [9], and adaptive virtual queues [103]. Some schemes based on fuzzy logic have been proposed as well [42, 55, 177]. This alternative will be addressed in chapter 5. However, little evidence, whether through simulation or experimentation, has shown the practical applicability of these alternatives in real networks.

1.4.2.1 QoS Provisioning Architectures

Traffic control and management in packet switched networks, i.e., the efficient distribution of bandwidth and other network resources in order to provide quality of service to end-to-end flows, poses many challenges due to the lack of a statistical characterization of these environments. In particular, it is very difficult to model and predict the behavior of traffic flows and flow aggregates [150]. Only incomplete information about traffic is usually available in routing equipment. Thus, static bandwidth distribution schemes are inefficient.

Two levels of quality of service can be distinguished in general. The first consists in providing better performance to elastic applications. This type of quality of service finds applications in best-effort networks. The second level requires providing well-specified performance bounds (or specialized management) to in-elastic applications as opposed to elastic ones. Besides, the number of classes of quality of service can be unbound in principle.

In most cases, elastic applications can perform well in best-effort networks as far as significant congestion does not occur. On the contrary, in-elastic applications may be often totally unusable when some quality of service parameters are not bound, such as delay too high or bandwidth too low. In this context, QoS guarantees can be provided and enforced within different administrative realms: end-to-end, edge-to-edge, edge-to-middle, middle-to-middle, edge-to-campus, etc.

IP networks have evolved throughout the years from a model that provided only *best-effort* services, i.e., the best possible service is provided yet with no guarantees, to a model where multiple types of services with different characteristics and QoS requirements are provided. Although a large majority of traffic in the current Internet is still best-effort, as new services are deployed users start to push for QoS guarantees and differentiated services.

More than fifteen years after the first standards were developed, quality of service (QoS) provisioning in Internet is still a highly debated topic. Among the proposed architectures, IntServ [17] was designed for providing individualized quality of service to application sessions. This architecture is based on the reservation of bandwidth on the end-to-end path by means of the RSVP signaling protocol [18, 11, 17].

Due to scalability and heterogeneity issues, IntServ has had little practical acceptance and has not been deployed [99]. In particular, it is not feasible for current routers to process the large number of per-flow states the IntServ architecture requires. In addition, the use of IntServ in an end-to-end path requires all the nodes involved to implemented the same reservation protocol, RSVP. As a result, currently there is little deployment of QoS technologies at the network layer.

Currently, it is accepted that it is not feasible to implement any individualized quality of service scheme in general in Internet routers. Consequently, the analysis of flow aggregates has become more important during the last years, and the technologies for QoS provisioning that have the highest possibilities of eventual large-scale deployment, DiffServ and MPLS, are designed to support flow aggregates.

As an alternative to IntServ, the IETF has proposed the architecture for differentiated QoS provisioning DiffServ [14, 89]. The approach of DiffServ emphasizes progressive deployment as an evolution of the current Internet and thus does not require significant structural changes. For better scalability, the DiffServ architecture defines specialized services at a more general scale than IntServ. In DiffServ, specialization can be performed on a per node or per flow aggregate basis.

The development of DiffServ was initially motivated by the requirements of voice and video applications. The QoS differentiation possible with DiffServ is relative or qualitative, i.e., of the type high bandwidth, low delay, low packet loss rate, etc. This is due to the nature of the reservation mechanisms of DiffServ where resources are allocated by mechanisms such as reservation of more bandwidth and lower packet dropping probability for preferential flow aggregates. That is, DiffServ does not allow for provisioning quantitative QoS guarantees [150].

The DiffServ working group of the IETF has defined two classes of service in addition to the best-effort class. The three classes are defined for every hop in a network, that is, in every router. The overall characteristics of these classes can be summarized as follows:

Expedited forwarding   (EF per-hop behavior [40]). The definition of EF includes guarantees of low packet loss rate, low latency and latency variation, as well as an end-to-end communication service with guaranteed bandwidth. The arrival rate of packets belonging to EF flows has to be lower or equal to the retransmission rate of such packets in every router. Therefore, implementing this class of service requires all the routers to reserve the proper resources in advance. This class of service has operating and structural implications and in practice requires the definition of service level agreements (SLA) between providers. Also, routers have to implement some traffic control mechanism for EF flows in order to provide the end-to-end bandwidth guarantee. The concrete implementation of this mechanisms is not addressed in the standard definition of the EF class. However, it is specified that *leaky-bucket* queues, see [30], have to be used.

Assured forwarding   (AF per-hop behavior [79]). The AF group provides 4 independent classes of IP packet delivery. Within each class, 2 o 3 levels of preference
are used regarding the packet loss rate. The objective of this group is to differentiate (with preferential drops in case of congestion) packets of best-effort type
and packets that exceed the subscribed information rate. As an example, AF can
be used to implement the so-called Olympic service. Three classes of service are
defined within this service: gold, silver and bronze. Packets in the gold class experience lighter load than silver packets. The same relationship applies to silver
and bronze packets.

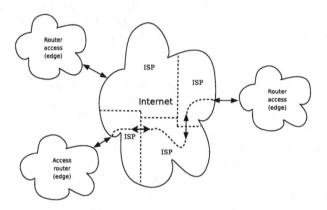

Fig. 1.3 Scheme of the DiffServ architecture

An scheme of a DiffServ network is shown in figure 1.3. Priorities or groups are
set by the end nodes, which is scalable and reduces the deployment complexity .
However, no concrete measures have been defined for keeping guarantees of service once packets leave edge routers towards inner routers. In conclusion, DiffServ
aims at providing QoS guarantees by means of congestion control algorithms that
consider service differentiation. RED is the most firm proposal to this end [14, 33].
The combined use of the mechanisms designed in the DiffServ architecture and the
network management systems based on MPLS [51] gives rise to an scalable model.
While MPLS allows for controling data paths, DiffServ allows for differentiating
QoS.

AQM With Differentiated Services

The RED AQM algorithm [61] is the most firm proposal both for best-effort and
DiffServ networks. It is thus the algorithm that has more probabilities of being
widely deployed. RED is based on the definition of two threshold values, minimum
and maximum, on a router packet queue. Within these two values packet are discarded to a varying probability, p. When the network is overloaded, i.e., the queue

size, q increases above the minimum threshold, some packets are discarded with a probability that increases with the queue size until the maximum probability is reached for the maximum threshold. This scheme is shown in figure 1.4(a).

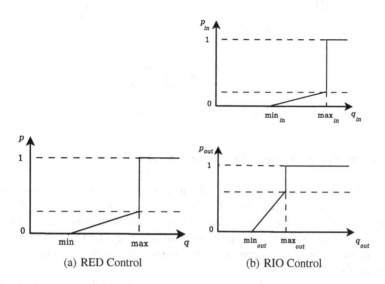

(a) RED Control   (b) RIO Control

Fig. 1.4 Packet drop rate versus queue size

The initial proposal for implementing RED within the DiffServ architecture adds the definition of different threshold values for each class of service. The two lowest threshold values are assigned to the best-effort class. Also, there is the possibility of reclassifying AF packets into best-effort packets when the subscribed rate is exceeded.

Figure 1.5 shows an scheme of the way a DiffServ router implements RED. First, a leaky-bucket type filter is applied on the incoming flows in order to enforce bandwidth constraints. Packets of class EF not compliant with these constraints are discarded, while non compliant AF packets are reclassified as best-effort packets. Packets belonging to the AF and best-effort classes are managed by the RIO (RED In/Out) scheme [39] which distinguishes between compliant and non compliant packets. Packets belonging to the EF class are managed by an independent high priority FIFO scheme.

The best-effort and EF classes are differentiated by means of different thresholds. RIO queues work in a similar manner to RED queues with the exception that two set of parameters are configured for compliant and non compliant packets. Parameters for non compliant packets are set in a particularly restrictive manner, as shown in figure 1.4(b).

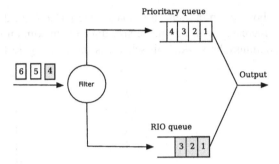

Fig. 1.5 Scheme of the RED algorithm as applied in DiffServ routers

1.5 Time Series Models for Network Traffic

It is well known that network traffic exhibits complex nonlinear behavior. Many classes of dynamical behavior have been described, including regular predictable and unpredictable behavior, transient and intermittent chaos, narrow-band and broadband chaos, pseudo-randomness and superposition of several basic patterns [96]. Most of them can be seen in network traffic load series (see chapter 3 for extensive experimental evidence).

In many cases network traffic shows patterns that suggest that regarding dynamical behavior traffic series can be properly classified within regular predictable phenomena. In this aspect, the analysis of network traffic can be addressed from the time series analysis viewpoint. In these cases, the theory of nonlinear dynamics provides a proper framework for the analysis, identification and prediction of network traffic time series.

Network traffic prediction finds applications in a variety of fields, including congestion and admission control, adaptive applications and network management. The essential idea in network traffic prediction is to predict traffic for the next control or action period based on on-line or off-line traffic measurements.

Services such as the network weather service (http://nws.cs.ucsb.edu) are lately becoming more important for adaptive applications. In particular, Grid Computing systems often rely on the availability of measurements and predictions of network conditions in order to optimize performance. This has motivated the development of grid oriented services for predicting TCP/IP end-to-end throughput and latency [110].

Predicting traffic load at low time scales (of the order of seconds and minutes) finds applications for dynamic resource allocation, whereas prediction at longer time scales (of the order of days and months) is of interest for higher level planning and dimensioning, as depicted in figure 1.6. Works within the first class are far more common than for the second class.

As in any field of application of time series prediction techniques, one of the major objectives is prediction accuracy. Hence, predictability of IP traffic, or the possibility to satisfy a certain prediction accuracy constraint, is a major issue that

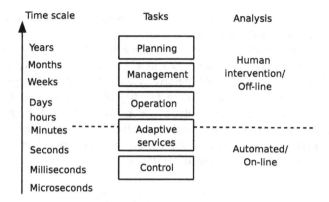

Fig. 1.6 Traffic load prediction at different time scales

is still essentially open. The first known attempt at long-term prediction of traffic in an IP network was performed in the NSFNET backbone in the mid-1990s [74]. However, the expression long-term should not be misunderstood as it is used to refer to the yearly scale and not for multiple steps ahead prediction. In the aforementioned work a single value for the next year is predicted.

Papagiannaki et al. in [126, 127] use ARIMA models in order to predict traffic in a Tier-1 backbone and achieve satisfactory results for up to 6 months ahead prediction on a 12 hours time scale (i.e. approximately 50 values are predicted within reasonable bounds). They apply the Box-Jenkins technique to two separated components of the traffic series, the long term trend and the fluctuations at the 12 hours scale, which are identified as contributing to 98% of the energy and 90% of the variance by means of the wavelet multiresolution method for multiscale analysis. Also, they are interested mostly in the overall long-term trend and target their analysis at a particular time scale (for instance, the standard deviation is computed as a weekly average of the daily standard deviations, and then the signal is approximated by the long term trend ± 3 times the standard deviation (also, the long-term trend is simplified with a weekly average)). In addition, no automated procedure is defined to select the combination of trend and detailed signals, and the methodology as applied in the paper requires human intervention. Thus, this method does not seem applicable for lower time scales.

Krithikaivasan et al. [102] use ARCH models in order to predict network traffic for one application case: an Internet link that connects the University of Missouri-Kansas City to MOREnet. Their focus is on nonstationarity.

Some authors have attempted at developing time series models at smaller time scales for limited data sets [12] as well as nonlinear time-series models for Ethernet traffic [27]. You and Chandra have analyzed stationarity issues and proposed threshold auto-regressive models for campus area traffic [179].

Some additional works have been published on the subject. These will be cited in chapter 3. However, except for the latter two papers [27, 179], no trace from these works is publicly available at present. Thus, these results are not only particular to specific applications but also difficult if not impossible to reproduce and compare.

All the aforementioned models, some of which will be described in what follows, are of the black-box kind. In chapter 3 we will address the prediction of traffic load in network links by means of autoregressive systems that can be interpreted from a linguistic viewpoint. Let us define a few building blocks to this end. Consider a discrete time series as a sequence of values, $X_t = X_0, X_1, \ldots, X_{n-1}$, that represents an ordered set of values, where t is the number of values in the series. The problem of predicting one future value, X_n, using a general autoregressive model (autoregressor) with no exogenous inputs can be stated as follows:

$$\hat{X}_n = f_r(X_{n-1}, X_{n-2}, \ldots, X_{n-M}),$$

where $\hat{X}_t$ is the prediction of model f_r for the prediction horizon 1 and M is the number of inputs to the regressor, i.e., the regressor size.

Very simple time-series models can be used as starting points. Among these, let us consider the following three:

- Mean (long-term mean of the series).
- Last value (or naive), where the last observed value is taken as prediction.
- BM(N): average over a history window of optimal size N.

The results that can be expected from these techniques are clearly not useful for dealing with traffic load. The next natural step is to use traditional autoregressive models. That is, Box-Jenkins and derived models, both for short-memory and long-memory stochastic models, including AR, MA, ARMA and ARIMA-like models.

In what follows, we overview a set of models that have been applied to network traffic as well as models that will be applied in the next chapter of this monograph. The former include short-memory and long-memory stochastic models and mean square error predictors. The latter include a set of computational intelligence based models. We do not aim at performing a complete overview of stochastic or computational intelligence predictive models. Instead, our focus is twofold. First, we discuss models that have been proposed for predicting traffic load. Second, we briefly discuss some computational intelligence methods that will be used in chapters 2 and 3. For a more detailed discussion of stochastic models we refer the interested reader to some of the many books that provide a more general treatment of this field [28, 16, 50].

1.5.1 Short-Memory Stochastic Models

A wide set of well established stochastic time series models, such as Markov and regression models [16], can only capture short-range dependencies. We refer to these models as short-memory models.

1.5.1.1 Markov Models

In a Markov model, the behavior of a system is modeled by defining a finite number of states. In general, increasing the number of states leads to more accurate models at the expense of higher computational complexity. The Markovian property, i.e., the next state of the system depends only on the current state is the common characteristic of these models. Markov models often have a complicated structure and many parameters to adjust when used to model a long-range dependent or a mixed process [6].

1.5.1.2 Regression Models

Regression models compute the next random variable in the sequence of a time series from previous ones within a specified time window and a moving average of white noise. These models are based on the lag operator, B, and the difference operator, Δ. B is defined as $BX_t = X_{t-1}$, where $B^sX_t = X_{t-s}$. Δ is defined as $\Delta X_t = X_t - X_{t-1}$, and analogously $\Delta^d = (1 - B)^d$, which can be expressed as follows using the binomial expansion:

$$(1 - B)^d = \sum b_{k=0}^{\infty} \binom{d}{k}(-1)^k B^k,$$

where

$$\binom{d}{k} = \frac{d!}{k!(d-k)!} = \frac{\Gamma(d+1)}{\Gamma(k+1)\Gamma(d-k+1)}.$$

Two additional polynomials, ϕ and θ are defined as well:

$$\phi(B) = (1 - \phi_1 B - \ldots - \phi_p B^p)$$

$$\theta(B) = (1 - \theta_1 B - \ldots - \theta_p B^q).$$

1.5.1.3 Autoregressive (AR) Models

An autoregressive model of order p, $AR(p)$, has the following general form:

$$\phi(B)X_t = \varepsilon_t,$$

where ε_t is an error component assumed to be white noise (independent identically distributed random variables with zero mean and variance σ^2). The variable X_t is regressed on its previous values:

$$X_t = \phi_1 X_{t-1} + \ldots + \phi_p X_{t-p} + \varepsilon_t.$$

AR models can model stationary time series. If all the root of $\phi(B)$ lie outside the unit circle, then it is invertible, i.e., can be rewritten in the form $X_t = \phi^{-1}(B)\varepsilon_t$. The autocorrelation of $AR(p)$ is expressed as follows:

$$\rho_k = A_1 G_1^k + \ldots + A_p G_p^k,$$

where $\frac{1}{G_i}, i = 1, \ldots, p$ are the roots of $\phi(B)$.

1.5.1.4 Autoregressive Moving Average (ARMA) Models

An ARMA model, $ARMA(p,q)$, has the following general form:

$$\phi(B)X_t = \theta(B)\varepsilon_t,$$

where $\theta(B)\varepsilon_t$ is the moving average component of the model. In an equivalent form, ARMA models can be expressed as follows:

$$X_t = \phi_t X_{t-1} + \ldots + \phi_p X_{t-p} + \varepsilon_t - \ldots - \theta_q \varepsilon t - q,$$

These models exhibit a high modeling flexibility but are also restricted to stationary time series. In practice, proper models for time series can be built with p and q equal to or even lower than 2.

1.5.1.5 Autoregressive Integrated Moving Average (ARIMA) Models

ARIMA models, $ARIMA(p,d,q)$, are defined as an extension to ARMA models where the polynomial $\phi(B)$ is allowed to have d roots equal to 1. The other roots lie outside the unit circle. An ARIMA model has the following general form:

$$\phi(B)\Delta^d X_t = \theta(B)\varepsilon_t.$$

ARIMA models can predict non-stationary processes. Note that

$$\Delta^d X_t = (1-B)^d X_t = \phi^{-1}(B)\theta(B)\varepsilon_t,$$

and,

$$X_t = (1+B+B^2+\ldots)^d \phi^{-1}(B)\theta(B)\varepsilon_t.$$

Thus, X_t is regressed to a sum (or integration) of infinite noise variables. By including the difference operator Δ, ARIMA models address those cases where the original series X_t is non-stationary but the increments $X_t - X_{t-d} = (1-B)^d X_t$ are stationary.

1.5.1.6 Nonparametric Autoregressive Models

Since nonlinear time series can have any possible form in general, a natural approach to modeling them is to adopt a nonparametric model form. In general, it can be assumed that:

$$X_t = f(X_{t-1}, \ldots, X_{t-p}) + \sigma(X_t - 1, \ldots, Xt - p)\varepsilon_t,$$

where f and σ are unknown functions and $\{\varepsilon_t\} \sim \text{IID}(0,1)$. Thus, no particular form is imposed on f and σ. Instead, some qualitative assumptions are usually made, such as that f and σ are smooth.

These models are referred to as nonparametric autoregressive conditional heteroscedastic (NARCH) or nonparametric autoregressive (NAR), if σ is a constant. NARCH and NAR models are manifestly very general, as they make very few assumptions on the process that generates the data. Such models are normally useful for a small value of p. For moderately large values of p, the functions in such a saturated form are difficult to estimate. This intrinsic limitation is commonly referred to as the curse of dimensionality in the nonparametric regression literature and other fields. A number of simplified models between parametric models and fully general nonparametric models have been proposed that can substantially ease the curse of dimensionality issue at the expense of restricting the model form [50].

A particular class of NAR models includes computational intelligence methods. In the next chapters, some computational intelligence methods will be used to develop nonparametric autoregressive models for network traffic load time series. Ordered Weighted Averaging-Induced nearest neighbor models, Least Squares Support Vector Machines and Optimally-Pruned Extreme Learning Machines will be briefly described later on in this section. In addition, a methodology to design nonlinear regressive models by means of fuzzy inference systems is described in chapter 2. These methods will be applied to a wide set of traffic load time series.

1.5.2 Long-Memory Stochastic Models

Long-memory models are more recent models which are capable of capturing long-range dependencies. We overview some models that have been widely used in both theory and practice.

1.5.2.1 Fractional Brownian Motion (fBm)

Brownian motion [44, 13, 73] is a stochastic process, $\text{Bm}_t, t \geq 0$, characterized by the property that increments $\text{Bm}_{t_0+t} - \text{Bm}_{t_0}$ are normally distributed with zero mean and variance $\sigma^2 t$. The fractional Brownian motion fBm_t is a self-similar process with Hurst parameter $1/2 \leq H \geq 1$, and with variance $\sigma^2 t^{2H}$. Fractional Brownian motion is non-stationary.

1.5.2.2 Fractional Gaussian Noise

Fractional Gaussian noise, fGn_t, is the increment process of a fractional Brownian motion for a finite increment τ:

$$fGn_t = fBm_{t\tau} - fBm_{(t-1)\tau},$$

As opposite to fractional Brownian motion, fractional Gaussian noise is stationary. The autocorrelation function of fGn_t has the following form:

$$\rho_k = 1/2[(k+1)^{2H} - 2k^{2H} + (k+1)^{2H}],$$

which, as $k \to \infty$, turns into:

$$\rho_k = H(2H-1)k^{2H-2}.$$

Fractional Gaussian noise processes can be used to generate synthetic self-similar traffic patterns [130, 157]. There is a direct predictor for fGn [73], although it is fairly complicated and computationally intensive. In addition, fGn processes are pure long-memory. Thus, fGn based predictors usuarlly require large regressor sizes (of the order of hundreds and thousands of values).

1.5.2.3 Fractional ARIMA (FARIMA)

Fractional ARIMA models, FARIMA, proposed in 1980 [85], are the natural extension to the ARIMA models when the parameter d of the difference operator can have a real value. X_t is a stationary invertible FARIMA(p,d,q) process if:

$$\phi(B)\Delta^d X_t = \theta(B)\varepsilon_t,$$

where d is a real number $(-1/2 < d < 1/2)$, and where $\phi(B)$ and $\theta(B)$ are stationary AR and invertible MA polynomials, respectively. The equation $H = d + \frac{1}{2}$ holds between d and H. Thus, X_t is a long-memory process if $(0 < d < 1/2)$ and a short-memory process if $d = 0$. These models have been used for network traffic modeling [148]. FARIMA$(0,d,0)$ is the basic form of these models:

$$\Delta^d X_t = \varepsilon_t.$$

For which,

$$\rho_k = \frac{(-d)(k+d+1)!}{(d-1)(k-d)!},$$

and, as $k \to \infty$,

$$\rho_k = \frac{(-d)!}{(d-1)!} k^{2d-1}.$$

1.5.2.4 Generalized ARMA (GARMA)

GARMA models are the generalization of all the regression models and can be applied to model both short-range and long-range dependencies. These models can also model cyclical patterns with fewer parameters than ARMA models. A GARMA(p,q) model of a process X_t is defined as follows:

$$\phi(B)(1 - 2\eta B + B^2)^d X_t = \theta(B)\varepsilon_t,$$

where $(-1/2 < d < 1/2)$ and $(1 < \eta < 1)$, and the term $(1 - 1\eta B + B^2)^d$ is the Gegenbauer polynomial, which can be expanded using the power series expansion.

1.5.2.5 Fractional Predictors

Let X_t be an invertible FARIMA(p,d,q) process:

$$\phi(B)\Delta^d X_t = \theta(B)\varepsilon_t.$$

Considering invertibility, it can be written:

$$\varepsilon_t = \sum_{j=0}^{\infty} \pi_j X_{t-j},$$

where

$$\sum_{j=0}^{\infty} \pi_j X_{t-j} = \phi(B)\theta^{-1}(B)(1-B)^d.$$

From the theorems of linear prediction, a one step ahead predictor of a FARIMA process can be defined as:

$$\hat{X}_t = -\sum_{j=1}^{\infty} \phi_j X_{t-j+1}.$$

GARMA models are very similar to FARIMA models. Thus, the method can be extended to a GARMA predictor. The extension lies in the computation of the π_j coefficients which are given as follows:

$$\sum_{j=0}^{\infty} \pi_j B^j = \phi(B)\theta^{-1}(B)(1 - 2\eta B + B^2)^d.$$

1.5.3 Mean Square Error Predictors

Let $\{X_t\}$ be a linear stochastic process and let us assume that the next value of $\{X_t\}$ can be expressed as a linear combination of previous observations:

$$X_{t+1} = w_n X_t + \ldots + w_1 X_{t-m+1} + \varepsilon_t,$$

where m is the order of the regression. The same predictor can be expressed in matrix form:

$$X_{t+1} = \mathbf{W}\mathbf{X}' + \varepsilon_t.$$

It can be seen that this is the case for all the regression models above, and particularly for the FARIMA and GARMA models. In many practical applications, such as network measurement and control, on-line prediction is needed and no prior knowledge about the underlying dynamics of the series is available. However, it is possible to estimate the weights w_i as follows. Let $\hat{\mathbf{W}}$ be the estimated weight vector, then:

$$\hat{X}_{t+1} = \hat{\mathbf{W}}\mathbf{X}' + \varepsilon_t,$$

where $\hat{X}_{t+1}$ is the predicted value of X_{t+1}. In what follows, we detail two solutions for this estimation problem. The first solution is based on minimum mean square error. The second approach is based on recursive linear regression. The former requires a matrix inversion and autocorrelation computations whereas the latter avoids these computationally intensive operations at the expense of lower accuracy.

1.5.3.1 Minimum Mean Square Error Predictors

One simple solution to the estimation problem above is the minimum square error (MMSE) method. In MMSE, the optimal weight vector is computed by minimizing the expected value of squared errors:

$$\varepsilon_t = X_{t+1} - \hat{X}_{t+1},$$

and the corresponding expected value is:

$$E|\varepsilon_t^2| = E|(X_{t+1} - \hat{X}_{t+1})^2|$$

That is, a minimization problems has to be solved. By using the derivative equation of the expression above, the following solution can be found:

$$\hat{\mathbf{W}} = \mathbf{\Gamma}\mathbf{G}^{-1},$$

where $\mathbf{G}$ is the autocorrelation matrix and $\mathbf{\Gamma}$ is an autocorrelation vector starting at lag m defined as follows:

$$\mathbf{G} = \begin{bmatrix} \rho_0 & \rho_1 & \cdots & \rho_{m-1} \\ \rho_1 & \rho_0 & \cdots & \rho_{m-2} \\ \vdots & \vdots & \ddots & \vdots \\ \rho_{m-1} & \rho_{m-2} & \cdots & \rho_0 \end{bmatrix}$$

and

$$\mathbf{\Gamma} = [\rho_m \ldots \rho_1].$$

The autocorrelation coefficients ρ_k can be computed by the following equation:

$$\rho_k = \frac{1}{m} \sum_{t=k+1}^{m} X_t X_{t-k},$$

where m is the order of the MMSE predictor.

MMSE predictors have the advantage that there is no need to know the underlying structure of the time series and thus can be used for on-line prediction. In addition, these models do not assume stationarity. They are also very simple to implement as they only require a few matrix manipulations which can be efficiently implemented as software and hardware. In addition, some approximation techniques have been proposed for computing the weight vector $\hat{\mathbf{W}}$ that eliminate the need for matrix inversion and autocorrelation computations [3]. However, for some traffic series it has been shown that the performance of MMSE for one step ahead prediction is not significantly better than that of a naive model [70].

1.5.3.2   Normalized Minimum Mean Square Error Predictors

The normalized MMSE method [77] uses an adaptive and recursive approach to compute the weight vectors for the MMSE method. It is sometimes referred to as normalized recursive linear regression [3]. It does not require prior knowledge of the correlation structure of the time series. Thus, it can be applied as an on-line prediction algorithm. The recursive linear estimator for the weight vector is defined as follows:

$$\hat{\mathbf{W}}_{t+1} = \hat{\mathbf{W}}_t + \mu \frac{\hat{\mathbf{X}}}{||\hat{\mathbf{X}}||^2} \varepsilon_t,$$

where μ is the adaptation constant and determines the convergence speed. NMMSE is convergent with respect to the mean square error if the adaptation constant satisfies the following relations: $0 < \mu < 2$.

The NNMSE method eliminates the need for computing a matrix inversion and autocorrelations as compared to MMSE. Thus, it is a very fast option. It has been claimed that NMMSE can attain a satisfactory accuracy for VBR video traffic prediction [1]. However, other authors have found this technique to provide very limited predictive capabilities for different traces [70]. In addition, it has been shown that NMMSE performs poorly for some traffic series, yielding worse results than a naive predictor for one step ahead prediction [70].

1.5.4 OWA-Induced Nearest Neighbor Models

Nearest neighbor pattern classification schemes have been applied in a variety of fields for several decades [34]. These schemes can provide good results in time series prediction applications [151] as well. A specific variant of the nearest neighbor scheme is the fuzzy k-nearest neighbor algorithm [100].

An Ordered Weighted Averaging (OWA) operator [173] of order m is a mapping $\mathscr{F}_W : \mathbb{R}^m \to \mathbb{R}$ characterized by an m-dimensional vector W, the weighting vector, whose elements, w_j, lie in the unit interval and sum to one. The mapping is defined as follows:

$$\mathscr{F}_W(a_1,\ldots,a_m) = \sum_{j=1}^m w_j b_j,$$

with b_j being the jth largest element within the a_i. If we call B the m dimensional vector whose jth element is b_j, then the mapping can be expressed as $\mathscr{F}_W(a_1,\ldots,a_n) = W^T B$. B is called the ordered augmented vector.

Induced OWA operators [175, 176] are a more general type of OWA operator that take as their argument pairs, called OWA pairs. One component of the pair is used to induce an ordering over the second components which are then aggregated.

IOWA operators are used to aggregate tuples of the form (v_i, a_i). In these pairs, v_i is called the order inducing value and a_i is called the argument value. The following procedure for performing IOWA aggregations has been proposed [176]:

$$\mathscr{F}_W(\langle v_1, a_1 \rangle, \ldots, \langle v_m, a_m \rangle) = W^T B_v,$$

where W is an OWA weighting vector of dimension m as before. However, the ordered augmented vector, B_v is such that its jth element is the argument value of the pair having the jth largest value for the order inducing variable, v_i. Thus, if we let $v - index$ be an index function such that $v - index(j)$ is the index of the argument pair with the jth largest order inducing value, then the mapping F_W for IOWA operators can be expressed as follows:

$$\mathscr{F}_W(\langle v_1, a_1 \rangle, \langle v_2, a_2 \rangle, \ldots, \langle v_m, a_m \rangle) = \sum_{j=1}^n w_j a_{v-index(j)}.$$

Among other methods, fuzzy methods based on IOWA operators have been proposed to model nearest neighbor rules [174].

1.5.5 Least Squares Support Vector Machines

Support Vector Machines (SVM) [144] are a type of learning algorithm developed in the 1990s based on results from statistical learning theory. The development of SVMs led to the definition of a new class of learning machines that use a central

concept of SVMs: kernels. Kernel machines or kernel based methods are by definition a modular framework that can be adapted to different tasks and domains by
choosing the kernel function and the base algorithm.

LS-SVM [159] are defined as a least square modification to SVMs. SVMs have
been shown to be a powerful method in classification and regression applications
due to its generalization capability and robustness against high dimensional problems. LS-SVM are regularized supervised approximators. Let us consider a set of
training samples given in the form of multiple input-single output pairs, $(\mathbf{x}_j, y_j)$,
with $j = 1, \dots, N$, where $\mathbf{x}_j \in \mathbb{R}^n$ are the inputs and $y_j \in \mathbb{R}$ are the corresponding
outputs. In short, an LS-SVM model is defined in its primal weight space as follows:

$$\hat{y} = \boldsymbol{\omega}^T \boldsymbol{\varphi}(\mathbf{x}) + b,$$

where $\hat{y} \in \mathbb{R}$ is the output of the model ($\hat{X}_{t+1}$ in the notation of previous sections),
$\mathbf{x} \in \mathbb{R}^n$ is an input vector (consisting of previous, known values of a time series for
the case of regression with no exogenous inputs), $\boldsymbol{\varphi} : \mathbb{R}^n \to \mathbb{R}^{n_\varphi}$ is a nonlinear feature map that transforms the original input space into a higher dimensional feature
space of dimension n_φ, $b \in \mathbb{R}$ is a bias term and $\boldsymbol{\omega} \in \mathbb{R}^{n_\varphi}$ is an unknown vector of
coefficients. The constrained optimization problem with a regularized cost function
for LS-SVM models is formulated as follows:

$$\min_{\boldsymbol{\omega}, b, e_j} = \frac{1}{2} \boldsymbol{\omega}^T \boldsymbol{\omega} + \gamma \frac{1}{2} \sum_{j=1}^{N} e_j^2,$$

where $y_j = \boldsymbol{\omega}^T \boldsymbol{\varphi}(\mathbf{x}_j) + b + e_j$, with e_j being the model errors (defined as $y_j - \hat{y}_j$),
and $\gamma > 0$ is a regularization parameter that balances the flatness-accuracy trade-
off for the regression function. By Mercer's theorem applied to the kernel matrix
$\boldsymbol{\omega}_{ij} = K(\mathbf{x}_i, \mathbf{x}_j) = \boldsymbol{\varphi}(\mathbf{x}_i)^T \boldsymbol{\varphi}(\mathbf{x}_j), i, j = 1, \dots, N$, it is not necessary to compute the
nonlinear mapping $\boldsymbol{\varphi}(\cdot)$ nor anything in the higher dimensional space, but this can
be done in an implicit manner using positive definite kernel functions K [144]. For
the kernel functions $K(\mathbf{x}_i, \mathbf{x}_j)$ there are three common choices:

- linear, $K(\mathbf{x}_i, \mathbf{x}_j) = \mathbf{x}_i^T \mathbf{x}_j$,
- polynomial of degree d, $K(\mathbf{x}_i, \mathbf{x}_j) = (\mathbf{x}_i^T \mathbf{x}_j + c)^d$, with $c \geq 0$ being a tuning
  parameter,
- and radial basis function (RBF) kernels, in particular Gaussian kernels, $K(\mathbf{x}_i, \mathbf{x}_j)$
  $= exp(-||\mathbf{x}_i - \mathbf{x}_j||_2^2 / \sigma^2)$, where σ is a tuning parameter.

The constrained optimization problem can be solved using Lagrangian multipliers.
This way, the final expression in dual form for estimating a regression function is as
follows:

$$\hat{f}(\mathbf{x}) = \sum_{j=1}^{N} \alpha_j K(\mathbf{x}_j, \mathbf{x}) + b,$$

where α_j are the Lagrangian multipliers. In practice, the training process involves
the selection of the regularization parameter and the kernel parameters. Given a set

of inputs, a proper choice of these parameters is key for obtaining a good regression model. To this end, it is common to use cross-validation approaches.

Whereas the training process for a SVM consists in a quadratic programming task that guarantees optimality, the optimization of an LS-SVM is simplified into a linear programming task. This implies that LS-SVM are significantly faster to optimize while training optimality is also guaranteed, i.e., local minima are avoided.

1.5.6   Extreme Learning Machine

The Extreme Learning Machine (ELM) [87] is a simple yet effective learning algorithm for training single-hidden-layer feed-forward artificial neural networks (SLFNs) with random hidden nodes. In ELM, the hidden neuron parameters are randomly assigned whereas the output weights are analytically determined. ELM is a unified framework of generalized SLFNs that has the universal approximation capability for a wide range of hidden node types. The training process for ELM can be several orders of magnitude faster than traditional learning algorithms for feed-forward neural network, while attaining similar or even better approximation capabilities.

In particular, in this monograph we will apply and assess the performance of Optimally Pruned Extreme Learning Machines (OP-ELM) models [115, 152] for traffic load. OP-ELM models are build in three stages and use Gaussian, sigmoid and linear kernels in general. First, an ELM is constructed, then, an exact ranking of the neurons in the hidden layer is obtained, and finally the decision on how many neurons are pruned is made based on an exact leave-one-out error estimation method. These stages are performed by means of fast methods and lead to extremely fast yet accurate models.

The accuracy of OP-ELM models has been shown to be comparable to or even better than that of other (much more computationally intensive) computational intelligence methods [115, 117], such as Least Squares Support Vector Machines [159] and Multilayer Perceptrons [76].

1.5.7   Prediction Performance Metrics

In regression problems, performance metrics are important not only because they allow for assessing and comparing different modeling methods but also because the goodness of fit metric is key for driving the tuning process. A number of metrics have been proposed in order to address different applications.

In the next chapter, a methodology for the long-term prediction of time series by means of fuzzy inference systems will be described. To this end, we use the Xfuzzy design environment for fuzzy inference systems [122]. The common mean square error (MSE) definition in the Xfuzzy environment, the MSE normalized with respect to the square of the range of the series:

$$\text{NMSE}_{range} = \frac{1}{N}\frac{1}{M}\sum_{i,j}\left(\frac{X_{ij}-\bar{X}_{ij}}{r_j}\right)^2, i=1,\ldots,M,$$

where N is the number of training data, M is the number of output variables of the system X_{ij} is the j-th output generated by the system for the i-th data, $\bar{X}_{ij}$ is the correct output given in the training data, and r_j is the range of the j-th output (used to normalize the deviations). By restricting the system to one output variable, with range r, we get the following simplified expression for the MSE normalized with respect to the square of the range of the series:

$$\text{NMSE}_{range} = \sum_{t=1}^{N} \left(\frac{X_i - \hat{X}_i}{r} \right)^2.$$

Alternatively, for single output systems, the normalized MSE with respect to the variance of the series is defined as follows:

$$\text{NMSE}_{var} = \frac{1}{var(y)} \sum_{t=1}^{N} (y_i - \hat{y}_i).$$

This variant will be used in chapter 3 for assessing regressive models for network traffic load.

An alternative metric for the goodness of a regressive model is the reverse of signal to noise ratio [1]:

$$\text{SNR}^{-1} = \frac{\sum \varepsilon^2}{\sum X^2}.$$

As before, the smaller the ratio, the better prediction accuracy.

An additional metric is the signal-to-error ratio proposed in [29], defined as follows:

$$\text{SRR} = 10 \cdot log_{10} \frac{E[X_t^2]}{E[\varepsilon_t^2]}.$$

The metrics described so far are based on squared errors. Regarding non-squared errors, the percentage error (PE) can be seen as a measure of the error magnitude, but also indicates the direction of error and identifies outliers. The absolute percentage error (APE) is computed without regard to the direction of errors. The mean absolute percentage error (MAPE) is an statistical measure that summarizes the distribution of the APE. APE and MAPE are defined as follows:

$$\text{APE}_t = \frac{|X_t - \hat{X}_t|}{X_t} \times 100.$$

$$\text{MAPE} = \sum_{t=1}^{n} \text{APE}_t.$$

It has been argued that the MAPE is not symmetric, i.e., it treats predictions errors above the actual values differently from errors below this value. In order to overcome this limitation, the the symmetric mean absolute percent error (SMAPE) has been proposed. The SMAPE is a measure of the symmetric absolute error in percent between the actual values X_t and the predictions $\hat{X}_t$ across all observations t of the test set of size n. The SMAPE is defined as follows:

$$\text{SMAPE} = \frac{1}{n}\sum_{t=1}^{n}\frac{|X_t - \hat{X}_t|}{(X_t + \hat{X}_t)}\cdot 100.$$

SMAPE is a mean percentage metric and thus accounts for different numbers of observations in the training and test subsamples as well as different scales among different time series. However, SMAPE is not really a symmetrical measure, specially when the errors have large absolute values [72].

It should be noted that all the metrics considered here assume no specific decision problem of the prediction task and hence assume errors have a symmetric cost.

In analysis of variance [94], the total sum of squares (SST), measures the variation of the observed x_i values about their expected value, and represents the total variation of the dependent variable X. Thus, SST is defined as follows:

$$SST = \sum_{i=1}^{M}(X_i - \bar{X})^2.$$

Can a significant part of this total variation be explained or attributed to a certain model? Analysis of Variance (ANOVA) provides information about levels of variability in a regression model and can be used as the basis for tests of significance. Using ANOVA, we test how representative a model is in terms of the proportion of variance it explains [94].

The ANOVA methodology is based on partitioning the SST into two sums: one due to the model relationship between inputs and outputs (referred to as the sum of squares due to regression, or SSR), and the residual variation that is not explained by the model (SSE). Then, SSR can be defined as the difference between SST and SSE,

$$SSR = SST - SSE,$$

or, equivalently, the sum of squares (SS) explained by the regression model is equal to the difference between the total SS and the unexplained SS. The basic regression line concept, where data are given a fit function plus a residual component, is reformulated this way. Thus, the distance $X_i - \bar{X}$ is divided into two components:

$$X_i - \bar{X} = (\hat{Y}_i - \bar{Y}) + (X_i - \hat{X}_i),$$

where the first term is the total variation in the signal, the second term is the variation in mean response of the model and the third term is the residual value. Or, in terms of sums of squares,

$$\sum_{i=1}^{M}(X_i - \bar{X})^2 = \sum_{i=1}^{M}(\hat{Y}_i - \bar{Y})^2 + \sum_{i=1}^{M}(X_i - \hat{X}_i)^2.$$

The ratio SSR/SST, called the *coefficient of determination*, R^2, tells the proportion of the variation of X that can be explained and determines the goodness of the model. That is, the square of the sample correlation is equal to the ratio of the regression model sum of squares to the total sum of squares: $R^2 = SSR/SST$. This formalizes the interpretation of R as explaining the fraction of variability in the data explained by the regression model. So if the R^2 term is for instance 0.78, it indicates that 78% of the variability in the response is explained by the model. The *coefficient of multiple correlation* is defined as the square root of the coefficient of determination.

Thus, within this framework, the mean square error of a predictive model normalized against the variance of the series, as defined above, is equivalent to the coefficient of determination and thus indicates the extent to which the variability of a series is explained by the model. We will apply a basic ANOVA technique as proposed in [127] in chapter 3 in order to assess a number of regressive models for network traffic load time series.

1.6 Conclusions

We have given a brief overview of a number of issues related to Internet Science, i.e., the study of laws and patterns in Internet structure. The following aspects have been briefly defined: measurement systems and infrastructures (both active and passive), performance metrics (with an emphasis on end-to-end metrics), models for network traffic, network modeling and simulation issues, relevant protocols and control mechanisms at the network and transport layers, quality of service and differentiated services.

Time series models for network traffic load were also reviewed. Both short-memory and long-memory models have been considered as well as parametric and nonparametric models. First, we focused on stochastic techniques that have been previously proposed for modeling Internet traffic. Then, we briefly described a set of computational intelligence techniques (including OWA-induced nearest neighbor models, LS-SVM and OP-ELM) that will be applied in the next chapters. Finally, we described some generic performance metrics for time series models.

References

[1] Adas, A.M.: Using Adaptive Linear Prediction to Support Real-Time VBR Video under RCBR Network Service Model. IEEE/ACM Trans. Networking 6(5), 635–644 (1998)

[2] Alberi, J.L., McIntosh, A., Pucci, M., Raleigh, T.: Overcoming Precision Limitations in Adaptive Bandwidth Measurements. In: 3rd New York Metro Area Networking Workshop (NYMAN 2003), New York (2003)

[3] Albert, A.E., Gardner, L.A.: Stochastic Approximation and Nonlinear Regression. MIT Press, Cambridge (1967)

[4] Albert, R., Jeong, H., Barabási, A.L.: Diameter of the world wide web. Nature 401, 130–131 (1999)

[5] Allman, M.: Measuring End-to-End Bulk Transfer Capacity. In: ACM SIGCOMM Internet Measurement Workshop, San Francisco, CA, USA, pp. 139–143 (2001)

[6] Andersen, A.T., Nielsen, B.F.: A Markovian Approach for Modeling Packet Traffic with Long-Range Dependence. IEEE Journal on Selected Areas in Communications 16(5), 719–732 (1998)

[7] Andrew, L., Floyd, S.: Common TCP Evaluation Suite. Tech. rep., Internet Engineering Task Force, Network Working Group, Internet Draft. Intended status: Best Current Practice (2008)

[8] Andrew, L., Marcondes, C., Floyd, S., Dunn, L., Guillier, R., Gang, W., Eggert, L., Ha, S., Rhee, I.: Towards a Common TCP Evauation Suite. In: 6th International Workshop on Protocols for Fast Long-Distance Networks, Manchester, UK (2008)

[9] Athuraliya, S., Li, V., Low, S., Yin, K.: REM: Active queue management. IEEE Network 15, 48–53 (2001)

[10] Atkinson, R., Floyd, S.: IAB Concerns and Recommendations Regarding Internet Research and Evolution. RFC 3869, Internet Engineering Task Force, Network Working Group, category: Informational (2004)

[11] Baker, F., Weinrib, A., Braden, B., Zhang, L., Bradner, S., Romanow, A., O'Dell, M., Mankin, A.: Resource ReSerVation Protocol (RSVP). Version 1 Applicability Statement. Some Guidelines on Deployment. RFC 2208, Internet Engineering Task Force, Network Working Group, category: Informational (1997)

[12] Basu, S., Mukherjee, A.: Time Series Models for Internet Traffic. In: Fifteenth Annual Joint Conference of the IEEE Computer and Communications Societies (Proceedings IEEE INFOCOM 1996). Networking the Next Generation, San Francisco, CA, USA, vol. 2, pp. 611–620 (1996)

[13] Biagini, F., Hu, Y., Øksendal, B., Zhang, T.: Stochastic Calculus for Fractional Brownian Motion and Applications. Springer, London (2008), ISBN: 978-1-85233-996-8

[14] Blake, S., Black, D.L., Carlson, M.A., Davies, E., Wang, Z., Weiss, W.: An Architecture for Differentiated Services. RFC 2475, Internet Engineering Task Force, Network Working Group, category: Informational (1998)

[15] Le Boudec, J.-Y., Thiran, P.: Network Calculus: A Theory of Deterministic Queuing Systems for the Internet. In: Thiran, P., Le Boudec, J.-Y. (eds.) Network Calculus. LNCS, vol. 2050, p. 3. Springer, Heidelberg (2001)

[16] Box, G., Jenkins, G.M., Reinsel, G.: Time Series Analysis: Forecasting & Control, 3rd edn. Prentice Hall, Englewood Cliffs (1994) ISBN: 0130607746

[17] Braden, B., Clark, D., Shenker, S.: Integrated Services in the Internet Architecture: an Overview. RFC 1633, Internet Engineering Task Force, Network Working Group, category: Informational (1994)

[18] Braden, B., Zhang, L., Berson, S., Herzog, S., Jamin, S.: Resource ReSerVation Protocol (RSVP). Version 1 Functional Specification. RFC 2206, Internet Engineering Task Force, Network Working Group, status: Proposed Standard (1997)

[19] Braden, B., Clark, D.D., Crowcroft, J., Davie, B., Deering, S., Estrin, D., Floyd, S., Jacobson, V., Minshall, G., Partridge, C., Peterson, L., Ramakrishnan, K.K., Shenker, S., Wroclawski, J., Zhang, L.: Recommendations on Queue Management and Congestion Avoidance in the Internet. RFC 2309, Internet Engineering Task Force, Network Working Group, category: Informational (1998)

[20] Brakmo, L., Peterson, L.: TCP Vegas: End to End Congestion Avoidance on a Global Internet. IEEE Journal on Selected Areas in Communications 13(8), 1465–1480 (1995)

[21] Broido, A., Hyun, Y., Gao, R., Claffy, K.C.: Their Share: Diversity and Disparity in IP Traffic. In: 5th Passive and Active Measurement Workshop (PAM), Antibes Juan-Les-Pins, France, pp. 113–125 (2004)

[22] Calyam, P., Krymskiy, D., Sridharan, M., Schopis, P.: Active and Passive Measurements on Campus, Regional and National Network Backbone Paths. In: 14th IEEE International Conference on Computer Communications and Networks (ICCCN 2005), San Diego, California USA, pp. 537–542 (2005)

[23] Cao, J., Cleveland, W., Lin, D., Sun, D.: Internet Traffic Tends Toward Poisson and Independent as the Load Increases. In: Nonlinear Estimation and Classification. Springer, New York (2002)

[24] Cappe, O., Moulines, E., Pesquet, J.C., Petropulu, A.P., Yang, X.: Long-range dependence and heavy-tail modeling for teletraffic data. IEEE Signal Processing Magazine 19(3), 14–27 (2002)

[25] Carson, M., Santay, D.: NIST Net Network Emulation Package. NIST Internetworking Technology Group. National Institute of Standards and Technology (2008), http://www-x.antd.nist.gov/nistnet/

[26] Casetti, C., Meo, M.: A New Approach to Model the Stationary Behavior of TCP Connections. In: Annual Joint Conference of the IEEE Computer and Communications Societies (IEEE INFOCOM), Tel Aviv, Israel, pp. 367–375 (2000)

[27] Chandra, K., You, C., Olowoyeye, G., Thompson, C.: Non-Linear Time-Series Models of Ethernet Traffic. Tech. rep., Center for Advanced Computation and Telecommunication, University of Massachusetts Lowell (1998)

[28] Chatfield, C.: The Analysis of Time Series. An Introduction, 6th edn., pp. 1–58488. CRC Press, Boca Raton (2003) ISBN: 1-58488-317-0

[29] Chen, B.S., Peng, S.C.: Traffic Modeling, Prediction, and Congestion Control for High-Speed Networks: A Fuzzy AR Approach. IEEE Transactions on Fuzzy Systems 8(5), 491–508 (2000)

[30] Cho, K.: Managing Traffic With ALTQ. In: 1999 USENIX Annual Technical Conference, FREENIX Track, Monterrey, CA, USA (1999), http://www.usenix.org/publications/library/proceedings/usenix%99/

[31] Chrysostomou, C., Pitsillides, A., Rossides, L., Sekercioglu, A.: Fuzzy Logic Controlled RED: Congestion Control in TCP/IP Differentiated Services Networks. Soft Computing Journal 8(20), 79–92 (2003)

[32] Claffy, K.C.: Measuring the Internet. IEEE Internet Computing 4(1), 73–75 (2000)

[33] Clark, D., Fang, W.: Explicit Allocation of Best Effor Packet Delivery Service. IEEE/ACM Transactions on Networking 6(4), 362–373 (1998)

[34] Cover, T.M., Hart, P.E.: Nearest neighbor pattern classification. IEEE Transactions on Information Theory 13, 21–27 (1967)

[35] Crodder, S., Elloumi, O., Pauwels, K.: RED Behavior with Different Packet Sizes. In: Fifth International Symposium on Computers and Communications (ISCC 2009), pp. 793–799 (2000)

[36] Crovella, M.E., Bestavros, A.: Self-Similarity in World Wide Web Traffic: Evidence and Possible Causes. IEEE/ACM Transactions on Networking 5(6), 835–846 (1997)

[37] Crovella, M.E., Krishnamurthy, B.: Internet Measurement: Infrastructure, Traffic and Applications. Wiley, Chichester (2006) ISBN: 978-0470014615

[38] DatCat, Datcat: Internet measurement data catalog (2008), http://www.datcat.org

[39] Clark, D.D., Fang, W.: Explicit Allocation of Best-Effort Packet Delivery Service. EEE/ACM Transactions on Networking 6(4), 362–373 (1998)

[40] Davie, B., Charny, A., Bennett, J., Benson, K., Boudec, J.Y.L., Courtney, B., Davari, S., Firoiu, V., Stiliadis, D.: An Expedited Forwarding PHB. RFC 3246, Internet Engineering Task Force, Network Working Group, status: Proposed Standard (2002)

[41] Devetsikiotis, M., da Fonseca, N.(eds.): Long Range Dependent Traffic. Computer Networks Journal 48(3), 289–488 (2005)

[42] Di Fatta, G., Lo Re, G., Urso, A.: A Fuzzy Approach for the Network Congestion Problem. In: Sloot, P.M.A., Tan, C.J.K., Dongarra, J., Hoekstra, A.G. (eds.) ICCS-ComputSci 2002. LNCS, vol. 2329, pp. 286–295. Springer, Heidelberg (2002)

[43] Donnelly, S.F.: High precision timing in passive measurements of data networks. PhD thesis, University of Waikato, Hamilton, New Zealand (2002)

[44] Doukhan, P., Oppenheim, G., Taqqu, M.S. (eds.): Theory and Applications of Long-Range Dependence, 1st edn. Birkhäuser, Basel (2002)

[45] Dovrolis, C., Ramanathan, P., Moore, D.: What Do Packet Dispersion Techniques Measure? In: Annual Joint Conference of the IEEE Computer and Communications Societies (IEEE INFOCOM), pp. 905–914 (2001)

[46] Dovrolis, C., Ramanathan, P., Moore, D.: Packet Dispersion Techniques and a Capacity Estimation Methodology. IEEE/ACM Transactions on Networking 12(6), 963–977 (2004)

[47] Endace Limited, DAG Network Monitoring Cards (2008),
http://www.endace.com/our-products/
dag-network-monitoring-cards/

[48] Estan, C., Savage, S., Varghese, G.: Automatically Inferring Patterns of Resource Consumption in Network Traffic. In: ACM SIGCOMM 2003, Karlsruhe, Germany, pp. 137–148 (2003)

[49] Faloutsos, M., Faloutsos, P., Faloutsos, C.: On Power-Law Relationships of the Internet Topology. Computer Communication Review 29(4), 251–262 (1999)

[50] Fan, J., Yao, Q.: Nonlinear Time Series: Nonparametric and Parametric Methods. Springer Series in Statistics. Springer, New York (2003)

[51] Faucheur, F.L., et al.: Multi-Protocol Label Switching (MPLS) Support of Differentiated Services. RFC 3270, Internet Engineering Task Force, Network Working Group (2002)

[52] Feldmeier, D., Biersack, E.: Comparison of Error Control Protocols for High Bandwidth-Delay Product networks. In: Conference on Protocols for High-Speed Networks II, Palo Alto, CA, USA, pp. 271–295 (1990)

[53] Feng, W., Kandlur, D., Saba, D.: A Self-configuring RED Gateway. In: Annual Joint Conference of the IEEE Computer and Communications Societies (IEEE INFOCOM), New York, USA, pp. 1320–1328 (1999)

[54] Feng, W., Shin, K., Kandlur, D., Saba, D.: The BLUE Active Queue Management Algorithms. IEEE/ACM Transactions on Networking 10, 513–528 (2002)

[55] Fengyuan, R., Xiunming, S.: Design of a Fuzzy controller for Active Queue Management. Computer Communications 25(9), 874–883 (2002)

[56] Fengyuan, R., Chuang, L., Xunhe, Y., Xiuning, S., Fobao, W.: A Robust Active Queue Management Algorithm Based on Sliding Mode Variable Structure Control. In: Annual Joint Conference of the IEEE Computer and Communications Societies (IEEE INFOCOM), New York, USA, pp. 13–20 (2002)

[57] Flannagan, M., Froom, R., Turek, K.: Cisco Catalyst Qos. Quality of Service in Campus Networks. Cisco Systems, Inc. (2003) ISBN: 1-58705-120-6

[58] Floyd, S.: Congestion Control Principles. RFC 2914, Internet Engineering Task Force, Network Working Group, category: Best Current Practice (2000)

[59] Floyd, S., Fall, K.: Promoting the Use of End-to-End Congestion Control in the Internet. IEEE/ACM Transactions on Networking 7(4), 458–472 (1999)

[60] Floyd, S., Jacobson, V.: On Traffic Phase Effects in Packet-Switched Gateways. Internetworking: Research and Experience 3(3), 115–156 (1992)

[61] Floyd, S., Jacobson, V.: Random Early Detection Gateways for Congestion Avoidance. IEEE/ACM Transactions on Networking 1(4), 397–413 (1993)

[62] Floyd, S., Jacobson, V.: The synchronization of periodic routing messages. IEEE/ACM Transactions on Networking 2(2), 122–136 (1994)

[63] Floyd, S., Kempf, J.: IAB Concerns Regarding Congestion Control for Voice Traffic in the Internet. RFC 3714, Internet Engineering Task Force, Network Working Group, category: Informational (2004)

[64] Floyd, S., Kohler, E.: Internet Research Needs Better Models. ACM SIGCOMM Computer Communication Review 33(1), 29–34 (2003)

[65] Floyd, S., Paxson, V.: Difficulties in Simulating the Internet. IEEE/ACM Transactions on Networking 9(4), 392–403 (2001)

[66] Floyd, S., Gummadi, R., Shenker, S.: Adaptive RED: An Algorithm for Increasing the Robustness of RED. Tech. rep., ICIR, ICSI Center for Internet Research (2001), http://www.icir.org/floyd/adaptivered/

[67] Floyd, S., et al.: Metrics for the Evaluation of Congestion Control Mechanisms. RFC 5166, Internet Engineering Task Force, Network Working Group, category: Informational (2008)

[68] Flux Research Group: Emulab – Network Emulation Testbed, University of Utah (2005), http://www.emulab.net/

[69] Fowler, H.J., Leland, W.E.: Local Area Network Traffic Characteristics, with Implications for Broadband Network Congestion Management. IEEE Journal on Selected Areas in Communication 9(7), 1139–1149 (1991)

[70] Ghaderi, M.: On the Relevance of Self-Similarity in Network Traffic Prediction. Tech. Rep. CS-2003-28, School of Computer Science, University of Waterloo (2003)

[71] Gong, W., Liu, Y., Misra, V., Towsley, D.F.: Self-similarity and long range dependence on the Internet: a second look at the evidence, origins and implications. Computer Networks 48(3), 377–399 (2005)

[72] Goodwin, P., Lawton, R.: On the asymmetry of the symmetric MAPE. International Journal of Forecasting 15(4), 405–408 (1999)

[73] Gripenberg, G., Norros, I.: On the prediction of fractional brownian motion. Journal of Applied Probability 33(2), 400–410 (1996)

[74] Groschwitz, N., Polyzos, G.: A Time Series Model of Long-term NSFNET Backbone Traffic. In: IEEE International Conference on Communications (SUPERCOMM/ICC 1994), New Orleans, LA, USA, vol. 3, pp. 1400–1404 (1994)

[75] Handley, M., Floyd, S., Pahdye, J., Widmer, J.: TCP Friendly Rate Control (TFRC): Protocol Specification. RFC 3448, Internet Engineering Task Force, Network Working Group, status: Proposed Standard (2003)

[76] Haykin, S.S.: Neural Networks: A Comprehensive Foundation, 2nd edn. Prentice-Hall, Englewood Cliffs (1998) ISBN: 978-0132733502

[77] Haykin, S.S.: Adaptive Filter Theory, 4th edn. Prentice-Hall, Englewood Cliffs (2001) ISBN: 978-0130901262

[78] He, Q., Dovrolis, C., Ammar, M.: On the predictability of large transfer TCP throughput. Computer Networks 51(14), 3959–3977 (2007)

[79] Heinamen, J., Baker, F., Weiss, W., Wroclawski, J.: An Assured Forwarding PHB Group. RFC 2597, Internet Engineering Task Force, Network Working Group (1999)

[80] Hemminger, S., Kuznetsov, A.N.: iproute2: Ip routing utilities (2008), http://developer.osdl.org/dev/iproute2/

[81] Heyde, A.A.: Investigating the performance of Endace DAG monitoring hardware and Intel NICs in the context of Lawful Interception. Tech. Rep. 080222A, Centre for Advanced Internet Architectures (CAIA), Swinburne University of Technology (2008), http://caia.swin.edu.au/reports/

[82] Hollot, C., Misra, V., Towsley, D., Gong, W.: A Control Theoretic Analysis of RED. In: Annual Joint Conference of the IEEE Computer and Communications Societies (IEEE INFOCOM), Anchorage, UK, pp. 1510–1519 (2001)

[83] Hollot, C., Misra, V., Towsley, D., Gong, W.: Analysis and Design of Controllers for RED Routers Supporting TCP Flows. IEEE Transactions on Automatic Control 47, 945–959 (2002)

[84] Hollot, C.V., Misra, V., Towsley, D.F., Gong, W.: On designing improved controllers for AQM routers supporting TCP flows. In: Annual Joint Conference of the IEEE Computer and Communications Societies (IEEE INFOCOM), Anchorage, UK, pp. 1726–1734 (2001)

[85] Hosking, J.R.M.: Fractional Differencing. Biometrika 83(1), 165–176 (1981)

[86] Hu, N., Steenkiste, P.: Evaluation and Characterization of Available Bandwidth Probing Techniques. IEEE Journal on Selected Areas in Communication 21(6), 879–894 (2003)

[87] Huang, G.B., Zhu, Q.Y., Siew, C.K.: Extreme learning machine: Theory and applications. Neurocomputing 70(1-3), 489–501 (2006)

[88] Information Sciences Institute University of Southern California, Viterbi School of Engineering, The Network Simulator – ns-2 (2008), http://www.isi.edu/nsnam/ns/

[89] Internet Engineering Task Force, Transport Area, Differentiated Services (diffserv) Charter, Internet Engineering Task Force, Transport Area (2008), http://www.ietf.org/html.charters/diffserv-charter.html

[90] Iren, S., Amer, P.D., Conrad, P.T.: The transport layer: Tutorial and survey. ACM Computing Surveys 31(4), 360–405 (1999)

[91] ITU-T X.214: Recommendation X.214. Information Technology, Open Systems Interconnection, Transport Service Definition International Telecommunications Union – Telecomunications Standardization Sector, 4th edn. (1995)

[92] Jacobson, V., Karels, M.J.: Congestion Avoidance and Control. In: ACM Computer Communication Review SIGCOMM 1988 Symposium: Communications Architectures and Protocols, vol. 18(4), pp. 314–329 (1988)

[93] Jacobson, V., Braden, B., Borman, D.: TCP Extensions for High Performance. RFC 1323, Internet Engineering Task Force, Network Working Group, status: Proposed Standard (1992)

[94] Jain, R.: The Art of Computer Systems Performance Analysis: Techniques for Experimental Design, Measurement, Simulation and Modeling. Wiley Interscience, Hoboken (1991) ISBN: 0471503361

[95] Jin, G., Tierney, B.L.: System Capability Effects on Algorithms for Network Bandwidth Measurement. In: Internet Measurement Conference, Miami Beach, FL, USA, pp. 27–38 (2003)

[96] Kantz, H., Schreiber, T.: Nonlinear Time Series Analysis, 2nd edn. Cambridge University Press, Cambridge (2004)

[97] Karagiannis, T., Molle, M., Faloutsos, M., Broido, A.: A Nonstationary Poisson View of Internet Traffic. In: 23th Annual Joint Conference of the IEEE Computer and Communications Societies (IEEE INFOCOM), Hong Kong, vol. 3, pp. 1558–1569 (2004)

[98] Karn, P., Partridge, C.: Improving round-trip time estimates in reliable transport protocols. ACM Transactions on Computer Systems 9(4), 364–373 (1991)

[99] Karsten, M., Cheriton, D.R.: Collected experience from implementing RSVP. IEEE/ACM Transactions on Networking 14(4), 767–778 (2006)

[100] Keller, J.C., Gray, M.R., Givens, J.A.: A fuzzy k-nearest neighbor algorithm. IEEE Transactions on Systems, Man and Cybernetics 15(4), 580–585 (1985)

[101] Krioukov, D., Chung, F., Claffy, K.C., Fomenkov, M., Vespignani, A.: Willinger The Workshop on Internet Topology (WIT) Report. ACM SIGCOMM Computer Communication Review 37(1), 69–73 (2007)

[102] Krithikaivasan, B., Zeng, Y., Deka, K., Medhi, D.: ARCH-based Traffic Forecasting and Dynamic Bandwidth Provisioning for Periodically Measured Nonstationary Traffic. IEEE/ACM Transactions on Networking 15(3), 683–696 (2007)

[103] Kunniyur, S., Srikant, R.: Analysis and Design of an Adaptive Virtual Queue (AVQ) Algorithm for Active Queue Management. ACM SIGCOMM Computer Communication Review 31(4), 123–134 (2001)

[104] Lakhina, A., Byers, J.W., Crovella, M.E., Matta, I.: On the Geographic Location of Internet Resources. IEEE J-SAC, Special Issue on Internet and WWW Measurement, Mapping and Modeling 21(6), 934–948 (2003)

[105] Lakshman, T., Madhow, U.: The Performance of TCP/IP for Networks With High Bandwidth Delay Products and Random Losses. IEEE/ACM Transactions on Networking 5(5), 336–350 (1997)

[106] Leland, W., Taqqu, M.S., Willinger, W., Wilson, D.: On the Self Similar Nature of Ethernet Traffic (Extended Version). IEEE/ACM Transactions on Networking 2(1), 1–15 (1994)

[107] Low, S., Paganini, F., Wang, J., Adlakha, S., Doyle, J.: Dynamics of TCP/RED and Scalable Control. In: Annual Joint Conference of the IEEE Computer and Communications Societies (IEEE INFOCOM), New York City, USA, pp. 239–248 (2002)

[108] Low, S., Paganini, F., Doyle, J.: Internet Congestion Control. IEEE Control Systems Magazine 22, 28–43 (2002)

[109] Low, S.H.: A Duality Model of TCP and Queue Management Algorithms. IEEE/ACM Transactions on Networking 11(4), 525–536 (2003)

[110] Lowekamp, B.B.: Combining active and passive network measurements to build scalable monitoring systems on the grid. ACM SIGMETRICS Performance Evaluation Review Special issue on grid computing 30(4), 19–26 (2003)

[111] Martin, J., Nilsson, A.: The Evolution of Congestion Control in TCP/IP: from Reactive Windows to Preventive Flow Control. Tech. Rep. TR 97/11, CACC, North Carolina State University (1997)

[112] Mathis, M., Semke, J., Madhavi, J.: The macroscopic behavior of the tcp congestion avoidance algorithm. ACM Computer Communications Review 27(3), 67–82 (1997)

[113] May, M., Bolot, J., Diot, C., Lyes, B.: Reasons not to Deploy RED. In: International Workshop on Quality of Service (IWQoS), London, United Kingdom, pp. 260–262 (1999)

[114] Medina, A., Allman, M., Floyd, S.: Measuring the Evolution of Transport Protocols in the Internet. ACM SIGCOMM Computer Communication Review 35(2), 37–52 (2005)

[115] Miche, Y., Sorjamaa, A., Lendasse, A.: OP-ELM: Theory, Experiments and a Toolbox.
In: Kůrková, V., Neruda, R., Koutník, J. (eds.) ICANN 2008, Part I. LNCS, vol. 5163,
pp. 145–154. Springer, Heidelberg (2008)

[116] Montesino-Pouzols, F.: Comparative Analysis of Active Bandwidth Estimation Tools.
In: Barakat, C., Pratt, I. (eds.) PAM 2004. LNCS, vol. 3015, pp. 175–184. Springer,
Heidelberg (2004)

[117] Montesino Pouzols, F., Lendasse, A.: Evolving fuzzy optimally pruned extreme learning machine for regression problems. Evolving Systems 1(1), 43–58 (2010)

[118] Montesino-Pouzols, F., Barriga, A., Lopez, D.R., Sánchez-Solano, S.: Performance
Analysis of Computer Networks. In: Encyclopedia of Networked and Virtual Orga-
nizations, Information Science Reference (an imprint of IGI Global), Hershey, New
York, USA, pp. 1216–1222 (2008) ISBN: 978-1-59904-885-7

[119] Montesino-Pouzols, F., Barriga, A., Lopez, D.R., Sánchez-Solano, S.: Performance
Analysis of Peer-to-Peer Traffic. In: Encyclopedia of Networked and Virtual Orga-
nizations, Information Science Reference (an imprint of IGI Global), Hershey, New
York, USA, vol. II, pp. 1210–1215 (2008) ISBN: 978-1-59904-885-7

[120] Montesino-Pouzols, F., Lopez, D.R., Barriga, A., Sánchez-Solano, S.: Performance
Analysis and Models of Web Traffic. In: Encyclopedia of Networked and Virtual Or-
ganizations, Information Science Reference (an imprint of IGI Global), Hershey, New
York, USA, vol. II, pp. 1196–1203 (2008) ISBN: 978-1-59904-885-7

[121] Montesino-Pouzols, F., Lopez, D.R., Barriga, A., Sánchez-Solano, S.: Performance
Analysis of Multimedia Traffic. In: Encyclopedia of Networked and Virtual Organiza-
tions, Information Science Reference (an imprint of IGI Global), Hershey, New York,
USA, vol. II, pp. 1204–1209 (2008) ISBN: 978-1-59904-885-7

[122] Moreno-Velo, F.J., Baturone, I., Sánchez-Solano, S., Barriga, A.: Rapid Design of
Fuzzy Systems With Xfuzzy. In: 12th IEEE International Conference on Fuzzy Systems (FUZZ-IEEE 2003), St. Louis, MO, USA, pp. 342–347 (2003)

[123] Murray, M., Claffy, K.C.: Measuring the Immeasurable: Global Internet Measurement
Infrastructure. In: 2nd Passive and Active Network Measurement Workshop (PAM),
pp. 159–167. Amsterdam, The Netherlands (2001)

[124] Murray, M., Claffy, K.C.: Measuring the immeasurable: Global internet measurement
infrastructure. In: Passive and Active Network Measurement (PAM) Workshop, Amsterdam (2001)

[125] Ott, T.J., Lakshman, T.V., Wong, L.: SRED: Stabilized RED. In: Annual Joint Conference of the IEEE Computer and Communications Societies (IEEE INFOCOM), New
York, USA, vol. 3, pp. 1346–1355 (1999)

[126] Papagiannaki, K., Taft, N., Zhang, Z.L., Diot, C.: Long-Term Forecasting of Internet
Backbone Traffic: Observations and Initial Models. In: Twenty-Second Annual Joint
Conference of the IEEE Computer and Communications Societies (IEEE INFOCOM
2003), San Francisco, USA, vol. 2, pp. 1178–1188 (2003)

[127] Papagiannaki, K., Taft, N., Zhang, Z.L., Diot, C.: Long-Term Forecasting of Internet
Backbone Traffic. IEEE Transactions on Neural Networks 16(5), 110–1124 (2005)

[128] Park, K., Willinger, W. (eds.): Self-Similar Network Traffic and Performance Evaluation. Wiley Interscience, New York (2000) ISBN: 0-471-31974-0

[129] Paxson, V.: Empirically-Derived Analytic Models of Wide-Area TCP Connections.
IEEE/ACM Transactions on Networking 2(4), 316–336 (1994)

[130] Paxson, V.: Fast, approximate synthesis of fractional Gaussian noise for generating
self-similar network traffic. Computer Communications Review 27(1), 5–18 (1997)

[131] Paxson, V.: Measurements and Analysis of End-to-End Internet Dynamics. PhD thesis. Computer Science Division, University of California (1997)

[132] Paxson, V.: End-to-End Internet Packet Dynamics. IEEE/ACM Transactions on Networking 7(3), 277–292 (1999)

[133] Paxson, V., Allman, M.: Computing TCP's Retransmission Timer. RFC 2988, Internet Engineering Task Force, Network Working Group, status: Proposed Standard (2000)

[134] Paxson, V., Floyd, S.: Wide Area Traffic: The Failure of Poisson Modeling. IEEE/ACM Transactions on Networking 3(3), 226–244 (1995)

[135] PlanetLab Consortium, Planetlab: An open platform for developing, deploying and accessing planetary-scale services (2008), http://www.planet-lab.org

[136] Prasad, R.S., Murray, M., Dovrolis, C., Claffy, K.C.: Bandwidth estimation: metrics, measurement techniques, and tools. IEEE Network 17(6), 27–35 (2003)

[137] Ramakrishnan, K., Floyd, S.: A Proposal to Add Explicit Congestion Notification to IP. RFC 2481, Internet Engineering Task Force, Network Working Group, category: Experimental (1999)

[138] Rekhter, Y., Gross, P.: Application of the Border Gateway Protocol in the Internet. RFC 1772, Internet Engineering Task Force, Network Working Group, status: Draft Standard (1995)

[139] Rekhter, Y., Li, T.: A Border Gateway Protocol 4 (BGP-4). RFC 1771, Internet Engineering Task Force, Network Working Group, status: Draft Standard (1995)

[140] Riedi, R.H.: Multifractal Processes. In: Theory and Applications of Long-Range Dependence, 1st edn., pp. 625–715. Birkhäuser, Boston (2002)

[141] Riedi, R.H., Crouse, M.S., Ribeiro, V.J., Baraniuk, R.G.: A Multifractal Wavelet Model with Application to Network Traffic. IEEE Transactions on Information Theory 45(3), 992–1018 (1999)

[142] Rizzo, L.: Dummynet: a Flexible Tool for Testing Network Protocols. dipartimento di Ingegneria dell'Informazione – Univ. di Pisa (2005), http://info.iet.unipi.it/~luigi/ip_dummynet/

[143] Rose, O.: Statistical Properties of MPEG Video Traffic and Their Impact on Traffic Modeling in ATM Systems. In: 20th Annual Conference on Local Computer Networks, pp. 397–406 (1995)

[144] Schölkopf, B., Smola, A.J.: Learning with Kernels. Support Vector Machines, Regularization, Optimization, and Beyond. MIT Press, Cambridge (2002) ISBN: 0262194759

[145] Shalunov, S., Teitelbaum, B.: TCP Use and Performance on Internet2. In: ACM SIGCOMM Internet Measurement Workshop, San Francisco, CA, USA, pp. 147–160 (2001)

[146] Shalunov, S., Lutzmann, B., Montesino-Pouzols, F.: Reporting IP Performance Metrics to Users. Tech. rep., Internet Engineering Task Force, Network Working Group, IP Performance Metrics Working Group, Internet Draft, draft-ietf-ippm-reporting-01 (2007)

[147] Shriram, A., Murray, M., Hyun, Y., Brownlee, N., Broido, A., Fomenkov, M., Claffy, K.: Comparison of public end-to-end bandwidth estimation tools on high-speed links. In: Dovrolis, C. (ed.) PAM 2005. LNCS, vol. 3431, pp. 306–320. Springer, Heidelberg (2005)

[148] Shu, Y., Jin, Z., Wang, J., Yang, O.W.W.: Prediction-Based Admission Control Using FARIMA Models. In: IEEE International Conference on Communications, ICC 2000, New Orleans, LA, USA, vol. 3, pp. 1325–1329 (2000)

[149] Siripongwutikorn, P., Banerjee, S., Tipper, D.: Adaptive Bandwidth Control for Efficient Aggregate QoS Provisioning. ACM Computer Communication Review (CCR) 32(3), 19–24 (2002)

[150] Siripongwutikorn, P., Banerjee, S., Tipper, D.: A Survey of Adaptive Bandwidth Control Algorithms. IEEE Communications Surveys 5(1), 14–26 (2003)

[151] Sorjamaa, A., Hao, J., Reyhani, N., Ji, Y., Lendasse, A.: Methodology for Long-Term Prediction of Time Series. Neurocomputing 70(16-18), 2861–2869 (2007)

[152] Sorjamaa, A., Miche, Y., Weiss, R., Lendasse, A.: Long-Term Prediction of Time Series using NNE-based Projection and OP-ELM. In: 2008 International Joint Conference on Neural Networks (IJCNN 2008), IEEE World Congress on Computational Intelligence, Hong Kong, China, pp. 2675–2681 (2008)

[153] Srikant, R.: The Mathematics of Internet Congestion Control. Systems & Control: Foundations & Applications. Birkhauser, Boston (2003) ISBN: 0-8176-3227-1

[154] SSF Research Network Scalable Simulation Framework. A Public Domain Standard for Discrete-Event Simulation of Large, Complex Systems in Java and C++ (2008), http://www.ssfnet.org/homePage.html

[155] Stevens, W.R.: TCP/IP Illustrated. The Protocols, vol. 1. Addison-Wesley, Reading (1994)

[156] Stevens, W.R.: TCP Slow Start, Congestion Avoidance Fast Retransmit, and Fast Recovery Algorithms. RFC 2001, Internet Engineering Task Force, Network Working Group, status: Proposed Standard (1997)

[157] Stoksik, M.A., Lane, R.G., Nguyen, D.T.: Accurate synthesis of fractional Brownian motion using wavelets. Electronics Letters 30(5), 383–384 (1994)

[158] Strauss, J., Katabi, D., Kaashoek, F.: A Measurement Study of Available Bandwidth Estimation Tools. In: 3rd ACM SIGCOMM Conference on Internet Measurement, Miami Beach, FL, USA, pp. 39–44 (2003)

[159] Suykens, J.A.K., Van Gestel, T., De Brabanter, J., De Moor, B., Vandewalle, J.: Least Squares Support Vector Machines. World Scientific, Singapore (2002) ISBN: 981-238-151-1

[160] Tirumala, A., Qin, F., Dugan, J., Ferguson, J., Gibbs, K.: Iperf Version 2.0.2. NLANR Distributed Applications Support Team (2005), http://dast.nlanr.net/Projects/Iperf/

[161] Traina, P.: BGP-4 Protocol Analysis. RFC 1774, Internet Engineering Task Force, Network Working Group, category: Informational (1995)

[162] Tran-Gia, P., Vicari, N.: COST-257 Final Report. Impacts of New Services on the Architecture and Performance of Broadband Networks. Tech. rep., EU Cooperation in the Field of Scientific and Technical Research (COST)-257 Management Committee (2000), http://www-info3.informatik.uni-wuerzburg.de/cost/Final/index.html

[163] Varga, A.: OMNet++ Discrete Event Simulator System. omnets Global Inc. (2005), http://www.omnetpp.org

[164] Varghese, G., Estan, C.: The Measurement Manifesto. ACM Computer Computer Communications Review 34(1), 9–14 (2004)

[165] Wang, C., Li, B., Hou, Y.T., Sohraby, K., Lin, Y.: LRED: A Robust Active Queue Management Scheme Based on Packet Loss Ratio. In: Annual Joint Conference of the IEEE Computer and Communications Societies (IEEE INFOCOM 2004), Hong Kong, China, vol. 1, pp. 1–12 (2004)

[166] Wang, H., Shin, K.: Refined Design of Random Early Detection Gateways. In: IEEE GLOBECOM 1999, Rio de Janeiro, Brazil, pp. 769–775 (1999)

[167] Wang, Z., Crowcroft, J.: Analysis of shortest-path routing algorithms in a dynamic network environment. ACM SIGCOMM Computer Communication Review 22(2), 63–71 (1992)

[168] Widmer, J.: Equation-Based Congestion Control for Unicast and Multicast Data Streams. PhD thesis, University of Mannheim (2003)

[169] Widmer, J., Denda, R., Mauve, M.: A Survey on TCP-Friendly Congestion Control. Special Issue of the IEEE Network Magazine "Control of Best Effort Traffic" 15(3), 28–37 (2001)

[170] Willinger, W., Paxson, V.: Where Mathematics meets the Internet. Notices of the American Mathematical Society 45(8), 961–970 (1998)

[171] Willinger, W., Paxson, V., Taqqu, M.S.: Self-Similarity and Heavy Tails: Structural Modeling of Network Traffic. In: A Practical Guide to Heavy Tails: Statistical Techniques and Applications, pp. 27–53. Brikhauser, Boston (1998)

[172] Wroclawski, J.: The Use of RSVP with IETF Integrated Services. RFC 2210, Internet Engineering Task Force, Network Working Group, status: Proposed Standard (1997)

[173] Yager, R.R.: The Ordered Weighted Averaging Operators: Theory and Applications. Kluwer, Norwell (1997)

[174] Yager, R.R.: Using fuzzy methods to model nearest neighbor rules. IEEE Transactions on Systems, And, and Cybernetics–Part B: Cybernetics 32(4), 512–525 (2002)

[175] Yager, R.R.: Induced aggregation operators. Fuzzy Sets and Systems 137(1), 59–69 (2003)

[176] Yager, R.R., Filev, D.P.: Induced Ordered Weighted Averaging Operators. IEEE Transactions on Systems, And, and Cybernetics–Part B: Cybernetics 29(2), 141–150 (1999)

[177] Yanfei, F., Fengyuan, R., Chuang, L.: Design of an Active Queue Management Algorithm Based on Fuzzy Logic Decision. In: IEEE International Conference on Communication Technology, ICCT, pp. 285–289 (2003)

[178] Yang, C., Reddy, A.: A Taxonomy for Congestion Control Algorithms in Packet Switching Networks. IEEE Network Magazine 9(5), 41–48 (1995)

[179] You, C., Chandra, K.: Time Series Models for Internet Data Traffic. In: 24th Annual IEEE International Conference on Local Computer Networks (LCN 1999), Lowell, MA, USA, pp. 164–171 (1999)

[180] Zhang, Z., Ribeiro, V., Moon, S., Diot, C.: Small-Time Scaling Behaviours of Internet Backbone Traffic: an Empirical Study. In: 22nd Annual Joint Conference of the IEEE Computer and Communications Societies (INFOCOM), San Francisco, CA, USA, vol. 3, pp. 1826–1836 (2003)

Chapter 2
Modeling Time Series by Means of Fuzzy Inference Systems

Abstract. In this chapter, we focus on long-term modeling and prediction of univariate nonlinear time series. First, a method for long-term time series prediction
by means of fuzzy inference systems combined with residual variance estimation
techniques is developed and validated through a number of time series prediction
benchmarks. This method provides an automatic means of modeling and predicting network traffic load, and can thus be classified as a method for predictive data
mining. Although the primary focus in this section is to develop a methodology for
building simple and thus interpretable fuzzy inference systems, it will be shown that
they also outperform some of the most accurate and commonly used techniques in
the field of time series prediction.

2.1 Predictive Models for Time Series

Time series prediction and analysis in general is a recurrent problem in virtually
all areas of natural and social sciences as well as in engineering. In the time series
prediction field, prediction accuracy is not the only major goal. Understanding the
behavior of time series and gaining insight into their underlying dynamics is a highly
desired capability of time series prediction methods [39].

In the past, conventional statistical techniques such as AR and ARMA models
have been extensively used for forecasting [4]. However, these techniques have
limited capabilities for modeling time series data, and more advanced nonlinear
methods including artificial neural networks have been frequently applied with
success [6].

Fuzzy logic based modeling techniques are appealing because of their interpretability and potential to address a broad spectrum of problems. In particular,
fuzzy inference systems exhibit a combined description and prediction capability
as a consequence of their rule-based structure [37]. The application of fuzzy inference systems to time series modeling and prediction dates back to [38], in which
the authors develop the well known learn from examples identification algorithm
for fuzzy inference systems and use the Mackey-Glass time series as a validation

F.M. Pouzols et al.: Mining & Control of Network Traffic by Computational Intelligence, pp. 53–85.
springerlink.com   © Springer-Verlag Berlin Heidelberg 2011

case. Nevertheless, in spite of its good performance in terms of accuracy and interpretability, fuzzy inference systems have seen little application in the field of time
series prediction as compared to other nonlinear modeling techniques such as neural
networks and support vector machines.

The methodology proposed here is intended to apply to crisp time series, i.e.
those time series consisting of crisp values, as opposed to other kinds of values,
such as interval and fuzzy values. That is, we propose here a methodology framework to perform autoregressive prediction of crisp time series by means of fuzzy
inference systems [23, 25]. We will call fuzzy autoregressors those autoregressors
implemented as fuzzy inference systems. This is not to be confused with what is
usually called fuzzy regression [5] and fuzzy time series [33] in the literature.

When developing fuzzy inference systems for time series prediction, many questions remain still open: How to perform long-term prediction? How many and what
inputs to the inference system must be defined? To what extent the theoretical universal approximation capability of fuzzy systems is achieved with existing techniques? What are the best fuzzy methods for these tasks?

In practice, one finds two problems when building a fuzzy model for a time series:
choosing variables or inputs to the inference system, and identifying the structure
of the system (linguistic labels and rule base). Once these steps have been accomplished, the fuzzy model can be tuned through supervised learning techniques. We
propose an automatic methodology framework to address these two problems using
fuzzy techniques and nonparametric residual variance estimation techniques in an
intertwined manner.

The first problem can be addressed by means of a priori feature selection techniques based on nonparametric residual variance estimation, which also provide an
estimate of the error of the most accurate nonlinear model that can be built without overfitting. The second problem is addressed by techniques for identification of
fuzzy systems from numerical examples [10], such as the algorithm by Wang and
Mendel (W&M) [37, 38] and fuzzy identification algorithms based on clustering
techniques [22, 7].

In this section, we also address a relatively recent challenge in the field of time series prediction: long-term prediction (as a generalization to short-term prediction),
for which lack of information and accumulated errors pose additional difficulties.
Also, real world benchmarking time series, in addition to synthetic series (chaotic
but noise-free) are analyzed. The methodology proposed here will be compared
against Least Squares Support Vector Machines (LS-SVM) [35], a method that has
been shown to be highly accurate in the field of time series prediction.

The next section outlines a nonparametric residual variance estimation method
that will be used for both variable and proper model complexity selection. In section 2.3 we propose a methodology framework and one concrete implementation
that uses well known algorithms for identifying and optimizing fuzzy inference systems. Section 2.4 illustrates the methodology through a case study. Finally, sections 2.5 and 2.5.7 present and further discuss experimental results for a number of
time series benchmarks from diverse fields of application.

2.2   Nonparametric Residual Variance Estimation: Delta Test

Nonparametric residual variance estimation (or nonparametric noise estimation, NNE) is a well-known technique in statistics and machine learning, finding many applications in nonlinear modeling [15]. NNE methods can be applied to recurrent problems such as variable and model structure selection. These methods are not however in widespread use in the machine learning community as most work has been done to date within the statistics community.

Delta Test (DT), introduced for time series in 1994 [30], is a NNE method, i.e, it estimates the lowest mean square generalization error (MSE) that can be achieved by a model without overfitting the training set. Given N multiple input-single output pairs, $(\bar{x}_i, y_i) \in \mathbb{R}^M \times \mathbb{R}$, the theory behind the DT method considers that the mapping between $\bar{x}_i$ and y_i is given by the following expression:

$$y_i = f(\bar{x}_i) + r_i,$$

where f is an unknown perfect fitting model and r_i is the noise. DT is based on hypothesis coming from the continuity of the regression function. When two inputs x and x' are close, the continuity of the regression function implies that the corresponding outputs, $f(x)$ and $f(x')$, will be close enough. When this implication does not hold, it is due to the influence of the noise.

Let us denote the first nearest neighbor of the point $\bar{x}_i$ in the set $\{\bar{x}_1, \ldots, \bar{x}_N\}$ by $\bar{x}_{NN}$. Then the DT, δ, is defined as follows:

$$\delta = \frac{1}{2N} \sum_{i=1}^{N} |y_{NN(i)} - y_i|^2,$$

where $y_{NN(i)}$ is the output corresponding to $\bar{x}_{NN(i)}$. For a proof of convergence, we refer to [18, 19]. DT is an unbiased and asymptotically perfect estimator with a relatively fast convergence [19] and is useful for evaluating nonlinear correlations between two random variables, namely, input-output pairs. DT can be seen as part of a more general NNE framework known as the Gamma Test [15]. Despite the simplicity of DT, it has been shown to be a robust method in real world applications [18]. This method will be used in the next sections for both model complexity selection and a priori input selection.

2.3   Methodology Framework for Time Series Prediction with Fuzzy Inference Systems

Consider a discrete time series as a vector, $\bar{y} = y_1, y_2, \ldots, y_{t-1}, y_t$ that represents an ordered set of values, where t is the number of values in the series. The problem of predicting one future value, y_{t+1}, using an autoregressive model (autoregressor) with no exogenous inputs can be stated as follows:

$$\hat{y}_{t+1} = f_r(y_t, y_{t-1}, \ldots, y_{t-M+1})$$

Where $\hat{y}_{t+1}$ is the prediction of model f_r and M is the number of inputs to the regressor.

Predicting the first unknown value requires building a model, f_r, that maps regressor inputs (known values) into regressor outputs (predictions). When a prediction horizon higher than 1 is considered, the unknown values can be predicted following two main strategies: recursive and direct prediction.

The recursive strategy applies the same model recursively, using predictions as known data to predict the next unknown values. For instance, the third unknown value is predicted as follows:

$$\hat{y}_{t+3} = f_r\left(\hat{y}_{t+2}, \hat{y}_{t+1}, y_t, y_{t-1}, \ldots, y_{t-M+3}\right)$$

Recursive prediction is the most simple and intuitive strategy and does not require any additional modeling after an autoregressor for one step ahead prediction is built. However, recursive prediction suffers from accumulation of errors. The longer the prediction term is, the more predictions are used as inputs. In particular, for prediction horizons greater than the regressor size, all inputs to the model are predictions.

Direct prediction requires that the process of building an autoregressor be applied for each unknown future value. Thus, for a maximum prediction horizon H, H direct models are built, one for each prediction horizon h:

$$\hat{y}_{t+h} = f_h(y_t, y_{t-1}, \ldots, y_{t-M+1}), \text{ with } 1 \le h \le H$$

While building a prediction system through direct prediction is more computationally intensive (as many times as values are to be predicted) it is also straightforward to parallelize. As opposed to recursive prediction, direct prediction does not suffer from accumulation of prediction errors.

We follow the direct prediction strategy. In order to build each autoregressor, a fuzzy inference system is defined as a mapping between a vector of crisp inputs, and a crisp output. This way, assuming all (M) inputs are used, the fuzzy autoregressor for prediction horizon h can be expressed as a set of N fuzzy rules:

$$R_i^h : \text{IF } y_t \text{ is } L_1^{i,h} \text{ AND } y_{t-1} \text{ is } L_2^{i,h} \text{ AND } \ldots$$
$$\ldots \text{ AND } y_{t-M+1} \text{ is } L_M^{i,h} \text{ THEN } \hat{y}_{t+h} \leftarrow \mu_{R_i^h}$$

Where the fuzzy sets $L_j^{i,h} \in \{L_j^{i,h,1}, L_j^{i,h,2}, \ldots, L_j^{i,h,n_j}\}$, with n_j being the number of linguistic labels (membership functions) defined for the input variable j. For example, in a system with two inputs, if $L_1^{i,h}$ is renamed LOW_1 and $L_2^{i,h}$ is renamed $HIGH_2$, the rule i for horizon 1, R_i^1, would have the following form:

$$\text{IF } y_t \text{ was } LOW_1 \text{ AND } y_{t-1} \text{ was } HIGH_2 \text{ THEN } \hat{y}_{t+h} \leftarrow \mu_{R_i^h}$$

Depending on the fuzzy operators, inference model and type of membership functions employed, the mapping between inputs and outputs can have different formulations. In principle, the methodology proposed here can be applied for any combination of membership functions, operators and inference model, but the selection can have a significant impact on practical results.

As a concrete implementation, we use the minimum as T-norm for conjunction operations and implications, Gaussian membership functions for inputs, singleton outputs and fuzzy mean as defuzzification method. Therefore, in this particular case a fuzzy autoregressor for prediction horizon h can be formulated as follows:

$$\mathscr{F}_h(\bar{y}) = \frac{\displaystyle\sum_{l=1}^{N_h} \min\left(\mu_{R_l^h}, \min_{1\leq v\leq M} \mu_{L_l^{i,h}}(y_v) \right)}{\displaystyle\sum_{l=1}^{N_h} \min_{1\leq v\leq M} \mu_{L_l^{i,h}}(y_v)}$$

Where N_h is the number of rules in the rulebase for horizon h, $\mu_{L_l^{i,h}}$ are Gaussian membership functions and $\mu_{R_l^h}$ are singleton membership functions.

The problem of building a regressor can be precisely stated as that of defining a proper number and configuration of membership functions and building a fuzzy rulebase from a data set of t sample data from a time series such that the fuzzy systems $\mathscr{F}_h(\bar{y})$ closely predict the $h-$th next values of the time series. The error metric to be minimized is the mean squared error (MSE).

We propose a methodology framework in which a fuzzy inference system is defined for each prediction horizon throughout the stages shown in figure 2.1. These stages are detailed in the following subsections.

2.3.1 Variable Selection

In principle, the whole set of known past values of a time series may influence the unknown future values. However, using all known values as inputs to a time series autoregressor does not necessarily improve its accuracy. As the number of inputs increases, and the known data become more sparse in a high-dimensional space, building a model gets more and more complex. This is the well known "curse of dimensionality" problem [3].

A proper choice of input variables can provide a balance between considering all the relevant inputs versus the simplicity of building an accurate autoregressor. As first step in the methodology, DT estimates are employed so as to perform an a priori selection of the optimal subset of inputs from the initial set of M inputs, given a maximum regressor size M.

Variable selection requires a selection criterion. We use the result of the DT applied to a particular variable selection as a measure of the goodness of the selection. The input selection that minimizes the DT estimate is chosen for the next stages.

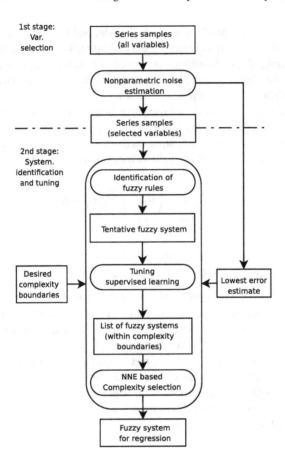

Fig. 2.1 Methodology framework for time series prediction

In addition, a selection procedure is required. For small (up to around 10-20) regressor sizes, an exhaustive evaluation of DT for all the possible selections (a total of $2^M - 1$) is feasible. We will call this procedure *exhaustive DT search*. Its main advantages is that the optimal selection is found. However, its algorithmic order is exponential and it is thus unfeasible for high regressor sizes.

For higher regressor sizes, forward-backward search of selections (FBS) [34] is employed. This procedure combines both forward and backward selection. FBS can be started alternatively from random selections or selections for lower regressor sizes performed by means of exhaustive search. While this procedure does not guarantee optimality of the chosen selection, it provides a convenient balance between performance and computational requirements.

NNE based selection can be classified into the set of input selection approaches which select a priori features, i.e., selection is based only on the dataset. Thus, the computational cost of DT-based selection is lower than that of the model dependent cases, in which input selection is addressed as a generalization error minimization problem, using leave-one-out, bootstrap or other resampling techniques.

2.3.2 System Identification and Tuning

This stage comprises three substages that are performed iteratively and in a coordinated manner. The whole process is driven by the third (complexity selection) substage, until a system that satisfies a training error condition derived from the DT estimate is constructed.

2.3.2.1 Stage 2.1: System Identification

In this substage, the structure of the inference system (linguistic labels and rule base) is defined by means of an automatic fuzzy systems identification algorithm. The set of inputs is fixed after the previous variable selection stage. Regardless of the identification algorithm used, one or more parameters are usually required that specify the potential complexity of the inference system. Thus, the desired boundaries of complexity for the systems being built are additional inputs.

The identification substage, as well as the next (tuning) substage are iteratively performed for increasing degrees of complexity. The concrete procedure used to explore different complexities depends on the identification and tuning algorithms applied.

For the concrete implementation analyzed here, identification is performed using the W&M algorithm driven by the DT estimate. The W&M algorithm is based on the "learn by example" principle and considers a fixed grid partition of the universes of discourse of the inputs, which are proper characteristics for modeling time series in an interpretable manner. Though a number of modifications and derived algorithms have been proposed, for the sake of simplicity and interpretability we adhere to the original specification of the algorithm for generating fuzzy inference rules directly from input-output data pairs [37] as implemented in version 3.2 of the Xfuzzy design environment [27].

In the case of the W&M algorithm, the number of membership functions per input must be specified a priori. Thus, the complexity boundaries would be specified as a maximum number of linguistic labels. Our approach is to explore the set of possible systems starting from the lowest possible number of linguistic labels. The same number of linguistic labels is used for each input.

2.3.2.2 Stage 2.2: System Tuning

We define an additional tuning step in the methodology as a substage separated from the identification substage. Note that in some cases (as for example in the algorithm by Higgins and Goodman [11]), these two substages can be integrated into a standalone algorithm. The tuning process is driven by one or more error metrics.

As concrete implementation, we apply the Levenberg-Marquardt second order optimization method [9, 2] for supervised learning driven by the normalized MSE (NMSE)[1]. A number of supervised learning and optimization methods implemented in the Xfuzzy environment were compared, including backpropagation and other gradient descent methods such as QuickProp and RProp, no derivative methods such as simulated annealing and downhill simplex, several second order methods as well as conjugate gradient methods. Matching previous experience in other fields of application [26], the best results in terms of accuracy were consistently achieved by the Levenberg-Marquardt method.

All the parameters of the membership functions of every input and output are adjusted so that the training error is minimized, i.e., self-tuning inference systems are defined. The learning algorithm applied is the Levenberg-Marquardt method as implemented in Xfuzzy [28].

2.3.3   Complexity Selection

As last step in the process of identifying and tuning fuzzy autoregressors, the proper complexity of the estimated best autoregressor is selected depending on the DT estimate. The iterative identification and tuning stage stops when a system is built such that its training error is equal to or lower than the DT estimate or a threshold based on the DT estimate. Since identification and tuning iterations are performed for increasing complexities, the simplest system that satisfies the DT-based error condition is selected.

For the particular implementation described in this section, the complexity of fuzzy systems is measured as the number of linguistic labels per input. Thus, this substage selects the system with the lowest number of labels per input that has a training error equal to or lower than an optimal error threshold based on the DT estimate.

We note that the DT estimate is an estimate of the lowest possible error, i.e, the error that an optimal model would achieve. Since the models we will apply are not likely to be optimal, we introduce a DT based threshold. The DT-based threshold, equal to or greater than the DT estimate, will be defined and validated experimentally in the next section.

Regarding the convergence and guarantee of finalization of this iterative process, neither the identification algorithm or the optimization method used here guarantee any error bound. However, it should be noted that fuzzy inference systems of the class being designed here, zero-order TSK models, are universal approximators [14, chapter 12]. Thus, for a sufficiently large number of membership functions and rules, any input-output mapping should be approximated with an arbitrary accuracy after the identification and optimization stage, i.e., the training error should be as small as required. In practice, it will be shown that the iterative identification and

---

[1] Normalization is performed against the squared range of the series.

tuning process proposed here converges fast and the number of membership functions required per input is in most cases below or around 5, with very few exceptions for which below or around 10 membership functions are required.

2.4 Case Study and Validation: ESTSP´07 Competition Dataset

For the purposes of validating and illustrating the proposed methodology framework and concrete algorithms and criteria, we analyze the data set from the ESTSP 2007 time series prediction competition [8]. This data set (see figure 2.2) consists of 875 samples of temperatures of the El Niño-Southern Oscillation phenomenon. Further illustration and examples can be found in [25].

In this section we analyze the original ESTSP 2007 series splitted into two subsets: a training set (first 475 samples) and a second set (last 400 samples) that will be used for validation. We will call this series ENSO. Though one of the major goals of the proposed methodology is to avoid the requirement of validation and test series, we define two subsets in order to validate the methodology with the residual noise estimator and algorithms being used.

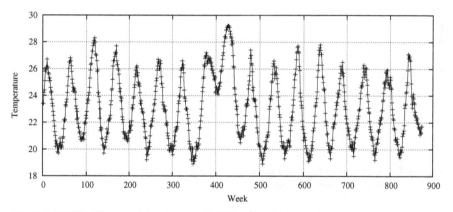

Fig. 2.2 ESTSP´07 competition data set (ENSO series, 875 samples)

A maximum regressor size of 10 and a prediction horizon of 50 are considered. As first stage within our methodology, DT is performed on the training set for all the possible variable selections ($2^{10} - 1$) and the one that leads to the lowest DT estimate is chosen. This process is performed independently for each prediction horizon. The number of selected variables is shown in figure 2.3. Between 3 and 5 variables are selected out of a maximum of 10. Thus, the employment of DT-based variable selection does not only increase accuracy but also leads to a significant decrease of the complexity of the fuzzy inference systems in terms of number of inputs. This fact, in turn, relieves the curse of dimensionality problem.

We should note that in principle an initial regressor size larger than 10 could be considered and it should be expected to improve the accuracy. However, for sizes above 15-20, an exhaustive search becomes too computationally expensive and finally unfeasible with current computational resources. A regressor size of 10 has thus been selected as a twofold heuristic compromise: first, the whole space of possible selections can be explored within a reasonable amount of time (around 1-6 hours approximately for 50 predictions, with current computers depending on the computer configuration). Second, after variable selection, the number of inputs is sufficiently small so as to avoid the curse of dimensionality in nonlinear models. Larger regressor sizes, for which the DT estimate is lower, usually lead however to little improvement or even poorer performance of the models. We thus, select 10 as an initial regressor size that additionally leads to fuzzy inference systems with a number of inputs sufficiently small so as to be easy to read. The effect that different initial regressor sizes can have on final performance will be illustrated through some examples in section 2.5.

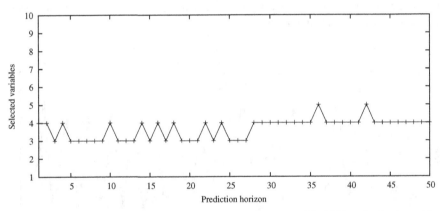

Fig. 2.3 ENSO: number of selected variables for horizon up to 50. DT-based selection with exhaustive search. Maximum regressor size 10.

As second stage, once input variables have been selected, the W&M algorithm is applied to the training set in order to identify fuzzy inference systems. These models are then tuned through supervised learning using the Levenberg-Marquardt algorithm over the training set. The process is repeated for increasing numbers of linguistic labels (membership functions) per input, starting from 2. Within this iterative process, the DT estimate is used to check whether the best possible approximation has been achieved, i.e., the right compromise between model complexity and training error has been found.

For the horizon 1 regressor, table 2.1 shows the number of rules identified for different numbers of linguistic labels per input (between 2 and 15). Training and validation errors are shown as well. The two columns labeled "before tuning" show the errors for the fuzzy systems as identified by means of the W&M algorithm, while the columns labeled "after tuning" show the errors for the systems tuned by means of supervised learning.

After the tuning substage, there is a considerable accuracy improvement. In particular, it can be seen that tuned systems with a low number of rules perform better than untuned systems with a much greater complexity. Thus the supervised learning substage also contributes to reducing model complexity.

Table 2.1 Number of membership functions and rules as well as errors for prediction horizon 1. Exhaustive DT-based selection of inputs. All errors as NMSE.

#MF	#Rules	Before tuning		After tuning	
		Training	Validation	Training	Validation
2	6	$2.833 \cdot 10^{-2}$	$2.899 \cdot 10^{-2}$	$1.479 \cdot 10^{-3}$	$1.705 \cdot 10^{-3}$
3	15	$8.813 \cdot 10^{-3}$	$1.016 \cdot 10^{-2}$	$1.250 \cdot 10^{-3}$	$1.558 \cdot 10^{-3}$
4	20	$4.190 \cdot 10^{-3}$	$4.884 \cdot 10^{-3}$	$1.189 \cdot 10^{-3}$	$1.580 \cdot 10^{-3}$
5	31	$2.709 \cdot 10^{-3}$	$3.113 \cdot 10^{-3}$	$1.082 \cdot 10^{-3}$	$1.616 \cdot 10^{-3}$
6	44	$1.986 \cdot 10^{-3}$	$2.466 \cdot 10^{-3}$	$1.009 \cdot 10^{-3}$	$1.738 \cdot 10^{-3}$
7	56	$1.868 \cdot 10^{-3}$	$2.617 \cdot 10^{-3}$	$9.228 \cdot 10^{-4}$	$1.794 \cdot 10^{-3}$
8	66	$1.453 \cdot 10^{-3}$	$1.978 \cdot 10^{-3}$	$9.509 \cdot 10^{-4}$	$1.869 \cdot 10^{-3}$
9	85	$1.289 \cdot 10^{-3}$	$1.915 \cdot 10^{-3}$	$8.676 \cdot 10^{-4}$	$1.979 \cdot 10^{-3}$
10	101	$1.229 \cdot 10^{-3}$	$1.920 \cdot 10^{-3}$	$7.509 \cdot 10^{-4}$	$2.153 \cdot 10^{-3}$
11	128	$1.130 \cdot 10^{-3}$	$2.043 \cdot 10^{-3}$	$6.104 \cdot 10^{-4}$	$2.602 \cdot 10^{-3}$
12	132	$1.114 \cdot 10^{-3}$	$2.113 \cdot 10^{-3}$	$5.848 \cdot 10^{-4}$	$2.491 \cdot 10^{-3}$
13	175	$1.121 \cdot 10^{-3}$	$2.139 \cdot 10^{-3}$	$4.902 \cdot 10^{-4}$	$2.816 \cdot 10^{-3}$
14	178	$1.006 \cdot 10^{-3}$	$2.194 \cdot 10^{-3}$	$4.426 \cdot 10^{-4}$	$3.455 \cdot 10^{-3}$
15	191	$9.713 \cdot 10^{-4}$	$2.126 \cdot 10^{-3}$	$4.793 \cdot 10^{-4}$	$2.865 \cdot 10^{-3}$

We also note that systems with a low number of linguistic labels per input (particularly between 2 and 5) are only very rough approximators before tuning. However, after the tuning substage their accuracy improve significantly while keeping the same rule base. This fact suggests that the rule bases correctly reflect the underlying dynamics of the series, though tuning the membership function parameters is no doubt required in order to build accurate models with such a low number of linguistic labels.

For selecting the proper complexity of the tuned fuzzy systems, a tolerance band above the DT estimate is considered. This band is defined by a threshold (DT-based threshold, $DTBT_h$) which increases with increasing horizons h according to equation 2.1, where DT_h is the DT estimate for horizon h:

$$DTBT_h = (1 + \min(0.90, 1.15 * h))DT_h. \tag{2.1}$$

For each horizon h, the simplest system that satisfies $MSE_h \leq DT_h$, where MSE_h is the training mean square error, is selected as the best autoregressor. This threshold has been defined on the basis of trial and error as a soft limit that favors simplicity to the detriment of accuracy. However, it was found to be robust for all the series analyzed. The definition is based on the following experimental observations:

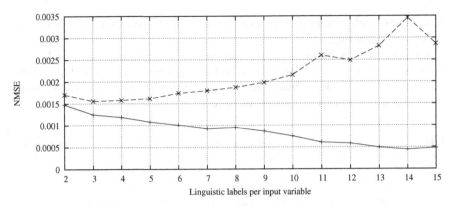

Fig. 2.4 ENSO: errors for horizon 1, exhaustive DT-based selection of inputs. Continuous line: training error. Dashed line: validation error.

- A tolerance approximately of 15% over the DT estimate for horizon 1 is appropriate.
- The best results can be achieved with tolerances increasing with the prediction horizon (particularly for the first 10 predictions approximately).
- A tolerance between 80%-100% over the DT estimates provides good results for long-term prediction.

We note though that the impact of the threshold is not determining for accuracy (the error increase is of the order of 10-20% at most for any prediction horizon). Similar results can be achieved by selecting a fixed adjustment factor of around 50%-75%. We chose the particular values in equation 2.1 so as to favor model simplicity to the detriment of accuracy.

For the ENSO series, $DTBT_1 \approx 1.26^{-3}$ and, as shown in figure 2.4, the fuzzy system with 3 linguistic labels per input is chosen as autoregressor for horizon 1.

Figure 2.5 shows the normalized DT NNE estimates (NDT-NNE) for prediction horizons up to 50 as well as the training and validation errors of the fuzzy autoregressors built and selected according to our methodology.

We note that besides the limitations of the fuzzy modeling techniques being employed, an additional source of error has been introduced in the proposed methodology: the DT-based selection of complexity does not guarantee optimal selection under all conditions. Although the fuzzy regressor for horizon 1 prediction that is chosen is the one with the lowest validation error, this is not the case for all horizons. In general, the deviation from the optimal selection depends on the time series being modeled and the prediction horizon.

By comparing the validation errors of the systems actually selected against the lowest validation errors that could have been achieved for any complexity we can know the order of magnitude of the error due to the DT-based selection of complexity. Figure 2.6 compares the NDT estimate (a robust estimation of the lowest training error that can be achieved without overfitting), the validation errors of the

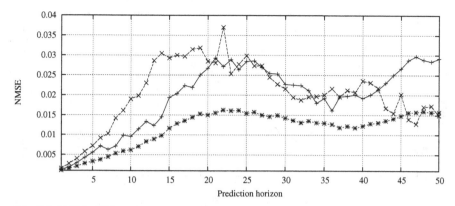

Fig. 2.5 ENSO: NDT estimates (∗), training (+) and validation (×) errors of fuzzy autore-gressors. Maximum regressor size 10. DT-based selection of inputs.

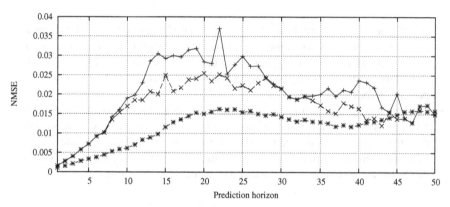

Fig. 2.6 ENSO: NDT estimates (∗), test errors for the selected fuzzy autoregressors (+), validation errors for the optimal complexity selections (×). Maximum regressor size 10. DT based selection of inputs.

fuzzy autoregressors selected according to the DT estimate, and the lowest possible validation errors for any number of linguistic labels.

Figure 2.7 shows the predictions for the first 50 values after the training set to-gether with a fragment of the actual time series.

Finally, we compare the accuracy of fuzzy models against LS-SVM models with the same autoregressor size and input selection. LS-SVMs were built for the same training subset selecting Radial Basis Function (RBF) kernels, grid search as opti-mization routine and cross-validation as cost function, see [35] for a detailed speci-fication of these and other options. Figure 2.8 shows the training and generalization errors for LS-SVM and fuzzy models. Averages errors are listed in table 2.2. Two main conclusions can be drawn from the comparison:

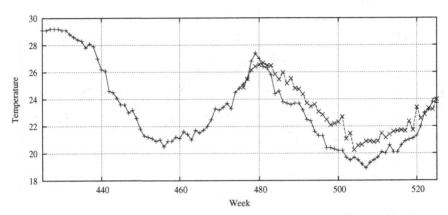

Fig. 2.7 ENSO: prediction of 50 values after the training set. Continuous line: actual time series. Dashed line: predictions.

- As for generalization capability, the performance of fuzzy autoregressors is clearly better than that of LS-SVM models. There are 4 exceptions: test errors of fuzzy autoregressors are slightly higher (less than 5%) for horizons 13 to 16. However, the overall superiority of fuzzy regressors is specially evident for long-term prediction (beyond horizon 25).
- Training and generalization errors are much closer for fuzzy models than for LS-SVM models. For long-term prediction, generalization errors may be even lower than training errors. Also, generalization errors are within approximately 200% of training errors for the worst cases. Thus, training errors of fuzzy models can be trusted as more realistic estimations of actual prediction errors.

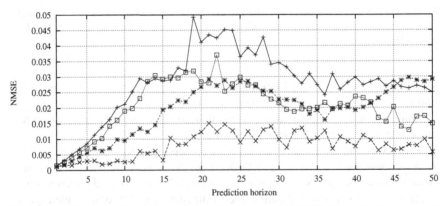

Fig. 2.8 ENSO: comparison of our methodology against LS-SVM. Generalization errors of LS-SVM models (+). Generalization errors of fuzzy models (□). Training errors of fuzzy models (∗). Training errors of LS-SVM models (×).

2.5 Experimental Results

In this section, the proposed concrete implementation of the methodology framework described is applied to a number of varied time series prediction problems from different fields of application, namely the Poland electricity time series prediction benchmark, the monthly averaged sunspot number, the daily averaged aggregated traffic in the Internet2 backbone network, the laser generated data set of the Santa Fe time series competition, the Mackey-Glass series and one of the series of the NN3 forecasting competition for neural networks and computational intelligence.

For every series, models are built to predict the next 50 values. Though one of the major goals of the methodology proposed here is to avoid the need for validation and test series, we will split the series into two subsets in order to assess the performance of the residual noise estimator and algorithms being used. Results will be compared against those of analogous LS-SVM models built using the same input selections, RBF kernels, grid search as optimization routine and cross-validation as cost function.

2.5.1 Poland Electricity Benchmark

This time series (PolElec henceforward) represents the normalized average daily electricity demand in Poland in the 1990s. The benchmark consists of a training set of 1400 samples, shown in figure 2.9, and a test set of 201 samples, shown in figure 2.10. It has been shown that the dynamics of this time series is nearly linear [17]. Besides the yearly periodicity, a clear weekly periodicity can be seen on smaller time scales (see figure 2.10).

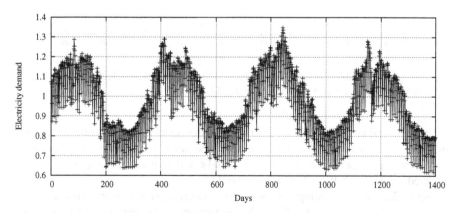

Fig. 2.9 PolElec: training series (1400 samples)

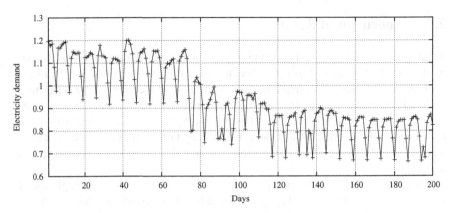

Fig. 2.10 PolElec: test series (201 samples)

We will show the results obtained for two different maximum regressor sizes: 7 and 14. In both cases, input selection was performed by exhaustive search of the lowest DT estimate. The number of selected variables is shown in figure 2.11.

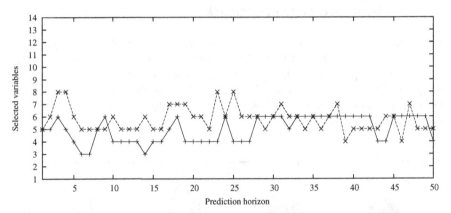

Fig. 2.11 PolElec: number of selected variables (exhaustive DT-based selection). Continuous line: regressor size 7. Dashed line: regressor size 14.

Training and test errors of a set of fuzzy autoregressors for horizon one are shown in figure 2.12 for different numbers of linguistic labels per input, in the case of a maximum regressor size of seven. The regressor with five membership functions is selected according to the DT-based threshold.

Figure 2.13 shows training and test errors of fuzzy regressors with different numbers of linguistic labels for prediction horizon seven (also in the case of a maximum regressor size of seven). The system with two linguistic labels is selected according to the DT-based threshold. However, the system with 3 linguistic labels achieves the lowest test error. This is an illustrative case in which a simpler and less accurate

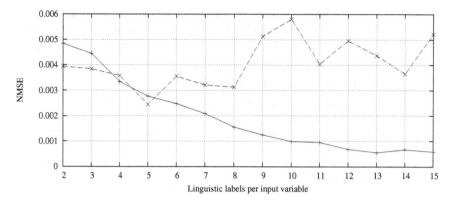

Fig. 2.12 PolElec: training (continuous line) and test (dashed line) errors against linguistic labels per input for horizon 1. Exhaustive DT-based selection of variables with regressor size 7.

model is selected because of the permissive nature of the DT-based threshold. Besides a lower number of linguistic labels, the system with two linguistic labels per input has eight rules, whereas the system with three linguistic labels per input has 15 rules.

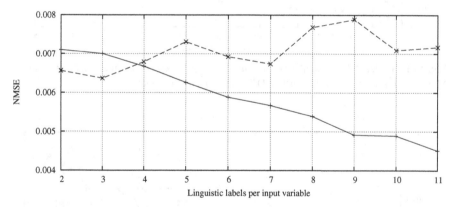

Fig. 2.13 PolElec: training (continuous line) and test (dashed line) errors against linguistic labels per input for horizon 7. Exhaustive DT-based selection of variables with regressor size 7.

For seven steps ahead prediction, using the notation for discrete time series introduced in section 2.3, three input variables are selected to predict y_{t+7}: y_t, y_{t-1} and y_{t-5}. As an example of the interpretability of the models developed, let us suppose that the last seven daily electricity demand measurements that are available correspond to the demand for a week from Monday through Sunday. Then, the fuzzy autoregressor predicts the demand for next Sunday based on the last known daily demand (Sunday), the demand of last Saturday and the demand of last Tuesday. The inputs can be called Sunday, Saturday and Tuesday. The two fuzzy sets defined for each input can be labeled as Low and High, and the output of the regressor can be called NextSunday. Considering this notation, a sample rule from the rule base of the regressor can be expressed as follows:

IF Tuesday was High AND Saturday was Low AND

Sunday was Low THEN NextSunday is "0.92"

Where "0.92" is used as linguistic label for a singleton output centered approximately at 0.92.

Figures 2.14 and 2.15 show the DT estimates as well as training and test errors for the two regressor sizes used.

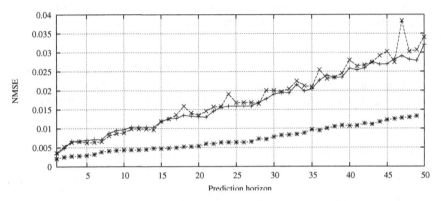

Fig. 2.14 PolElec: NDT estimates (∗), training (+) and test (×) errors of fuzzy autoregressors. Maximum regressor size 7. Exhaustive DT-based selection of inputs.

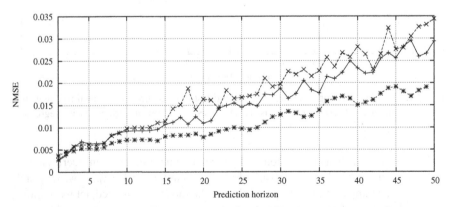

Fig. 2.15 PolElec: NDT estimates (∗), training (+) and test (×) errors of fuzzy autoregressors. Maximum regressor size 14. Exhaustive DT-based selection of inputs.

The average training and test error of LS-SVM models are shown together with the errors of fuzzy models in table 2.2. Fuzzy autoregressors achieve a greater approximation accuracy for the test subset. In this case, there are no exceptions for any prediction horizon, and the differences are higher than in the case of the ENSO series. We also note that for this series test errors are bounded within a range of 133% of training errors.

Table 2.2 Training and test errors of LS-SVM and fuzzy models averaged for horizons 1 through 50. All errors as NMSE. Maximum regressor size specified between parenthesis.

Series	LS-SVM Training	LS-SVM Test	Fuzzy inference Training	Fuzzy inference Test
ENSO (10)	$8.055 \cdot 10^{-3}$	$3.192 \cdot 10^{-2}$	$1.943 \cdot 10^{-2}$	$2.043 \cdot 10^{-2}$
PolElec (7)	$1.158 \cdot 10^{-2}$	$3.566 \cdot 10^{-2}$	$1.696 \cdot 10^{-2}$	$1.779 \cdot 10^{-2}$
PolElec (14)	$1.037 \cdot 10^{-2}$	$3.241 \cdot 10^{-2}$	$1.582 \cdot 10^{-2}$	$1.816 \cdot 10^{-2}$
Sunspots (9)	$1.338 \cdot 10^{-2}$	$3.284 \cdot 10^{-2}$	$1.691 \cdot 10^{-2}$	$2.623 \cdot 10^{-2}$
Sunspots (12)	$9.637 \cdot 10^{-3}$	$3.024 \cdot 10^{-2}$	$1.590 \cdot 10^{-2}$	$2.546 \cdot 10^{-2}$
AbileneI (7)	$8.587 \cdot 10^{-3}$	$2.476 \cdot 10^{-2}$	$1.448 \cdot 10^{-2}$	$1.732 \cdot 10^{-2}$
AbileneI (12)	$6.771 \cdot 10^{-3}$	$2.153 \cdot 10^{-2}$	$1.228 \cdot 10^{-2}$	$1.506 \cdot 10^{-2}$
SFL (10)	$1.481 \cdot 10^{-3}$	$6.578 \cdot 10^{-3}$	$1.020 \cdot 10^{-2}$	$1.285 \cdot 10^{-2}$
SFL (16)	$5.275 \cdot 10^{-4}$	$5.290 \cdot 10^{-3}$	$8.791 \cdot 10^{-3}$	$1.202 \cdot 10^{-2}$
MG (9)	$7.881 \cdot 10^{-4}$	$3.658 \cdot 10^{-3}$	$1.385 \cdot 10^{-2}$	$1.775 \cdot 10^{-2}$

2.5.2 Sunspot Numbers

The series of sunspot numbers is a periodic measure of the sunspot activity as a function of the number of spots visible on the face of the sun and the number of groups into which they cluster. Values from this series (Sunspots) are subject to uncertainty and noise, particularly during the past centuries. We analyze a series of monthly averaged sunspot numbers covering from January 1749 to December 2007, as provided by the National Geographical Data Center from the US National Oceanic and Atmospheric Administration[2]. The series is split into a set of 1000 values for training and a set of 2108 values for testing. The whole series is shown in figure 2.16.

Figure 2.17 shows the number of variables selected for the two maximum regressor sizes used for the Sunspots series: 9 and 12. Figures 2.18 and 2.19 show the DT estimates as well as training and test errors for the two regressor sizes chosen. The average training and test error of LS-SVM models are shown together with the errors of fuzzy models in table 2.2. For both regressor sizes, fuzzy autoregressors are more accurate with no exception for any of the prediction horizons.

[2] The series used here can be obtained from http://www.ngdc.noaa.gov/stp/SOLAR/ftpsunspotnumber.html. The International Sunspot Number is produced by the Solar Influence Data Analysis Center (SIDC) at the Royal Observatory of Belgium [36].

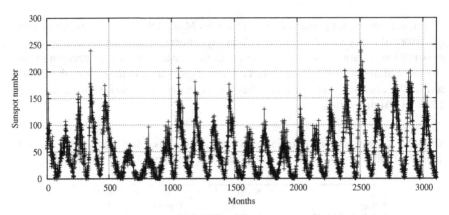

Fig. 2.16 Sunspots: training (first 1000 samples) and test (last 2098 samples) series

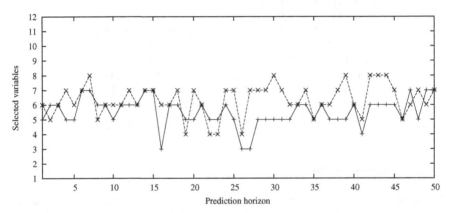

Fig. 2.17 Sunspots: number of selected variables (exhaustive DT-based selection). Continuous line: regressor size 9. Dashed line: regressor size 12.

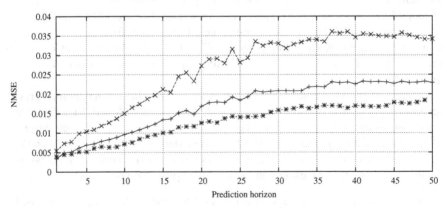

Fig. 2.18 Sunspots: NDT estimates (∗), training (+) and test (×) errors of fuzzy autoregressors. Maximum regressor size 9. Exhaustive DT-based selection of inputs.

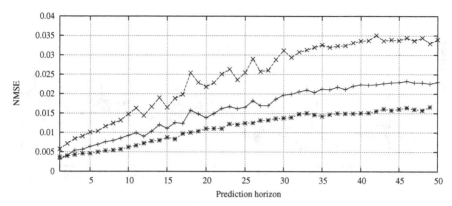

Fig. 2.19 Sunspots: NDT estimates (∗), training (+) and test (×) errors of fuzzy autoregressors. Maximum regressor size 12. Exhaustive DT-based selection of inputs.

2.5.3 Aggregated Incoming Traffic in the Internet2 Backbone Network

This series, Internet2 henceforward, represents the total amount of aggregated incoming traffic in the routers of the Abilene network, the Internet2 backbone, during several years. The Internet2 series consists of 1458 daily averages (in b/s), shown in figure 2.20 covering from the 4th of January of 2003 to the 31st of December of 2006. The data are available from the Abilene Observatory [12]. The daily averages for years 2003 and 2004 (the first 728 values) were selected as training set, whereas the daily averages for years 2005 and 2006 (the last 730 values) were selected as test set.

Figure 2.21 shows the number of variables selected for the two maximum regressor sizes considered for the Internet2 series: 7 and 12. Figures 2.22 and 2.23 show the DT estimates as well as training and test errors for the two regressor sizes chosen. The average training and test error of LS-SVM models are shown together with the errors of fuzzy models in table 2.2. Again, for both regressor sizes, fuzzy autoregressors are more accurate on the average and with no exception for any of the prediction horizons.

2.5.4 Santa Fe Time Series Competition: Laser Dataset

The laser data set of the Santa Fe Laser time series competition [32] (SFL) consists of 1000 training samples and 9000 test samples, as shown in figures 2.24 and 2.25, respectively. The series represents the intensity of a far-infrared-laser in a chaotic state, measured in a physics laboratory experiment. This time series is a cross-cut through periodic to chaotic pulsations of the laser. Chaotic pulsations can be closely modeled using the theoretical Lorenz model of a two level system [39].

This series is a good example of noise-free complicated behavior in a clean, stationary, low-dimensional physical system for which the underlying dynamics is well

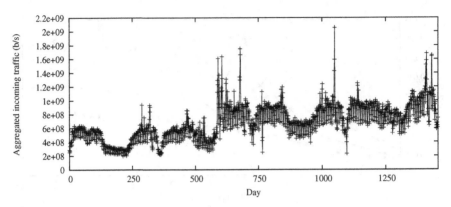

Fig. 2.20 Internet2: daily averaged aggregated incoming traffic in the Abilene backbone for 1458 days. Training series (first 728 values) and test series (last 730 values).

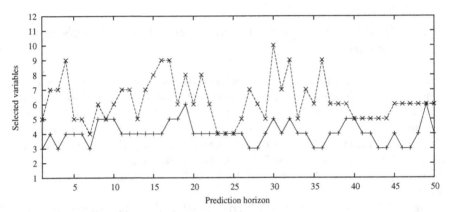

Fig. 2.21 Internet2: number of selected variables (exhaustive DT based selection). Continuous line: regressor size 7. Dashed line: regressor size 12.

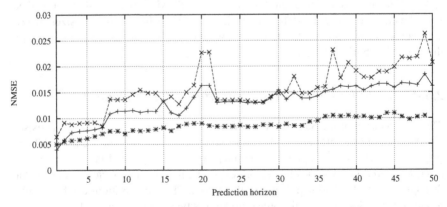

Fig. 2.22 Internet2: NDT estimates (∗), training (+) and test (×) errors of fuzzy autoregressors. Maximum regressor size 7. Exhaustive DT-based selection of inputs.

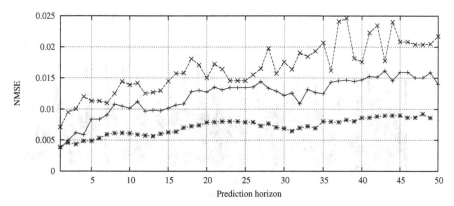

Fig. 2.23 Internet2: NDT estimates (∗), training (+) and test (×) errors of fuzzy autoregressors. Maximum regressor size 12. Exhaustive DT-based selection of inputs.

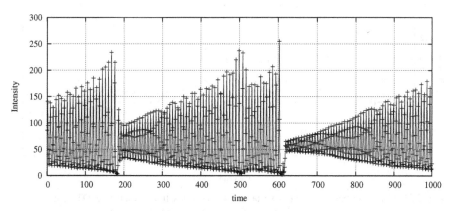

Fig. 2.24 SFL: training series (1000 samples)

understood. The data set is very predictable on short time scales because of the relatively simple oscillations. However, the rapid decay of the oscillations are events harder to predict.

In this case, we develop fuzzy autoregressors for two maximum sizes: 10 (for which exhaustive search of DT estimates is applied) and 16 (for which the exhaustive search is extended with a forward-backward search up to size 16). The number of variables selected for both cases is shown in figure 2.26.

Figure 2.27 shows training and test errors of fuzzy regressors with different numbers of linguistic labels for predicting horizon 1, for the case with maximum regressor size 10. The autoregressor with 8 linguistic labels is the first to fall within the DT-based threshold. However, the test error is nearly monotonically decreasing with the number of linguistic labels.

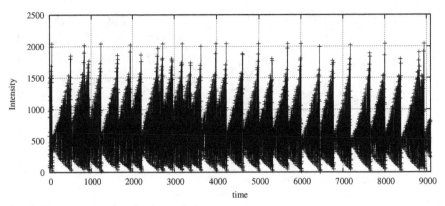

Fig. 2.25 SFL: test series (9093 samples)

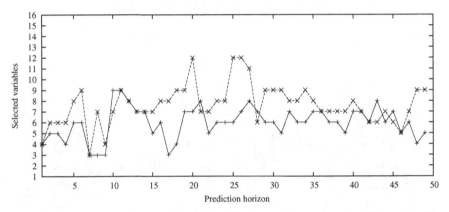

Fig. 2.26 SFL: number of selected variables. Continuous line: exhaustive DT search with maximum regressor size 10. Dashed line: forward-backward DT search with maximum regressor size 16.

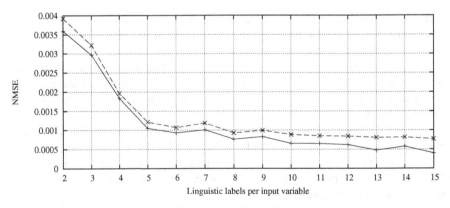

Fig. 2.27 SFL: training (continuous line) and test (dashed line) errors against linguistic labels per input. Horizon 1. Maximum regressor size 10. Exhaustive DT-based selection of inputs.

This fact suggests that the number of linguistic labels required to achieve the lowest possible test errors may be even higher than 15. This phenomenon occurs for short-term prediction horizons but is less evident for longer-term predictions. Figure 2.28 shows training and test errors for different numbers of linguistic labels for prediction horizon 35. A similar behavior can be observed for regressors with maximum size set to 16.

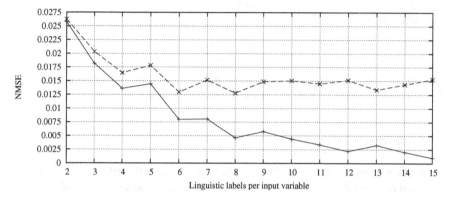

Fig. 2.28 SFL: training (continuous line) and test (dashed line) errors against linguistic labels per input. Horizon 35. Maximum regressor size 10. Exhaustive DT-based selection of inputs.

Figures 2.29 and 2.30 show the DT estimates as well as training and test errors for the two regressor sizes used. As shown in table 2.2, for this series LS-SVM based autoregressors clearly outperform their fuzzy counterpart.

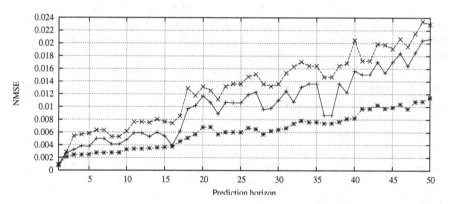

Fig. 2.29 SFL: NDT estimates (∗), training (+) and test (×) errors of fuzzy autoregressors. Maximum regressor size 10. Exhaustive DT-based selection of inputs.

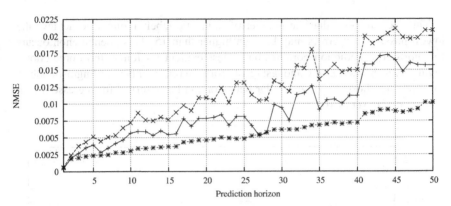

Fig. 2.30 SFL: NDT estimates (∗), training (+) and test (×) errors of fuzzy autoregressors. Maximum regressor size 16. Forward-backward DT-based selection of inputs.

2.5.5 Mackey-Glass Series

The Mackey-Glass time series [21] (MG henceforth) is fully deterministic and is generated numerically, as opposed to the time series analyzed so far. It is often used in the literature for evaluating fuzzy systems identification and prediction methods [1, 38, 16, 31, 13, 7, 25]. The time series is defined by the following differential equation:

$$\frac{dy(t)}{dt} = \frac{0.2y(t-\tau)}{1+y^{10}(t-\tau)} - 0.1y(t)$$

When $\tau > 17$, the series exhibits chaotic behavior. Higher values of τ yield higher dimensional chaos. In this section, a discrete time series is generated using the 4th order Runge-Kutta numerical integration method with $\tau = 30$.

A series of 1500 values (see figure 2.31) was generated and splitted into a set of 500 samples for training and a set comprising the remaining 1000 samples for test. As in [38], we use a maximum regressor size of 9. We note though that accurate long-term prediction would require greater regressor sizes.

Figure 2.32 shows the number of selected variables for horizons up to 50. Figure 2.33 shows the training and test errors together with the DT estimates. From table 2.2, it is evident that LS-SVM models achieve a greater accuracy averaged for horizons 1 through 50.

For comparison purposes with the literature about fuzzy modeling of the Mackey-Glass series, we examine the 1 step ahead autoregressor for the MG series. For a regressor size of 9, the inference system has only two inputs, both with 5 linguistic labels, and 13 rules. In spite of the simplicity of this system, its test error is approximately 9% lower than the DT estimate.

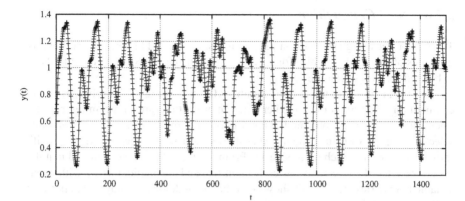

Fig. 2.31 MG: fragment of the Mackey-Glass series (1500 samples). The first 500 samples are selected as training set. The remaining 1000 samples are selected as test set.

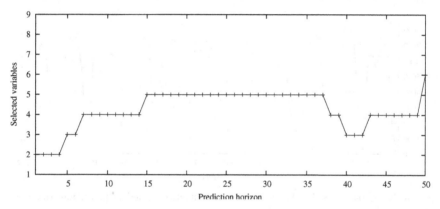

Fig. 2.32 MG: Number of selected variables for horizons up to 50. Exhaustive DT-based selection of inputs. Maximum regressor size 9.

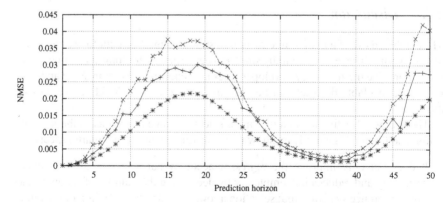

Fig. 2.33 MG: NDT estimates (*), training (+) and test (×) errors of fuzzy autoregressors. Exhaustive DT-based selection of inputs. Maximum regressor size 9.

2.5.6 NN3 Competition

The NN3 forecasting competition [29] comprises a set of 111 series with monthly measures of financial variables for several years. The next unknown 18 values have to be predicted for each time series.

Here we analyze the time series number 104, belonging to the subset B of the competition. The time series together with the predictions are shown in figure 2.34. These predictions were obtained using a maximum regressor size of 18. Variable selection was performed through exhaustive search up to size 12 extended with forward-backward search up to size 18. From the plot, it can be concluded that the cyclic behavior of the series is correctly identified and the predictions are within reasonable boundaries. This result shows that the methodology employed can perform well when the training series is small.

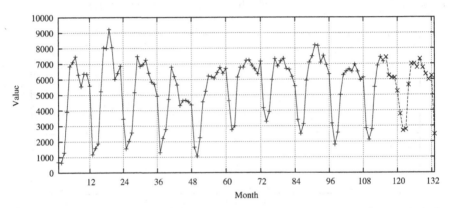

Fig. 2.34 NN3_104 series. 115 known values (continuous line) and 18 predictions (dashed line).

2.5.7 Discussion

The combined use of a nonparametric noise estimation method with fuzzy modeling techniques has been experimentally shown to perform well for long-term time series prediction. The methodology developed does not require a validation stage and thus the whole available data set can be used as training data to build autoregressive models.

The use of DT estimates in a first input selection stage as well as in the identification and tuning stage has been shown to be advantageous in two main aspects:

• It does not only improve the regressor accuracy but also increases its interpretability and reduces its complexity by decreasing the number of inputs to the fuzzy inference system. Input selection allows for a drastic reduction of the number of inputs. For instance, it is specially clear for the MG series, for which only

2 inputs out of 9 are selected for short-term prediction. This complexity reduction is also very clear in the case of the PolElec series with maximum regressor size 14.

• It has been shown to be a robust solution to the problem of selecting the proper system complexity.

A tolerant DT-based threshold has been defined. In general, the optimal threshold is dependent on the nonlinear approximation technique employed, i.e., fuzzy operators, membership functions, inference model as well as the identification and tuning methods. This threshold can be thus understood as a hint on what degree of accuracy can be expected from a particular fuzzy modeling technique. By using a tolerant DT-based threshold, we have favored simplicity to the detriment of accuracy.

All these factors contribute to a methodology for building fuzzy inference models that are both accurate and interpretable for both short-term and long-term prediction. In addition, fuzzy models have been shown to clearly outperform LS-SVM models in terms of prediction accuracy in the case of noisy time series for which there are no satisfactory deterministic models available.

A remarkable property of the fuzzy regressors developed with our methodology is their generalization capability. Test errors have been found to be very close to training errors. The difference between them is typically no more than 20-30% except in the case of the Sunspots series, where test errors are approximately 60% higher than training errors. While LS-SVM are usually praised for their good generalization performance, fuzzy autoregressors exhibit a much lower degree of overfitting.

On the other hand, It has been shown that LS-SVM models achieve a greater accuracy than fuzzy models for a specific type of series represented by the SFL and MG series. The MG series is purely deterministic, whereas the SFL series is nearly deterministic. Both are noise-free and stationary, can be predicted with relatively simple dynamical models and can be approximated with a very high accuracy. This fact leads us to conclude that in the absence of noise and perturbations, fuzzy inference based autoregression may not be a proper technique if the main objective is approximation accuracy and interpretability is a secondary objective. This type of series is however not common in real world applications. In addition, the higher accuracy of LS-SVM does not come at no cost. For the MG (9), SFL (10), and SFL (16) series, the construction and optimization of the LS-SVM model requires approximately 37, 35 and 103 times more run time respectively, as shown in table 2.3. Thus, the methodology proposed here, while clearly less accurate for this kind of series, is still significantly faster and exhibits less overfitting.

As far as computational requirements is concerned, the proposed methodology has a very low cost compared against the LS-SVM method. A software tool, xftsp [24], has been developed that implements the methodology as a whole and provides support for the identification and tuning algorithms included in the Xfuzzy development environment [40]. This Java based implementation of the methodology presented here is consistently between 1 and 2 orders of magnitude faster than the implementation of LS-SVM used for this study: the optimized C version of the LS-SVMlab1.5 Matlab/C toolbox [20]. Table 2.3 shows the time required to build time series models with both methods for a subset of the time series considered here. Memory consumption is also much lower for the fuzzy methodology, which enables it to be applied to large training series beyond the few thousand samples current practical limitation of LS-SVM models.

Table 2.3 Run time (in seconds) required to build models for prediction horizons 1-50. All tests were run on the same system, with no significant competing load. Maximum regressor size specified between parenthesis.

Series	LS-SVMlab1.5	Fuzzy inference
ENSO (10)	$3.45 \cdot 10^5$	$1.05 \cdot 10^4$
PolElec (7)	$3.04 \cdot 10^5$	$1.05 \cdot 10^4$
PolElec (14)	$9.91 \cdot 10^5$	$2.30 \cdot 10^4$
Sunspots (9)	$3.10 \cdot 10^5$	$1.04 \cdot 10^4$
Sunspots (12)	$2.42 \cdot 10^5$	$1.22 \cdot 10^4$
AbileneI (7)	$1.40 \cdot 10^5$	$1.75 \cdot 10^3$
AbileneI (12)	$1.27 \cdot 10^5$	$4.69 \cdot 10^3$
SFL (10)	$1.28 \cdot 10^6$	$3.49 \cdot 10^4$
SFL (16)	$1.61 \cdot 10^6$	$4.55 \cdot 10^4$
MG (9)	$3.64 \cdot 10^5$	$3.54 \cdot 10^3$

The fact that the test results are improved when a DT-based threshold higher than the DT estimate itself is introduced, leads us to two remarks on the performance of the identification and tuning stage:

- There is likely room for improvement of the identification and tuning procedures.
- The DT-based threshold can be seen as an aggressiveness index. 1 would be the most aggressive option, most often leading to overfitting and high complexity. Values in the range $[1.2, 2]DT_h$ are reasonable for the identification and learning techniques employed, most often leading to both low complexity and overfitting.

The fact that the impact of the DT-based threshold is very similar for all the series analyzed leads us to conclude that it is a factor eminently dependent on the identification and learning procedure and its inner limitations. Other methods for fuzzy inference systems identification, tuning and simplification exist, and the ones used here could be improved. This is an area of future research.

For complex and noisy time series, it is common that the most simple fuzzy system that can be built (the one with 2 linguistic labels per input) is comparable in accuracy to the LS-SVM model. For example, the fuzzy system with 2 linguistic labels per input for horizon 1 prediction of the PolElec series outperforms LS-SVM with the same input selection. In this case, the test error of the fuzzy regressor is approximately 35% lower.

In general, it can be concluded that fuzzy systems with the minimum number of linguistic terms, though not optimal in terms of accuracy, provide a reasonable approximation to the best system that can be built. Thus, it is easy to obtain very simple approximate models that ease the understanding of the time series dynamics.

We have developed an automatic methodology framework for long-term time series prediction by means of fuzzy inference systems. Experimental results for a concrete implementation of the methodology confirm good approximation accuracy and generalization capability. Linguistic interpretability and low computational requirements are two remarkable advantages over common time series prediction methods.

A fundamental advantage of autoregressive time series prediction with fuzzy inference systems is the fact that the models constructed consist of linguistic rules that can be interpreted by humans. For some time series, the most accurate rulebases have a low number of rules (below 10-15 rules), which makes it easy to extract a linguistic explanation for the system dynamics.

Several procedures have been shown to play a key role in achieving good approximation accuracy and low overfitting while keeping the complexity low: variable selection, application of a supervised learning method for tuning after identification, and using DT-NNE for selecting the proper number of linguistic labels per input. Also, when systems have a high number of rules and are thus not interpretable by humans in practice, there is still the possibility to build simpler, approximate models with a degree of accuracy of the same order.

2.6 Conclusions

A methodology for long-term time series prediction by means of fuzzy inference systems has been presented and validated through a number of benchmarks. In this methodology, the use of DT, a nonparametric residual variance estimation technique, provides two advantages: it is used for optimal a priori input selection and it provides a quantitative estimation that can be used for driving the model learning process.

This methodology has been shown to outperform LS-SVM models in terms of accuracy, interpretability and computational cost for an heterogeneous set of benchmarks. It has been extensively applied to a wide set of packet level traffic traces in order to extract linguistic models of traffic load at different time scales.

References

[1] Angelov, P.P., Filev, D.P.: An approach to online identification of takagi-sugeno fuzzy models. IEEE Transactions on Systems, Man and Cybernetics - Part B: Cybernetics 34(1), 484–498 (2004)
[2] Battiti, R.: First and Second Order Methods for Learning: Between Steepest Descent and Newton's Method. Neural Computation 4(2), 141–166 (1992)
[3] Bellman, R.E.: Dynamic Programming, 1957th edn. Dover Publications Inc., New York (2003) (republished) ISBN: 0486428095
[4] Box, G., Jenkins, G.M., Reinsel, G.: Time Series Analysis: Forecasting & Control, 3rd edn. Prentice Hall, Englewood Cliffs (1994) ISBN: 0130607746

[5] Chang, Y.H.O., Ayyub, B.M.: Fuzzy regression methods - a comparative assessment. Fuzzy Sets and Systems 119(2), 187–203 (2001)
[6] Chatfield, C.: The Analysis of Time Series. An Introduction, 6th edn. CRC Press, Boca Raton (2003) ISBN: 1-58488-317-0
[7] Chiu, S.L.: A Cluster Estimation Method with Extension to Fuzzy Model Identification. In: IEEE Conference on Fuzzy Systems, pp. 1240–1245. IEEE World Congress on Computational Intelligence, Orlando (1994)
[8] ESTSP07, ESTSP 2007 European Symposium on Time Series Prediction: Prediction Competition (2008), http://www.estsp.org
[9] Fletcher, R.: Practical Methods of Optimization, 2nd edn. Wiley Interscience, New York (2000) ISBN: 978-0-471-49463-8
[10] Guillaume, S.: Designing fuzzy inference systems from data: An interpretability-oriented review. IEEE Transactions on Fuzzy Systems 9(3), 426–443 (2001)
[11] Higgins, C.M., Goodman, R.M.: Fuzzy Rule-Based Networks for Control. IEEE Transactions on Fuzzy Systems 2(1), 82–88 (1994)
[12] internet2observatory, The Internet2 Observatory (2008), http://www.internet2.edu/observatory/
[13] Jang, J.S.R.: ANFIS: Adaptive-Network-Based Fuzzy Inference System. IEEE Transactions on Systems, Man and Cybernetics 23(3), 665–685 (1993)
[14] Jang, J.S.R., Sun, C.T., Mizutani, E.: Neuro-Fuzzy and Soft Computing A Computational Approach to Learning and Machine Intelligence. Prentice Hall, Upper Saddle River (1997) ISBN 0-13-61066-3
[15] Jones, A.J.: New Tools in Non-linear Modelling and Prediction. Computational Management Science 2(1), 109–149 (2004)
[16] Kasabov, N.K., Song, Q.: DENFIS: Dynamic Evolving Neural-Fuzzy Inference System and Its Application for Time-Series Prediction. IEEE Transactions on Fuzzy Systems 10(2), 144–154 (2002)
[17] Lendasse, A., Lee, J., Wertz, V., Verleysen, M.: Forecasting Electricity Consumption using Nonlinear Projection and Self-Organizing Maps. Neurocomputing 48(1), 299–311 (2002)
[18] Liitiäinen, E., Lendasse, A., Corona, F.: Non-parametric Residual Variance Estimation in Supervised Learning. In: Sandoval, F., Prieto, A.G., Cabestany, J., Graña, M. (eds.) IWANN 2007. LNCS, vol. 4507, pp. 63–71. Springer, Heidelberg (2007)
[19] Liitiäinen, E., Lendasse, A., Corona, F.: Bounds on the mean power-weighted nearest neighbour distance. Proceedings of the Royal Society A: Mathematical, Physical and Engineering Sciences 464(2097), 2293–2301 (2008)
[20] lssvmlab, Least Squares - Support Vector Machines Matlab/C Toolbox (2008), http://www.esat.kuleuven.ac.be/sista/lssvmlab
[21] Mackey, M.C., Glass, L.: Oscillations and Chaos in Physiological Control Systems. Science 197(4300), 287–289 (1977)
[22] Montesino Pouzols, F., Barriga Barros, A.: Automatic clustering-based identification of autoregressive fuzzy inference models for time series. Neurocomputing 73(10), 1937–1949 (2010)
[23] Montesino-Pouzols, F., Lendasse, A., Barriga, A.: Fuzzy Inference Based Autoregressors for Time Series Prediction Using Nonparametric Residual Variance Estimation. In: 17th IEEE International Conference on Fuzzy Systems (FUZZ-IEEE 2008), IEEE World Congress on Computational Intelligence, Hong Kong, China, pp. 613–618 (2008)

[24] Montesino-Pouzols, F., Lendasse, A., Barriga, A.: xftsp: a Tool for Time Series Prediction by Means of Fuzzy Inference Systems. In: 4th IEEE International Conference on Intelligent Systems (IS 2008), Varna, Bulgaria, pp. 2–2–2–7 (2008)

[25] Montesino Pouzols, F., Lendasse, A., Barros, A.B.: Autoregressive time series prediction by means of fuzzy inference systems using nonparametric residual variance estimation. Fuzzy Sets and Systems 161(4), 471–497 (2010)

[26] Moreno-Velo, F.J.: Un entorno de desarrollo para sistemas de inferencia complejos basados en lógica difusa. PhD thesis, University of Seville (2003)

[27] Moreno-Velo, F.J., Baturone, I., Sánchez-Solano, S., Barriga, A.: Rapid Design of Fuzzy Systems With Xfuzzy. In: 12th IEEE International Conference on Fuzzy Systems (FUZZ-IEEE 2003), St. Louis, MO, USA, pp. 342–347 (2003)

[28] Moreno-Velo, F.J., Baturone, I., Barriga, A., Sánchez-Solano, S.: Automatic Tuning of Complex Fuzzy Systems with Xfuzzy. Fuzzy Sets and Systems 158(18), 2026–2038 (2007)

[29] NN3Competition, NN3 Artificial Neural Network & Computational Intelligence Forecasting Competition (2008),
http://www.neural-forecasting-competition.com

[30] Pi, H., Peterson, C.: Finding the embedding dimension and variable dependencies in time series. Neural Computation 6(3), 509–520 (1994)

[31] Rong, H.J., Sundararajan, N., Huang, G.B., Saratchandran, P.: Sequential Adaptive Fuzzy Inference System (SAFIS) for nonlinear system identification and prediction. Fuzzy Sets and Systems 157(9), 1260–1275 (2006)

[32] SantaFeLaser, The Santa Fe Time Series Competition Data. Data Set A: Laser generated data (2008),
http://www-psych.stanford.edu/ andreas/
Time-Series/SantaFe.html

[33] Song, Q., Chissom, B.: Fuzzy time series and its models. Fuzzy Sets and Systems 54(3), 269–277 (1993)

[34] Sorjamaa, A., Hao, J., Reyhani, N., Ji, Y., Lendasse, A.: Methodology for Long-Term Prediction of Time Series. Neurocomputing 70(16-18), 2861–2869 (2007)

[35] Suykens, J.A.K., Van Gestel, T., De Brabanter, J., De Moor, B., Vandewalle, J.: Least Squares Support Vector Machines. World Scientific, Singapore (2002) ISBN: 981-238-151-1

[36] Van der Linden R.A.M., The SIDC Team: Online Catalogue of the Sunspot Index. RWC Belgium, World Data Center for the Sunspot Index, Royal Observatory of Belgium, years 1748-2007 (2008), http://sidc.oma.be/html/sunspot.html

[37] Wang, L.X.: The WM Method Completed: A Flexible System Approach to Data Mining. IEEE Transactions on Fuzzy Systems 11(6), 768–782 (2003)

[38] Wang, L.X., Mendel, J.M.: Generating Fuzzy Rules by Learning from Examples. IEEE Transactions on Systems, Man, and Cybernetics 22(4), 1414–1427 (1992)

[39] Weigend, A., Gershenfeld, N.: Times Series Prediction: Forecasting the Future and Understanding the Past. Addison-Wesley Publishing Company, Reading (1994) ISBN: 0201626020

[40] xfuzzy, Xfuzzy: Fuzzy Logic Design Tools (2008),
https://forja.rediris.es/projects/xfuzzy

Chapter 3
Predictive Models of Network Traffic Load

Abstract. Understanding the dynamics and performance of packet switched networks on the basis of measurements enables practitioners to optimize resources. As network measurement research further advances and new measurement tools and infrastructures are available, the task of network operation becomes more and more complex. In this chapter we apply the methodology developed in the previous chapter to time series concerning network traffic load. An extensive predictability analysis is performed using the same nonparametric residual variance estimation technique that is integrated into the prediction methodology. Based on the predictability results, fuzzy inference based models that are both interpretable and accurate are derived for a wide set of heterogeneous time series for network traffic.

3.1 Models for Network Traffic Load

Traffic load in packet switched networks is hard to model and predict. In general, most if not all classes of complex dynamical behavior can be observed in network traffic. Thus, traditional linear methods do not seem powerful enough so as to properly model and predict network traffic. Instead, nonlinear and nonparametric models have to be explored. Although some work along these lines has been done, much work remains.

Complex properties have been identified in traffic at different time scales. As outlined in section 1.5, models -and predictive models in particular- are sought at all time scales. Models at scales ranging from the order of the microseconds up to milliseconds are key to develop prediction-based control mechanisms. At scales of seconds, minutes and hours, predictive models can be used for network weather prediction services, useful for adaptive applications, network operators and users. At scales of weeks and months, predictive models are specially relevant for medium- and long-term capacity planning.

In this section, we will analyze time series for the traffic volume measured at certain network points for a given period. We will refer to these time series as network traffic time series. In principle, since models will be entirely derived from data, a

F.M. Pouzols et al.: Mining & Control of Network Traffic by Computational Intelligence, pp. 87–145.
springerlink.com   © Springer-Verlag Berlin Heidelberg 2011

direct physical interpretation is not possible. In order to address this issue and make it possible to interpret the models, we will apply the fuzzy inference based method developed in chapter 2.

Here we adopt a broad perspective on predictive models and analyze network traces at different time scales. In general, we will analyze 4 or 5 different scales differing by exactly one order of magnitude (ranging from a minimum of 1 ms to a maximum of 100 s and 1 day in a particular case). Not all traces are analyzed at all these time scales for different reasons, such as lack of enough measurements and low link speed that would render series at low time scales very sparse.

When developing a predictive model for a time series, multiple factors such as the features selected and the class of model have an impact on the final performance. The most direct approximation to the problem is to develop an autoregressive model. Many nonlinear modeling techniques exist. In this section we look into the possibility of building autoregressive nonlinear models for network traffic with satisfactory performance. We use the nonparametric residual variance estimation technique described above in order to analyze the predictability of network traffic load at different time scales.

Four different methods for predictive modeling are used: fuzzy inference systems with nonparametric residual variance estimation (NRVE-FIS), optimally pruned extreme learning machines (OP-ELM), induced OWA operators based nearest neighbors (IOWA-NN) and ARIMA models. Fuzzy inference models are built using the methodology described in the preceding section. The variable selection stage is performed not only for NRVE-FIS models but also for OP-ELM and IOWA-NN models. It is important to note that by using the ARIMA technique we do not aim at performing a thorough comparison between computational intelligence methods and stochastic methods. More elaborated linear and nonlinear stochastic methods that consider long-range dependence have been developed [10, 6, 4]. We used ARIMA models here as a reference method that is simple, feasible and well established.

Some previous works on predictive models for network traffic load can be found in the literature. These studies are however very specific regarding the time scale and kind of models taken into consideration. Long-term load prediction studies on the the NSFNET backbone [14], the Sprint Tier-1 backbone [28] and the Georgia Institute of Technology campus network [3], among others, reveal trends, periodicities and other predictable patterns and show that predictions can be made for long-term horizons within acceptable error bounds. Predictability analyses of network traffic with models assumptions have been performed as well. In particular, nonlinear threshold autoregressive models have been applied to filtered traffic traces [46]. Also, Markov models have been applied to modeling long-range dependence [2]. Also, Sang and Li [32] use of autoregressive moving average (ARMA) and Markov-modulated Poisson process (MMPP) models in order to assess predictability in a parametric manner. Qiao et al study the predictability of some network traces. However, the study is limited to one-step-ahead predictability and a small set of similar network traces. Yi et al. study one-step-ahead predictability as well using wavelet-based multiresolution analysis and a set of linear and nonlinear regressive models [45]. Recently, Cortez et al. [8] have attempted at predicting network traffic in several links belonging to the United Kingdom Education and Research Network (UKERNA) network using univariate and multivariate feed-forward neural network models. Two scales are taken into consideration for one step-ahead prediction in this work, 10 minutes and hourly intervals, where it is found that neural networks can perform better than some naive methods as well as Holt-Winters exponential smoothing models.

A related predictability problem arises in mobile environments is that of the predictability of mobility and its effects on bandwidth provisioning among other tasks [37].

Possible applications of traffic load prediction include overall weather services that provide predictions at a high scale, both temporal and spatial [41]. Predictors have been proposed as well for supporting real time video [1, 23], and admission control [36] among other applications. Many works have attempted at predicting video traffic, particularly in ATM scenarios. A concrete application of multiresolution learning neural networks to predicting VBR video traffic has been reported by Liang [23].

Predictability of traffic is important not only at network links but also from an end-to-end perspective [13]. As discussed in chapter 1, so-called formula based predictive models for the stationary behavior of TCP transfers have been available for some years. More recently, predictability of large TCP transfers has been addressed by He at al. [15], where both analytical models for TCP throughput and autoregressive models approaches are analyzed and it is found that autoregressive models can be quite accurate even for very short training samples (of the order of 10–20). However, the study is parametric and restricted to one-step ahead prediction as well as simple linear regressive models.

3.2 Analysis of Traffic Traces

In this chapter, we focus on aggregate traffic at network links. Links with different degrees of aggregation are studied. From a qualitative perspective, it is obvious that the amount of traffic is predictable to a certain degree for some networks. For instance, clear daily patterns can be identified in most links. Other weekly and monthly seasonal variations can be observed as well [28]. Current knowledge about network traffic and previous works seem to indicate that two factors have an important impact on predictability: the prediction horizon and the degree of traffic aggregation.

On the one hand, it can be expected that predictability will decrease as the prediction horizon increases, i.e., prediction errors increase as we try to predict further in the future. the other hand, it has been observed that traffic load in links with a high degree of aggregation is sometimes simpler to predict. A number of techniques for estimating complexity exist, including fractal dimension and Lyapunov exponents. Even if these invariants could be calculated accurately, they still do not provide enough information for the construction of a predictive model. Here we look into this issue from the point of view of time series analysis and prediction.

Some previous results [45] indicate that predictability does not necessarily increase monotonically with decreasing time resolution, i.e., smoother signals, that it is largely independent of the prediction mechanism and that simple models are sufficient while more complex non-stationary, nonlinear models are advantageous only at very coarse resolutions. As this section will show, time scale is a crucial parameter when looking at the questions above. Besides, both short-term and long-term prediction are addressed.

The approach followed here is to predict network traffic load in the form of the load series aggregated at a certain time interval. Thus, inter-arrival times are not modeled.

In order to apply a time series prediction technique, three general issues have to be addressed:

- choosing the input variables,
- defining the architecture and parameters of the model, and
- applying a criterion for selecting the best model.

Here we address the first issue using the delta test based a priori input selection scheme described in chapter 2 in the context of a methodology for time series prediction by means of fuzzy inference systems. This scheme is followed for all the modeling techniques used. In particular, the following modeling techniques are compared: fuzzy inference systems with the methodology described in chapter 2, optimally pruned extreme learning machines (OP-ELM) [24, 38] and nearest neighbors with induced OWA operators [43, 44] (IOWA-NN) as well as traditional autoregressive models. In particular, ARIMA models are analyzed as a representative case of traditional autoregressive models.

Each of these techniques define specific architectures, parameters and criteria for model selection, that were briefly described in chapter 1. More specifically, OP-ELM models are built using the following configuration options: a combination of linear, Gaussian and sigmoid kernels is used, a maximum of 100 neurons is allowed, and data are normalized before modeling. ARIMA models are built for maximum orders of 10. The model that minimizes the Akaike information criteria (AIC) is selected. IOWA-NN models are optimized by means of the simple iterative learning procedure proposed in [42]. These linear models have been included for comparison purposes and should be regarded as a lower bound on the acceptable performance of more complex models.

Models will be built for long-term prediction at all the time scales considered. In particular, the 30 next values will be modeled.

Besides, no preprocessing technique is applied. This includes smoothing techniques based on wavelets such as the one used in [28] which is an example of a filtering technique with a particular objective: approximate forecasting of long-term overall trend for provisioning. Thus, the training and test sets that will be used for building models in what follows consist in the aggregated traffic load as measured in a particular link. This way, we intend to assess the overall performance of the method presented here, independently of specific preprocessing steps which are usually dependent on the particular network and time scale of interest.

In general, in what follows we extract network traffic load time series from passive traces and then analyze them. The following overall principles apply:

- Some traces are available in the form of constant period series, such as the Bellcore traces analyzed in the next section. However, most of the traces analyzed here have been generated from packet level traces (which would allow for reconstructing the corresponding packet point process). In order to get a discrete time series from a packet level trace, a uniform time scale is needed. Time series have been generated by aggregating packet sizes at constant intervals. It is worthwhile to note that clock inaccuracies in packet arrival times can thus distort the final time series, specially for small periods. This will be pointed out for those series for which timestamping inaccuracies can be significant.

- For all series, at least 5 different time resolutions, as far as available, scaled by a factor of 10 are analyzed for the whole timespan of the trace (be it originally recorded as a constant period series or a packet level trace). The scales used range from 1 millisecond up to 1 day. This way, for example the WIDE-F-DITL-200701 trace is used to analyze a series of traffic load at 10ms, 100 ms, 1 s, 10 s and 100 s intervals.

- A particular case is that of the Internet2 Abilene backbone traffic measurements. For this trace, a packet level trace is not available. In addition, the time span depends on the time interval and the set of time intervals was fixed at measurement time as 1 day, 12 hours, 2 hours, 30 minutes and 5 minutes. The whole time span of the measurements carried out in the Abilene backbone is only available at the daily interval. For higher resolution measurements, the total time span is smaller.

- Load at the IP layer is considered, i.e., link header/trailer information is not included for accounting load. This is in line with the definition of MTU by the IETF, where the link layer headers are not summed up for computing the MTU, i.e, IP MTU as defined by the IETF is not the total frame size.

In general, a maximum regressor size of 10 was used as a reasonable bound that made it possible to perform the analysis outline above which took more than 400000 hours of computation. A regressor size of 10 is also a convenient choice from the point of view of the dimensionality of the nonlinear predictive models that are built after the input selection stage, as it was shown in chapter 2.

The symmetric absolute percent error (SAPE) is used instead of the absolute percent error (APE) in order to have more readable plots of an absolute error measure. As compared to the APE, the SAPE mitigates the effect of large prediction errors for small actual values, in which cases the APE measure has greatly oscillations that decrease the readability of figures.

For each network trace we perform a systematic analysis presented here by means of four figures, following the general scheme outlined below:

- In a first figure, a plot of the traffic load series at the different time scales considered is shown.

- Then, a second figure shows the residual variance estimates computed with the Delta Test. The intend of this figure is to give an overall idea of the predictability of the traffic load trace at different time scales. For each time scale two plots are shown.
  - In the first one, the normalized Delta Test estimates for prediction horizons 1, 2, 5 and 15 are shown for different subseries lengths (ranging from 200 up to 3000 or 3200 when available). The estimates shown are averages taken over 20 repetitions for randomly selected subseries. This plot serves two purposes: 1) the variance of the noise (or non-predictable component of the signal) at both short- and mid-term is depicted, and 2) the stability of the variance of the noise estimate with regards to the subseries length can be drawn from the variations in the NDT estimate.
  - The second plot shows a 3D representation of the NDT estimate against the regressor size (ranging from 1 to 10) and the prediction horizon (ranging from 1 to 30). This plot provides a visual representation of: 1) the effect of the maximum regressor size used on the lowest training error possible (that decreases as the maximum regressor size increases), and 2).
- A third figure shows for every time scale considered the training and test errors of the four predictive models considered here. The models are build using 1000 samples long (as far as available) training subseries and tested on 1000 samples long (as far as available) test subseries. The plots show average errors for up to 20 models built for each of up to 20 subseries (as far as available). The errors are shown for prediction horizons ranging from 1 to 30. Errors for NRVE-FIS and OP-ELM models are shown on the left, whereas errors for ARIMA and IOWA-NN models are shown on the right.
- In a fourth figure, an example of the results of the fuzzy inference predictive model is shown for one randomly selected training subseries of 1000 samples (as far as available) and a test subseries of 1000 samples (as far as available). This figure shows the SAPE for the whole training set (on the left side) as well as the prediction of the next 30 samples after the training set (on the right). The intend of the figure is to give an approximate idea of the order of the linear errors of the predictions on the test subseries as well as how close the short-term and medium-term predictions are to the real test subseries.

Some specific remarks will be made for individual time series throughout the next pages. However, for better readability the results will be discussed in detail in a unified manner in section 3.5, on page 130.

We will analyze traces made publicly available from a number of institutions and projects, including both traces of historic interest as well as more recent traces. Within the historic traces, we analyze the LBL traces, the BellCore traces and the DEC traces. The recent traces will be analyzed distributed among sections devoted to different classes of links: backbone links, exchange points, intercontinental links, access links and wireless links. Traces for two backbones, the Abilene network and the CAIDA OC48 measurement point will be analyzed. Regarding exchange traffic, traces recorded at the AMPATH exchange point will be analyzed. For intercontinental traffic, traces recorded at the measurement point of a link between Japan and the USA will be analyzed. Then, we will analyze traces for two access links corresponding to two research institutions. Finally, we analyze a trace recorded at a wireless access point.

Unless explicitly stated, traces collected by the Cooperative Association for Internet Data Analysis (CAIDA) can be obtained, with certain use and access restrictions, from the CAIDA Internet Data – Passive Data Sources site (http://www.caida.org/data/passive/) as well as from the Internet Measurement Data Catalog [33, 9] (http://datcat.org). Traces from the Internet Traffic Archive can be obtained from [20]. Additionally, traces originally collected by the National Laboratory for Network Research (NLANR) Passive Measurements and Analysis (PMA) project can be obtained from the PMA special traces archive [27]; traces collected by the MAWI working group of the WIDE project can be obtained from [40]; and traces belonging to the Community Resource for Archiving Wireless Data At Dartmouth (CRAWDAD) can be obtained from [19].

3.3 Series of the Internet Traffic Archive

The Internet Traffic Archive (ITA) [20] hosts a number of network traffic traces of high historical relevance. These include the the Bellcore traces and the traces taken at the Lawrence Berkeley National Laboratory. The first were the empirical basis for finding self-similarity and long-range dependence in Ethernet traffic [12, 21] whereas the second were instrumental in showing that the Poisson model can fail to capture the behavior of traffic in wide area networks [29]. These traces have been extensively analyzed throughout the literature and have been the basis for some remarkable developments in traffic modeling [22, 21, 12, 20].

3.3.1 LBL Traces

This set of traces includes the LBL-TCP-3 and LBL-PKT-4 and LBL-PKT-5 traces and correspond to the LBL-PKT-3, LBL-PKT-4 and LBL-PKT-5 traces used in [29]. The tracing was done on the Ethernet demilitarized zone network over which flows all traffic into or out of the Lawrence Berkeley Laboratory, located in Berkeley, California. All times were recorded in Pacific Standard Time and timestamps have microsecond precision. We should note that the LBL-CONN-7 trace available from ITA is not analyzed here since no information on the packet arrival process is available in the original trace. As the results obtained for the three LBL traces are similar to a significant extent, for better readability we only show here the results for the LBL-TCP-3 trace. The reader interested in details about other LBL traces is referred to [25].

3.3.1.1   LBL-TCP-3

This trace includes two hours of WAN TCP traffic between the Lawrence Berkeley Laboratory and the rest of the world. The trace ran from 14:10 to 16:10 on Thursday 20th of January, 1994, capturing 1.8 million TCP packets. About 0.0002% of these were dropped.

In figure 3.1, the traffic load series for the LBL-TCP-3 trace at 4 different time scales are shown. Figure 3.2 shows the DT based estimation of predictability of the series derived from the LBL-TCP-3 trace at different time scales. The column on the left shows for every time scale the averaged NDT for 100 randomly selected subseries of different lengths (horizons 1, 2, 5, 10 and 15, regressor size 10). The column on the right shows for every time scale the NDT for horizons 1-30 and regressor sizes 1-10 in the case of one randomly selected subseries of length 1000 (or less when not enough samples are available).

In figure 3.3, the training and test errors for the LBL-TCP-3 trace at different time scales are shown. Four models are used: a) fuzzy inference models built with the NRVE-FIS methodology and OP-ELM models (left column), and b) IOWA derived nearest neighbor models and ARIMA models (right). Prediction horizons 1 through 30 are considered.

Figure 3.4 shows an example of prediction made with fuzzy inference models for every time scale, for one random selection of a training set of up to 1000 training samples and up to 1000 test samples (as far as available). On the left, it is shown the test symmetric absolute percentage error (SAPE). On the right, it is shown the predictions of the next 30 values after the training set.

3.3.2   Bellcore Traces

The Bellcore dataset includes four traces each containing a million packet arrivals on an Ethernet link at the computing laboratory of the Bellcore Morristown Research and Engineering facility in the late 80s. These traces were the empirical basis for finding self-similarity and long-range dependence in Ethernet traffic [21]. The measurement techniques used to record these traces are discussed in detail in [12]. The first two traces, BC-pAug89 and BC-pOct89 contain packet arrivals for both local traffic and traffic to and from the Internet. The second two traces, BC-Oct89Ext and BC-Oct89Ext4 include only external packet arrivals, i.e., WAN traffic. Timestamps are provided in microsecond units but have a nominal precision of 4 microsecond. Further analysis by the authors of the traces indicate that the actual accuracy is limited to roughly 10 microseconds. All Ethernet packets were captured in these traces [20]. The interested reader is referred to [25], where a similar analysis to that shown above for the LBL-TCP-3 trace is detailed.

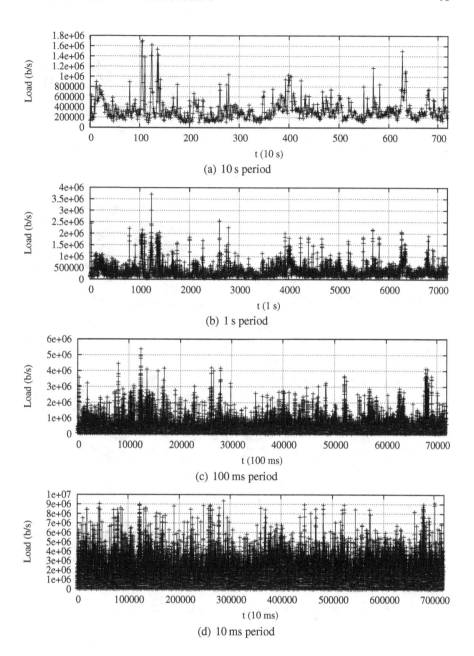

(a) 10 s period

(b) 1 s period

(c) 100 ms period

(d) 10 ms period

Fig. 3.1 LBL-TCP-3: traffic load series at different time scales

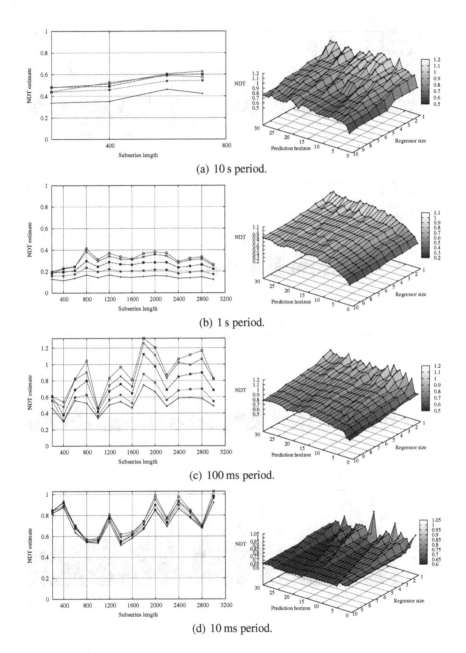

(a) 10 s period.

(b) 1 s period.

(c) 100 ms period.

(d) 10 ms period.

Fig. 3.2 LBL-TCP-3: predictability estimate for different time scales. NDT for prediction horizons 1, 2, 5, 10 and 15 are shown on the left for different subseries lengths (starting at random points, averaged for 100 repetitions). NDT for a randomly selected subseries of 1000 samples is shown on the right (horizons 1-30 and regressor sizes 1-10).

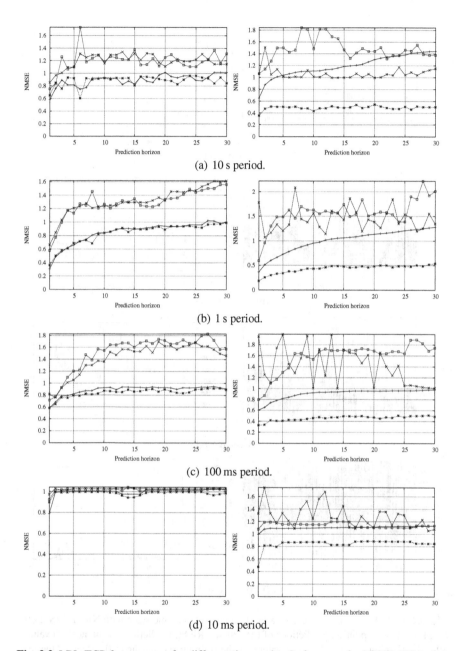

(a) 10 s period.

(b) 1 s period.

(c) 100 ms period.

(d) 10 ms period.

Fig. 3.3 LBL-TCP-3: test errors for different time scales. Left: errors for NRVE-FIS (+ for training, × for test) and OP-ELM (∗ for training, □ for test) models. Right: errors for ARIMA (+ for training, × for test) and IOWA-NN (∗ for training, □ for test) models.

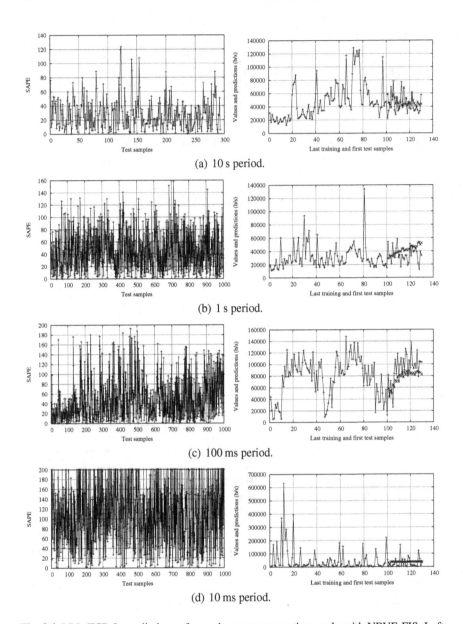

(a) 10 s period.

(b) 1 s period.

(c) 100 ms period.

(d) 10 ms period.

Fig. 3.4 LBL-TCP-3: predictions of a random test set per time scale with NRVE-FIS. Left: SAPE for one-step ahead predictions of the test set. Right: predictions of the next 30 values after the training set. The last 100 values of the training set, the first 30 values of the test set (both as +, continuous line) and the 30 corresponding predictions (×, dashed line) are plotted.

3.3.3   DEC Traces

Each of the four original DEC-PKT traces [20] recorded an hour of WAN traffic between Digital Equipment Corporation (DEC) and the rest of the world. The traces were recorded at the primary Internet access point of DEC, which is an Ethernet demilitarized zone (DMZ) network operated by DEC Palo Alto research groups. The traces correspond to the DEC-WRL-1 through DEC-WRL-4 traces used in [29]. All times were recorded in Pacific Standard Time and timestamps have millisecond precision.

The traces that we will call DEC-PKT-1 through DEC-PKT-4 correspond to the files dec-pkt-1.tcp through dec-pkt-4.tcp in the original traces. Only TCP traffic is thus taken into consideration here. UDP, encapsulated IP and other traffic is not analyzed since not enough information is included in the original traces in order to construct traffic load series (as packet sizes were recorded only for TCP packets). These traces and the associated time series that will be analyzed are thus transport protocol specific. Again, we refer the interested reader to [25].

3.4   Application to Recent Traffic Time Series

The more recent traffic time series will be analyzed in four groups: backbone traffic (for high capacity links internal to specific networks), exchange traffic (for normally lower capacity long distance links), access links and wireless traffic (including both campus range and event specific setups).

3.4.1   Backbone Traffic

In this section we analyze traffic traces recorded at two backbone links of high performance networks: the Internet2 Abilene network and the CAIDA OC-48 measurement point.

3.4.1.1   Aggregated Incoming and Outgoing Traffic in the Internet2 Backbone Network

This trace represents the total amount of aggregated incoming and outgoing traffic in the routers of the Abilene network, the Internet2 backbone, during several years. The data are available from the Abilene Observatory [16]. This trace is analyzed at the following time scales: 1 day, 12 hours, 2 hours, 30 minutes and 5 minutes. These are the time scales at which it is used to be plotted in the Internet2 observatory for online monitoring.

The daily series consists of 1458 daily averages covering from the 4th of January, 2003 to the 31st of December, 2006. Although measurements for some months of 2007 are also available, we have discarded all measurements taken on and after 1st of January, 2007. This is because some routers of the Abilene backbone started to be shutdown as part of the transition to the new Internet2 backbone in 2007 that took a few months. As a consequence of the transition, the topology of the Abilene network as well as the total amount of traffic traversing the Abilene routers were deeply affected. We thus prefer to exclude from our study those measurements taken during the transition.

Note as well that the series at 5 minutes, 30 minutes, 2 hours and 12 hours span from 4th of August, 2005 to 31st of December, 2006. Earlier values were not available at these time scales. Aggregated incoming traffic is analyzed in the subtrace Abilene-I whereas outgoing traffic is analyzed in the subtrace Abilene-O. Strong correlations between them can be observed in the plots of the two series at all the time scales examined. As a consequence of the clear symmetry between the two series, the short- and long-term predictability as well as the training and test errors for the different techniques applied are very similar. The interested reader can find the results for the Abilene-O trace in [25].

In figure 3.5, the traffic load series for the Abilene-I trace at 5 different time scales are shown. Figure 3.6 shows the DT based estimation of predictability of the series derived from the Abilene-I trace at different time scales. The column on the left shows for every time scale the averaged NDT for 100 randomly selected subseries of different lengths (horizons 1, 2, 5, 10 and 15, regressor size 10). The column on the right shows for every time scale the NDT for horizons 1-30 and regressor sizes 1-10 in the case of one randomly selected subseries of length 1000 (or less when not enough samples are available).

In figure 3.7, the training and test errors for the Abilene-I trace at different time scales are shown. Four models are used: a) fuzzy inference models built with the NRVE-FIS methodology and OP-ELM models (left column), and b) IOWA derived nearest neighbor models and ARIMA models (right). Prediction horizons 1 through 30 are considered.

Figure 3.8 shows an example of prediction made with fuzzy inference models for every time scale, for one random selection of a training set of up to 1000 training samples and up to 1000 test samples (as far as available). On the left, it is shown the test symmetric absolute percentage error (SAPE). On the right, it is shown the predictions of the next 30 values after the training set.

3.4.1.2   CAIDA OC48 Traces

Here we analyze the OC 48 measurements made in 2002 [34], for the OC48-20020814-0 and OC48-20020814-1 traces; January 2003, for the OC48-20030115-0 and OC48-20030115-1 traces; and March 2003 [35], for the OC48-20030424-0 and OC48-20030424-1 traces. All these traces belong to the CAIDA OC48 Traces Dataset. Extensive results are only shown for the OC48-20030424-1 trace. For an in depth analysis of all the traces, the interested reader is referred to [25].

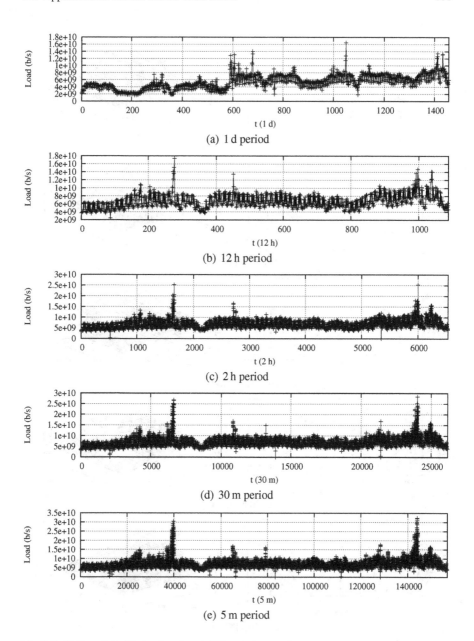

Fig. 3.5 Abilene-I: traffic load series at different time scales

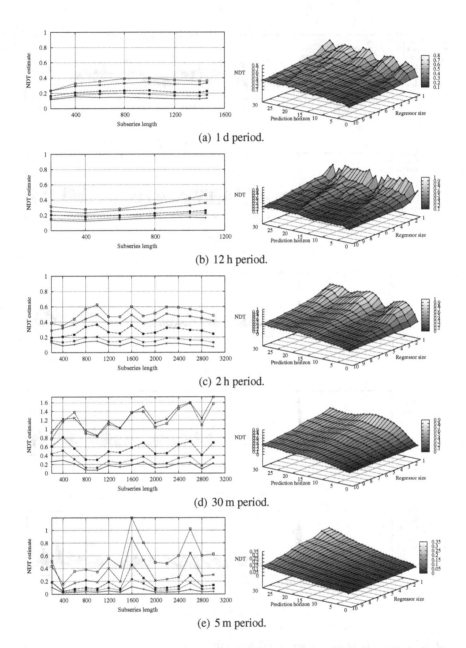

(a) 1 d period.

(b) 12 h period.

(c) 2 h period.

(d) 30 m period.

(e) 5 m period.

Fig. 3.6 Abilene-I: predictability estimate for different time scales. NDT for prediction horizons 1, 2, 5, 10 and 15 are shown on the left for different subseries lengths (starting at random points, averaged for 100 repetitions). NDT for a randomly selected subseries of 1000 samples is shown on the right (horizons 1-30 and regressor sizes 1-10).

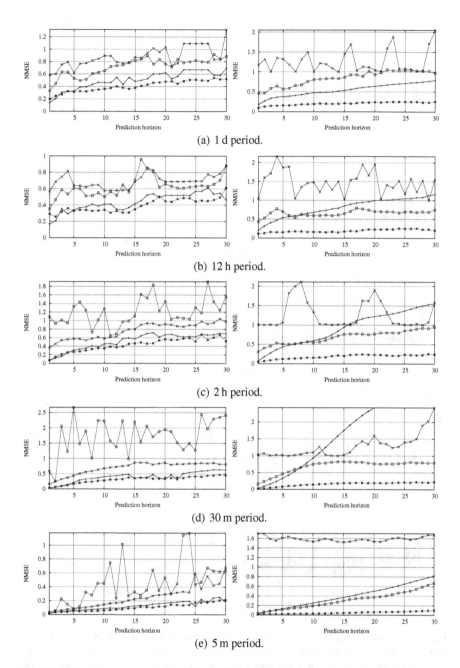

(a) 1 d period.

(b) 12 h period.

(c) 2 h period.

(d) 30 m period.

(e) 5 m period.

Fig. 3.7 Abilene-I: test errors for different time scales. Left: errors for NRVE-FIS (+ for training, × for test) and OP-ELM (∗ for training, □ for test) models. Right: errors for ARIMA (+ for training, × for test) and IOWA-NN (∗ for training, □ for test) models.

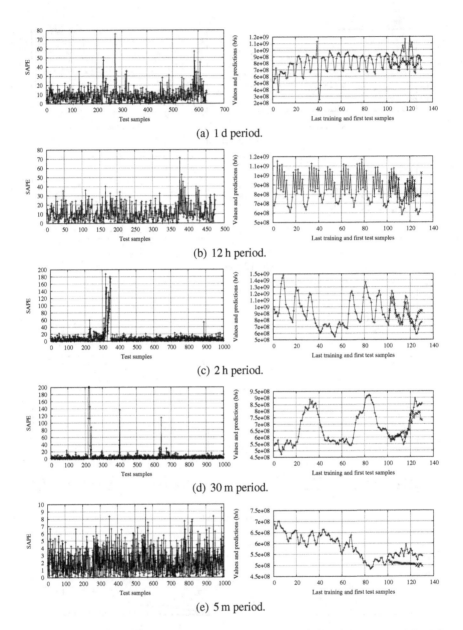

(a) 1 d period.

(b) 12 h period.

(c) 2 h period.

(d) 30 m period.

(e) 5 m period.

Fig. 3.8 Abilene-I: predictions of a random test set per time scale with NRVE-FIS. Left: SAPE for one-step ahead predictions of the test set. Right: predictions of the next 30 values after the training set. The last 100 values of the training set, the first 30 values of the test set (both as +, continuous line) and the 30 corresponding predictions (×, dashed line) are plotted.

3.4.1.3   CAIDA OC48-20030424-1

This trace is one of the two traces collected in both directions of an OC48 link on 24th of April, 2003 for 1 hour starting at 00:00. The measurement link is a USA West Coast peering link for a large ISP. The traces consist of 13 GB of data for 203 million packets and 96 GB of observed IP traffic [35]. Here we analyze the trace for direction 1. Results for the trace for the complementary direction, OC48-20030424-0, are detailed in [25].

In figure 3.9, the traffic load series for the OC48-20030424-1 trace at 5 different time scales are shown. Figure 3.10 shows the DT based estimation of predictability of the series derived from the OC48-20030424-1 trace at different time scales. The column on the left shows for every time scale the averaged NDT for 100 randomly selected subseries of different lengths (horizons 1, 2, 5, 10 and 15, regressor size 10). The column on the right shows for every time scale the NDT for horizons 1-30 and regressor sizes 1-10 in the case of one randomly selected subseries of length 1000 (or less when not enough samples are available).

In figure 3.11, the training and test errors for the OC48-20030424-1 trace at different time scales are shown. Four models are used: a) fuzzy inference models built with the NRVE-FIS methodology and OP-ELM models (left column), and b) IOWA derived nearest neighbor models and ARIMA models (right). Prediction horizons 1 through 30 are considered.

Figure 3.12 shows an example of prediction made with fuzzy inference models for every time scale, for one random selection of a training set of up to 1000 training samples and up to 1000 test samples (as far as available). On the left, it is shown the test symmetric absolute percentage error (SAPE). On the right, it is shown the predictions of the next 30 values after the training set.

3.4.1.4   WIDE-C-20080211

This trace was recorded on Monday 11th of February, 2008 at the measurement point C of the WIDE backbone. This measurement point is an IPv6 line connected to 6Bone. It started at 18:00:01 and ended at 21:16:37. 2 million packets were captured. This trace shows some specific properties. In particular, 100% of traffic is IPv6, and 25% of packets are related to DNS requests. The average transfer rate during the recording was 188.01 Kb/s.

In figure 3.13, the traffic load series for the WIDE-C-20080211 trace at 4 different time scales are shown. Figure 3.14 shows the DT based estimation of predictability of the series derived from the WIDE-C-20080211 trace at different time scales. The column on the left shows for every time scale the averaged NDT for 100 randomly selected subseries of different lengths (horizons 1, 2, 5, 10 and 15, regressor size 10). The column on the right shows for every time scale the NDT for horizons 1-30 and regressor sizes 1-10 in the case of one randomly selected subseries of length 1000 (or less when not enough samples are available).

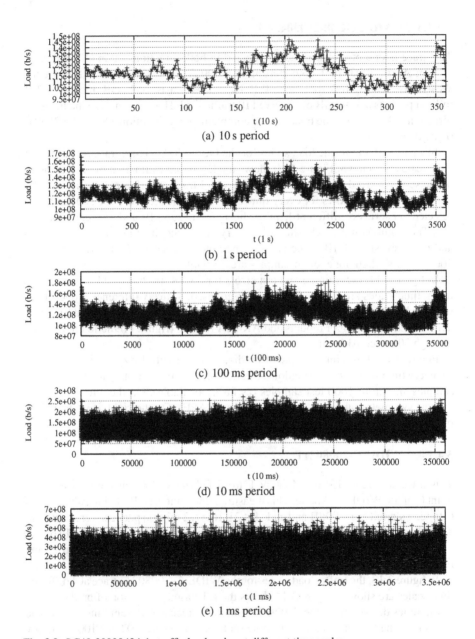

Fig. 3.9 OC48-20030424-1: traffic load series at different time scales

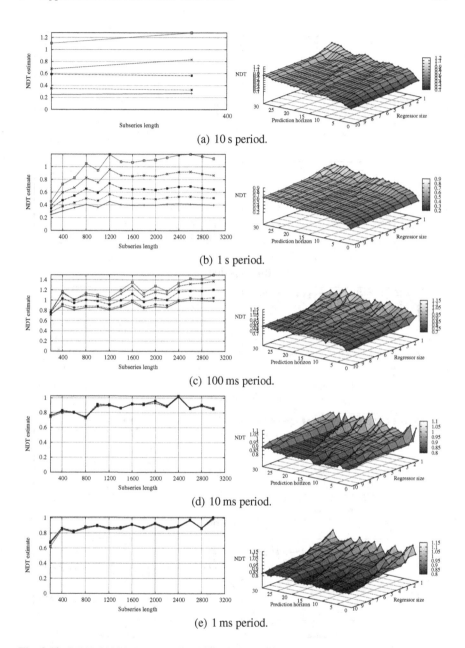

(a) 10 s period.

(b) 1 s period.

(c) 100 ms period.

(d) 10 ms period.

(e) 1 ms period.

Fig. 3.10 OC48-20030424-1: predictability estimate for different time scales. NDT for prediction horizons 1, 2, 5, 10 and 15 are shown on the left for different subseries lengths (starting at random points, averaged for 100 repetitions). NDT for a randomly selected subseries of 1000 samples is shown on the right (horizons 1-30 and regressor sizes 1-10).

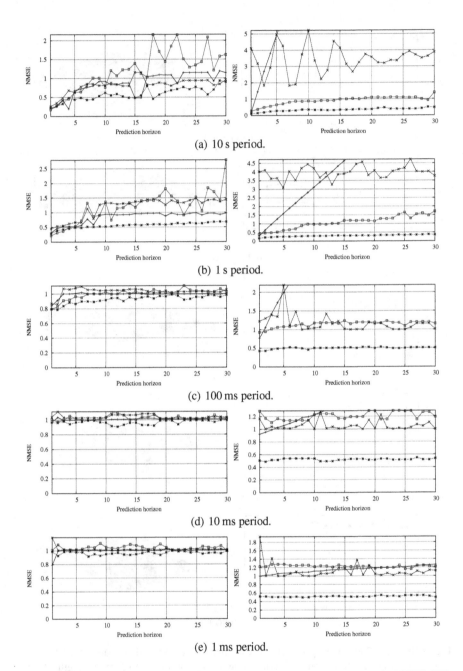

(a) 10 s period.

(b) 1 s period.

(c) 100 ms period.

(d) 10 ms period.

(e) 1 ms period.

Fig. 3.11 OC48-20030424-1: test errors for different time scales. Left: errors for NRVE-FIS (+ for training, × for test) and OP-ELM (∗ for training, □ for test) models. Right: errors for ARIMA (+ for training, × for test) and IOWA-NN (∗ for training, □ for test) models.

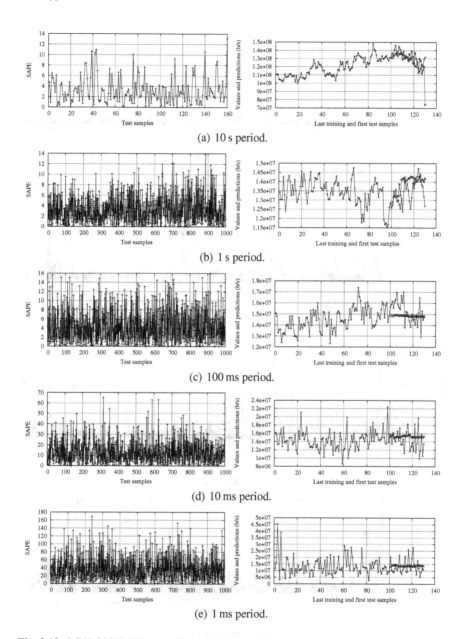

(a) 10 s period.

(b) 1 s period.

(c) 100 ms period.

(d) 10 ms period.

(e) 1 ms period.

Fig. 3.12 OC48-20030424-1: predictions of a random test set per time scale with NRVE-FIS. Left: SAPE for one-step ahead predictions of the test set. Right: predictions of the next 30 values after the training set. The last 100 values of the training set, the first 30 values of the test set (both as +, continuous line) and the 30 corresponding predictions (×, dashed line) are plotted.

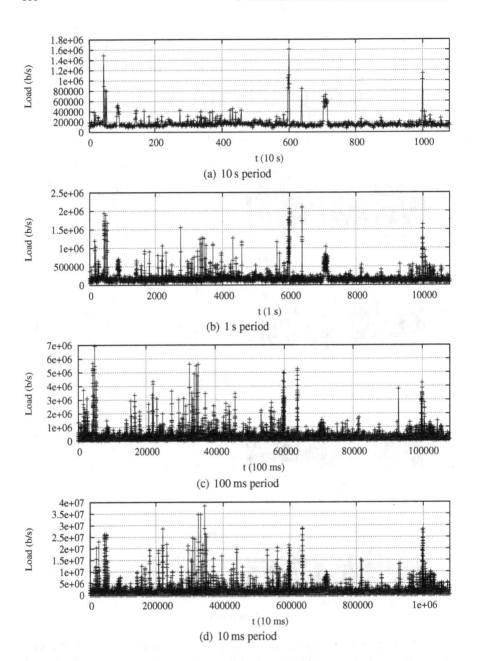

(a) 10 s period

(b) 1 s period

(c) 100 ms period

(d) 10 ms period

Fig. 3.13 WIDE-C-20080211: traffic load series at different time scales

In figure 3.15, the training and test errors for the WIDE-C-20080211 trace at different time scales are shown. Four models are used: a) fuzzy inference models built with the NRVE-FIS methodology and OP-ELM models (left column), and b) IOWA derived nearest neighbor models and ARIMA models (right). Prediction horizons 1 through 30 are considered.

Figure 3.16 shows an example of prediction made with fuzzy inference models for every time scale, for one random selection of a training set of up to 1000 training samples and up to 1000 test samples (as far as available). On the left, it is shown the test symmetric absolute percentage error (SAPE). On the right, it is shown the predictions of the next 30 values after the training set.

3.4.2 Exchange and Peering Traffic

As examples of time series for network traffic at exchange points we study the AMPATH OC12 traces recorded in the DITL 2007 event and the Equinix-Chicago traces recorded during the DITL 2008 event.

The AMPATH (AMericasPATH) traces, belonging to the CAIDA Anonymized 2007 Internet Traces, contains pcap packet header traces collected on both directions of an OC12 link at the AMPATH International Internet Exchange point located in Miami, Florida. This OC12 link carries traffic between U.S. research and education (R&E) networks and R&E networks in South and Central America. These traces were collected as part of the Day in the Life of the Internet (DITL) project in January, 2007. They cover the full 2 days of DITL-2007-01-09 which started midnight on 9th of January, 2007 (UTC) and ended midnight on 11th of January, 2007 (UTC). These traces consist of over 850 million IPv4 packet headers. The reader is referred to [25] for the omitted details.

3.4.2.1 Equinix-Chicago-DITL-2008

This collection, belonging to the CAIDA Anonymized 2008 Internet Traces, contains trace files for a single direction recorded at the CAIDA passive monitor equinix-chicago during the DITL event in 2008. The equinix-chicago Internet data collection monitor is located at an Equinix datacenter in Chicago, IL, and is connected to an OC192 backbone link (9953 Mb/s) of a Tier1 ISP between Chicago and Seattle, WA. It was recorded on direction B (from Chicago to Seattle). The trace started on 19th of March, 2008 at 18:59:08 UTC, and ended at 20:01:00 UTC. This trace was recorded as part of the Day in the Life of the Internet (DITL) 2008 measurement event (http://www.caida.org/projects/ditl/).

In figure 3.17, the traffic load series for the Equinix-Chicago-DITL-2008 trace at 5 different time scales are shown. Figure 3.18 shows the DT based estimation of predictability of the series derived from the Equinix-Chicago-DITL-2008 trace at different time scales. The column on the left shows for every time scale the averaged NDT for 100 randomly selected subseries of different lengths (horizons 1, 2, 5, 10 and 15, regressor size 10). The column on the right shows for every time scale the

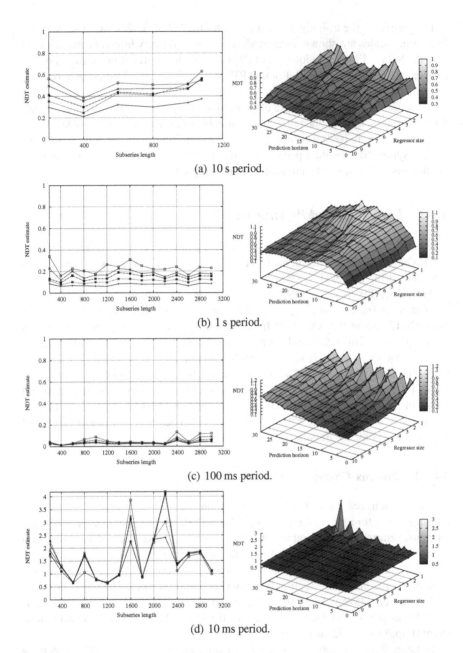

(a) 10 s period.

(b) 1 s period.

(c) 100 ms period.

(d) 10 ms period.

Fig. 3.14 WIDE-C-20080211: predictability estimate for different time scales. NDT for prediction horizons 1, 2, 5, 10 and 15 are shown on the left for different subseries lengths (starting at random points, averaged for 100 repetitions). NDT for a randomly selected subseries of 1000 samples is shown on the right (horizons 1-30 and regressor sizes 1-10).

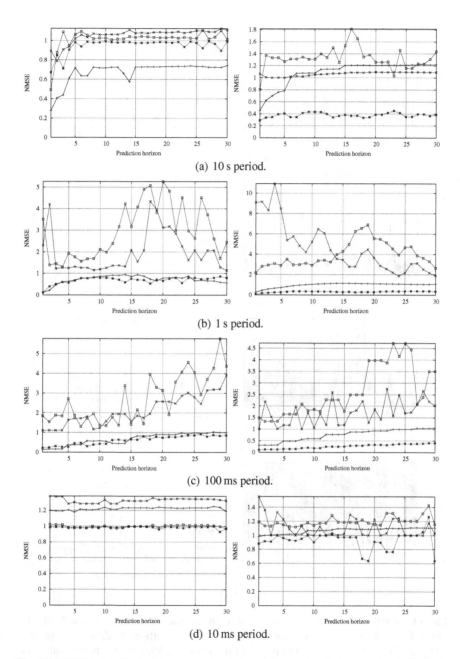

(a) 10 s period.

(b) 1 s period.

(c) 100 ms period.

(d) 10 ms period.

Fig. 3.15 WIDE-C-20080211: test errors for different time scales. Left: errors for NRVE-FIS (+ for training, × for test) and OP-ELM (∗ for training, □ for test) models. Right: errors for ARIMA (+ for training, × for test) and IOWA-NN (∗ for training, □ for test) models.

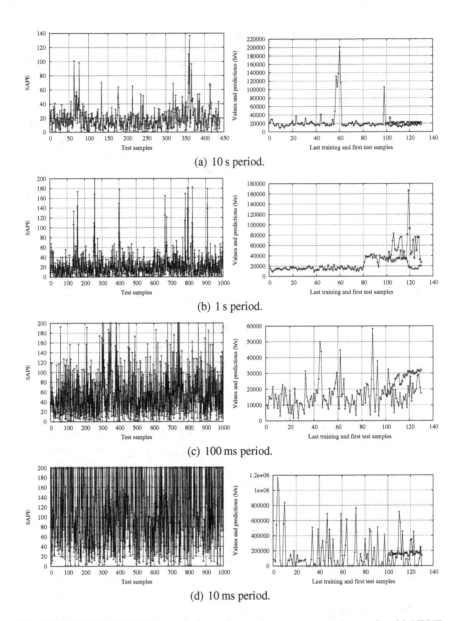

(a) 10 s period.

(b) 1 s period.

(c) 100 ms period.

(d) 10 ms period.

Fig. 3.16 WIDE-C-20080211: predictions of a random test set per time scale with NRVE-FIS. Left: SAPE for one-step ahead predictions of the test set. Right: predictions of the next 30 values after the training set. The last 100 values of the training set, the first 30 values of the test set (both as +, continuous line) and the 30 corresponding predictions (×, dashed line) are plotted.

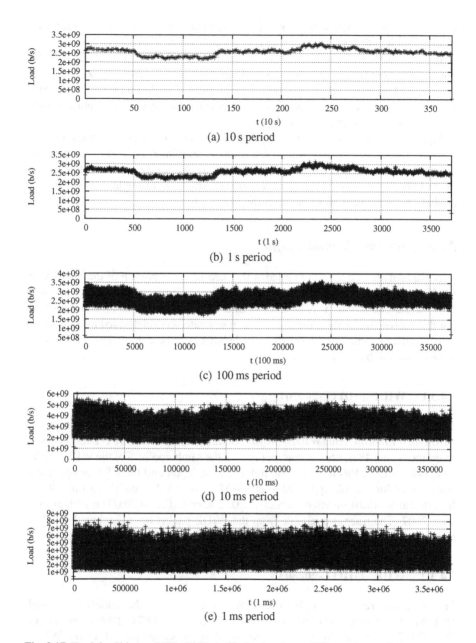

(a) 10 s period

(b) 1 s period

(c) 100 ms period

(d) 10 ms period

(e) 1 ms period

Fig. 3.17 Equinix-Chicago-DITL-2008: traffic load series at different time scales

NDT for horizons 1-30 and regressor sizes 1-10 in the case of one randomly selected subseries of length 1000 (or less when not enough samples are available).

In figure 3.19, the training and test errors for the Equinix-Chicago-DITL-2008 trace at different time scales are shown. Four models are used: a) fuzzy inference models built with the NRVE-FIS methodology and OP-ELM models (left column), and b) IOWA derived nearest neighbor models and ARIMA models (right). Prediction horizons 1 through 30 are considered.

Figure 3.20 shows an example of prediction made with fuzzy inference models for every time scale, for one random selection of a training set of up to 1000 training samples and up to 1000 test samples (as far as available). On the left, it is shown the test symmetric absolute percentage error (SAPE). On the right, it is shown the predictions of the next 30 values after the training set.

3.4.3 Intercontinental Traffic

Here we analyze two traces recorded at the measurement point F of the the MAWI Working Group of the Widely Integrated Distributed Environment (WIDE) [7, 40] This measurement point is is an intercontinental link between Japan and the USA. For this reason, we analyze these traces in a separate section devoted to transcontinental traffic as some differences in traffic behavior can be expected as compared to intracontinental backbone links.

3.4.3.1 WIDE-F-DITL-200701

This trace was recorded by the MAWI Working Group of the Widely Integrated Distributed Environment (WIDE) project as part of the DITL 2007 event. The trace spans more than 48 hours and was recorded at the samplepoint-F of the WIDE backbone [40], a 155 Mb/s trans-pacific link between Japan and the USA in operation since first of June, 2006, upgraded from 100 Mb/s to 150 Mb/s on first of June, 2007. It started at 07:45:01 on 9th of January, 2007 and ended at 10:00:00 on 11th of January, 2007. The trace is available from the WIDE packet traces archive [40]. The reader interested in the complete analysis of this trace is referred to [25].

3.4.3.2 WIDE-F-DITL-200803

This trace was recorded by the MAWI Working Group of the Widely Integrated Distributed Environment (WIDE) project as part of the DITL 2008 event. As in the DITL 2007 event, this trace was recorded at the samplepoint-F of the WIDE backbone [40], a 155 Mb/s trans-pacific link between Japan and the USA. It spans more than 72 hours, starting at 23:45:00 on 17th of March, 2008 and ending at 00:00:00 on 21st of March, 2008. The trace is available from the WIDE packet traces archive [40].

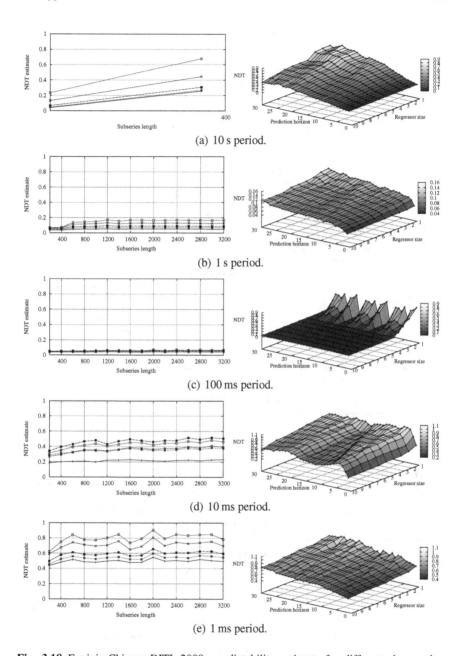

Fig. 3.18 Equinix-Chicago-DITL-2008: predictability estimate for different time scales. NDT for prediction horizons 1, 2, 5, 10 and 15 are shown on the left for different subseries lengths (starting at random points, averaged for 100 repetitions). NDT for a randomly selected subseries of 1000 samples is shown on the right (horizons 1-30 and regressor sizes 1-10).

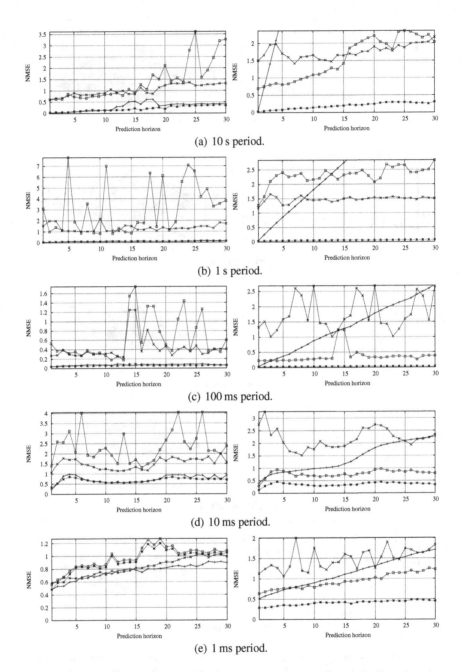

(a) 10 s period.

(b) 1 s period.

(c) 100 ms period.

(d) 10 ms period.

(e) 1 ms period.

Fig. 3.19 Equinix-Chicago-DITL-2008: test errors for different time scales. Left: errors for NRVE-FIS (+ for training, × for test) and OP-ELM (∗ for training, □ for test) models. Right: errors for ARIMA (+ for training, × for test) and IOWA-NN (∗ for training, □ for test) models.

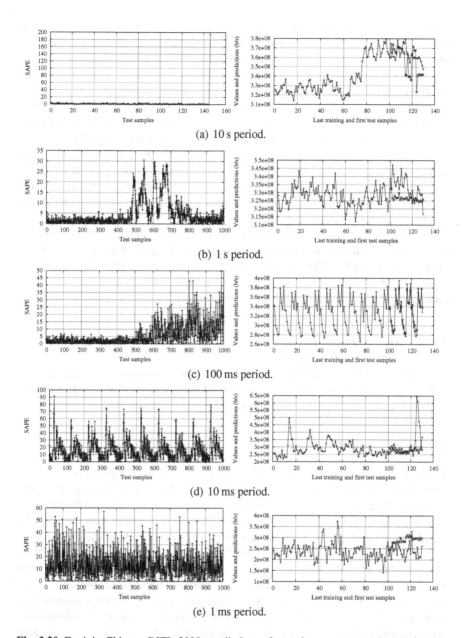

Fig. 3.20 Equinix-Chicago-DITL-2008: predictions of a random test set per time scale with NRVE-FIS. Left: SAPE for one-step ahead predictions of the test set. Right: predictions of the next 30 values after the training set. The last 100 values of the training set, the first 30 values of the test set (both as +, continuous line) and the 30 corresponding predictions (×, dashed line) are plotted.

In figure 3.21, the traffic load series for the WIDE-F-DITL-200803 trace at 5 different time scales are shown. Figure 3.22 shows the DT based estimation of predictability of the series derived from the WIDE-F-DITL-200803 trace at different time scales. The column on the left shows for every time scale the averaged NDT for 100 randomly selected subseries of different lengths (horizons 1, 2, 5, 10 and 15, regressor size 10). The column on the right shows for every time scale the NDT for horizons 1-30 and regressor sizes 1-10 in the case of one randomly selected subseries of length 1000 (or less when not enough samples are available).

In figure 3.23, the training and test errors for the WIDE-F-DITL-200803 trace at different time scales are shown. Four models are used: a) fuzzy inference models built with the NRVE-FIS methodology and OP-ELM models (left column), and b) IOWA derived nearest neighbor models and ARIMA models (right). Prediction horizons 1 through 30 are considered.

Figure 3.24 shows an example of prediction made with fuzzy inference models for every time scale, for one random selection of a training set of up to 1000 training samples and up to 1000 test samples (as far as available). On the left, it is shown the test symmetric absolute percentage error (SAPE). On the right, it is shown the predictions of the next 30 values after the training set.

3.4.4 Access Point Traffic

Here we analyze traffic traces recorded at access points where a certain organization connects to the Internet, such as the Bell Labs-I trace, measured at a 9 Mb/s link to the Internet, and the NCAR-I trace, measured at a Gb/s access link.

3.4.4.1 Bell Labs-I

The Bell Labs-I trace was captured jointly by the NLANR PMA project and the Internet Traffic Research group from 19th to 25th of May, 2002. It is a one week contiguous Internet access IP header trace collected from Sunday 19th through Saturday 25th at Bell Labs research, Murray Hill, NJ, USA. The data set was collected with a Dag3.2E 10/100 Mb/s Ethernet card at the outside of the firewall servicing researchers at Bell Labs via a 9 Mb/s link to the Internet. The Dag3.2E is sychronzed to a CDMA time receiver, with an absolute accuracy to UTC to within 10 microseconds, the frequency stability comparable to GPS-based solutions. Some overall characteristics of the traffic at this link can be summarized as follows: the local network hosts serve 400 people, mostly technical, and about 50 administrative staff, all HTTP traffic has clients inside the network and servers outside, and all flows are bidirectional whereas routing is fully symmetric. Again, the interested reader is referred to [25] for an exhaustive analysis of all the traces.

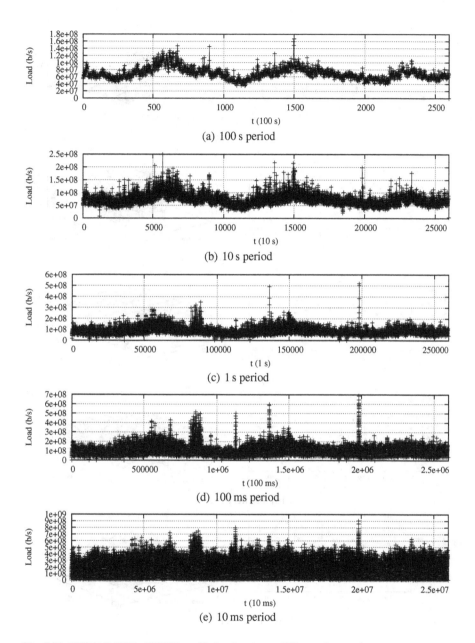

Fig. 3.21 WIDE-F-DITL-200803: traffic load series at different time scales

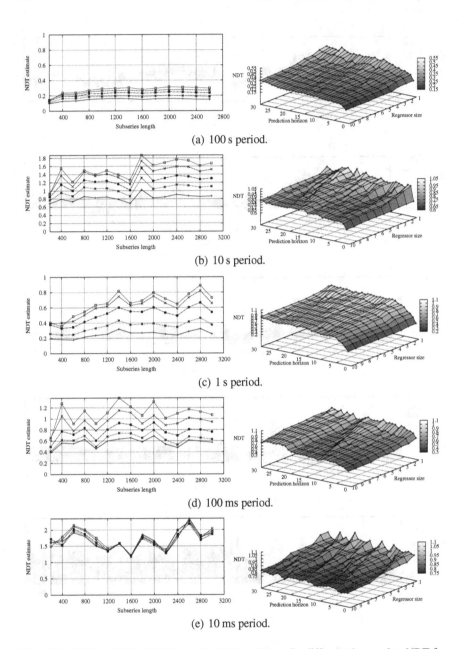

(a) 100 s period.

(b) 10 s period.

(c) 1 s period.

(d) 100 ms period.

(e) 10 ms period.

Fig. 3.22 WIDE-F-DITL-200803: predictability estimate for different time scales. NDT for prediction horizons 1, 2, 5, 10 and 15 are shown on the left for different subseries lengths (starting at random points, averaged for 100 repetitions). NDT for a randomly selected sub-series of 1000 samples is shown on the right (horizons 1-30 and regressor sizes 1-10).

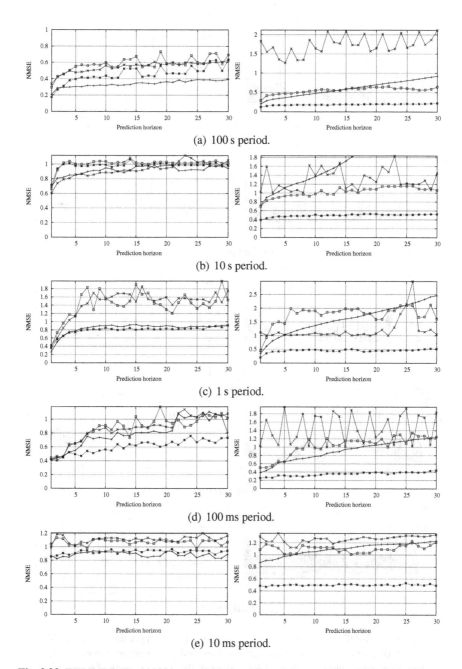

(a) 100 s period.

(b) 10 s period.

(c) 1 s period.

(d) 100 ms period.

(e) 10 ms period.

Fig. 3.23 WIDE-F-DITL-200803: test errors for different time scales. Left: errors for NRVE-FIS (+ for training, × for test) and OP-ELM (∗ for training, □ for test) models. Right: errors for ARIMA (+ for training, × for test) and IOWA-NN (∗ for training, □ for test) models.

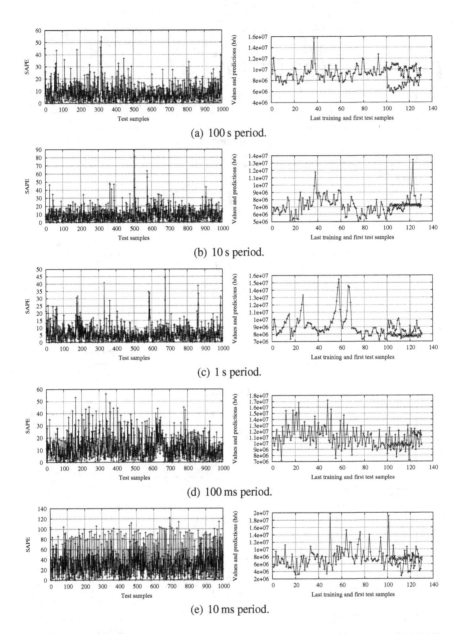

(a) 100 s period.

(b) 10 s period.

(c) 1 s period.

(d) 100 ms period.

(e) 10 ms period.

Fig. 3.24 WIDE-F-DITL-200803: predictions of a random test set per time scale with NRVE-FIS. Left: SAPE for one-step ahead predictions of the test set. Right: predictions of the next 30 values after the training set. The last 100 values of the training set, the first 30 values of the test set (both as +, continuous line) and the 30 corresponding predictions (×, dashed line) are plotted.

3.4.4.2 NCAR-I

The NCAR-I data set is a one hour long IP header trace captured by the NLANR PMA project using an Endace DAG4.2GE dual Gigabit Ethernet network measurement card at the end of December 2003. The measurement point is located at the access link of the National Center for Atmospheric Research, Boulder.

In figure 3.25, the traffic load series for the NCAR-I trace at 5 different time scales are shown. Figure 3.26 shows the DT based estimation of predictability of the series derived from the NCAR-I trace at different time scales. The column on the left shows for every time scale the averaged NDT for 100 randomly selected subseries of different lengths (horizons 1, 2, 5, 10 and 15, regressor size 10). The column on the right shows for every time scale the NDT for horizons 1-30 and regressor sizes 1-10 in the case of one randomly selected subseries of length 1000 (or less when not enough samples are available).

In figure 3.27, the training and test errors for the NCAR-I trace at different time scales are shown. Four models are used: a) fuzzy inference models built with the NRVE-FIS methodology and OP-ELM models (left column), and b) IOWA derived nearest neighbor models and ARIMA models (right). Prediction horizons 1 through 30 are considered.

Figure 3.28 shows an example of prediction made with fuzzy inference models for every time scale, for one random selection of a training set of up to 1000 training samples and up to 1000 test samples (as far as available). On the left, it is shown the test symmetric absolute percentage error (SAPE). On the right, it is shown the predictions of the next 30 values after the training set.

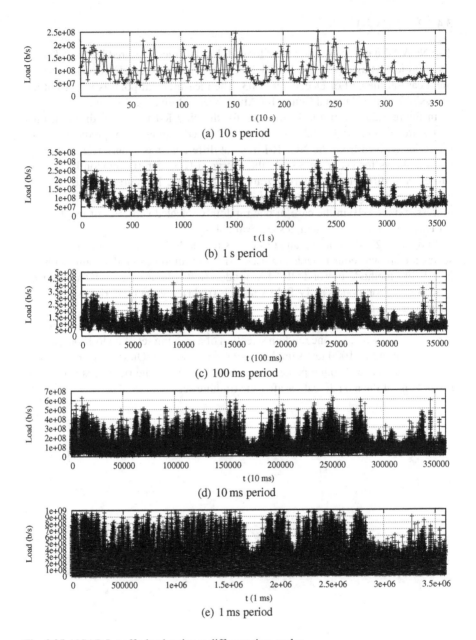

(a) 10 s period

(b) 1 s period

(c) 100 ms period

(d) 10 ms period

(e) 1 ms period

Fig. 3.25 NCAR-I: traffic load series at different time scales

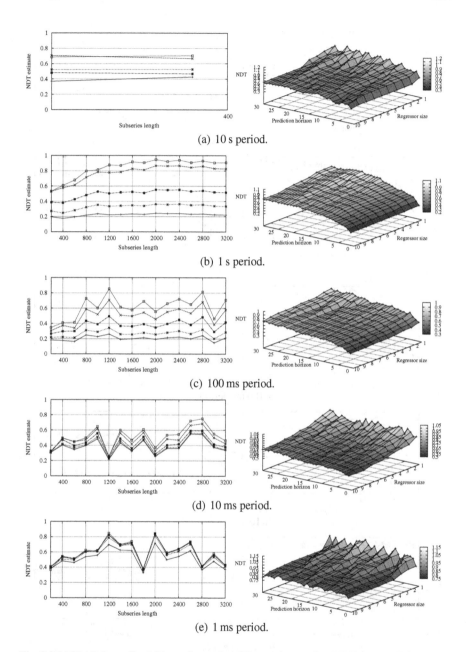

Fig. 3.26 NCAR-I: predictability estimate for different time scales. NDT for prediction horizons 1, 2, 5, 10 and 15 are shown on the left for different subseries lengths (starting at random points, averaged for 100 repetitions). NDT for a randomly selected subseries of 1000 samples is shown on the right (horizons 1-30 and regressor sizes 1-10).

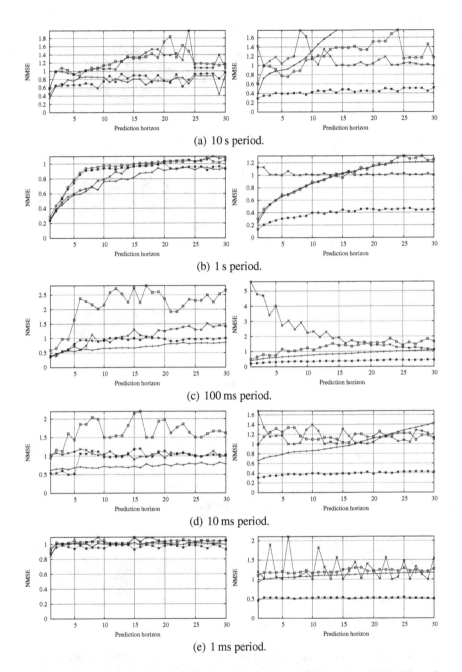

(a) 10 s period.

(b) 1 s period.

(c) 100 ms period.

(d) 10 ms period.

(e) 1 ms period.

Fig. 3.27 NCAR-I: test errors for different time scales. Left: errors for NRVE-FIS (+ for training, × for test) and OP-ELM (∗ for training, □ for test) models. Right: errors for ARIMA (+ for training, × for test) and IOWA-NN (∗ for training, □ for test) models.

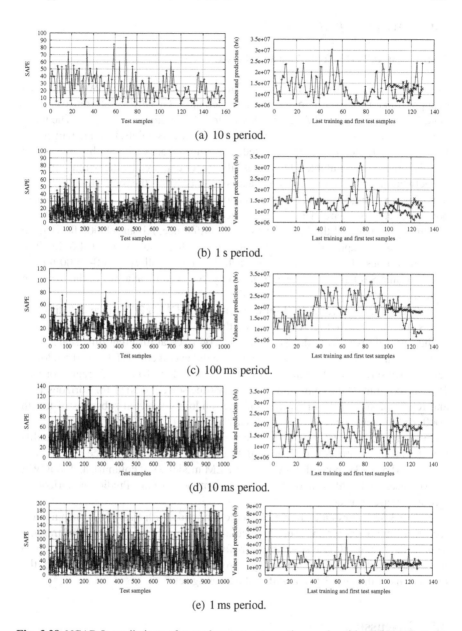

Fig. 3.28 NCAR-I: predictions of a random test set per time scale with NRVE-FIS. Left: SAPE for one-step ahead predictions of the test set. Right: predictions of the next 30 values after the training set. The last 100 values of the training set, the first 30 values of the test set (both as +, continuous line) and the 30 corresponding predictions (×, dashed line) are plotted.

3.4.5   Wireless Traffic

3.4.5.1   SIGCOMM 2004 Wireless Access

This trace was recorded at an access point of the open wireless network provided
to approximately 550 participants who attended the SIGCOMM conference in Portland, Oregon from 30th of August, 2004 to 3rd of September, 2004 [31]. The wireless network set up for the conference comprised 5 access points. Connectivity to the
Internet was provided by four separated DSL lines. Traces from both wireless and
wired monitors are available. Here we use the traces captured by the wired sniffer on
the network segment connected to the access points. Thus, the trace provides a view
of packets exchanged between the access points and the Internet. This measurement
point and setup is regarded as characteristic of a large hotspot setting.

This trace spans all the three days of the conference. About 300000 flows from
195 users were captured for an overall bandwidth consumption of 4.6 GB. The
scheduled times of the conference on each day are as follows: 8:00-18:00 for the
full days, and 8:00-14:00 on the half (last) day. An in depth analysis of the trace can
be found in [30].

In figure 3.29, the traffic load series for the SIGCOMM 2004 trace at 5 different
time scales are shown. Figure 3.30 shows the DT based estimation of predictability
of the series derived from the SIGCOMM 2004 trace at different time scales. The
column on the left shows for every time scale the averaged NDT for 100 randomly
selected subseries of different lengths (horizons 1, 2, 5, 10 and 15, regressor size
10). The column on the right shows for every time scale the NDT for horizons 1-30
and regressor sizes 1-10 in the case of one randomly selected subseries of length
1000 (or less when not enough samples are available).

In figure 3.31, the training and test errors for the SIGCOMM 2004 trace at different time scales are shown. Four models are used: a) fuzzy inference models built
with the NRVE-FIS methodology and OP-ELM models (left column), and b) IOWA
derived nearest neighbor models and ARIMA models (right). Prediction horizons 1
through 30 are considered.

Figure 3.32 shows an example of prediction made with fuzzy inference models
for every time scale, for one random selection of a training set of up to 1000 training
samples and up to 1000 test samples (as far as available). On the left, it is shown
the test symmetric absolute percentage error (SAPE). On the right, it is shown the
predictions of the next 30 values after the training set.

3.5   Discussion

Let us consider a first general summary of the analysis performed:

- The plots of the series at different time scale (first figure) confirm the often found
  fact that traffic load is self-similar for a wide range of time scales. Though there
  are notable exceptions, and the degree of self-similarity is diverse, which is in line
  with experimental results that show that the Hurst coefficient is usually not high

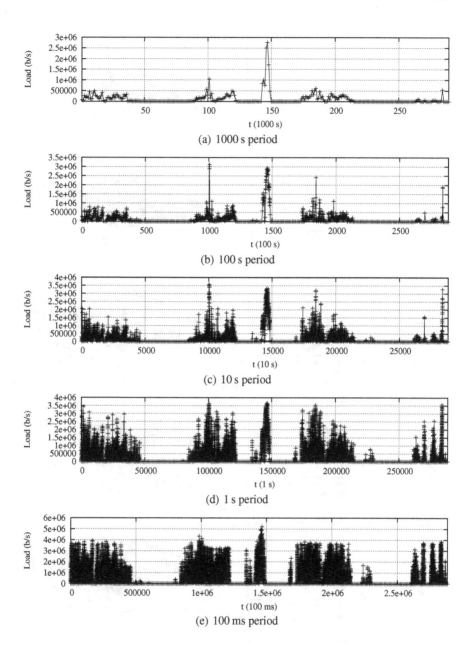

Fig. 3.29 SIGCOMM 2004: traffic load series at different time scales

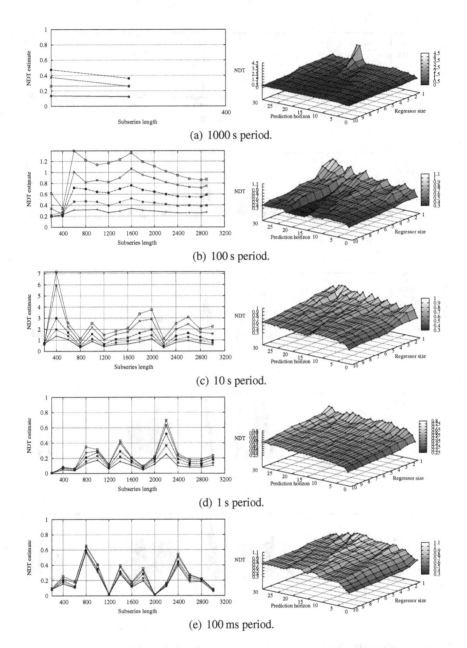

(a) 1000 s period.

(b) 100 s period.

(c) 10 s period.

(d) 1 s period.

(e) 100 ms period.

Fig. 3.30 SIGCOMM 2004: predictability estimate for different time scales. NDT for prediction horizons 1, 2, 5, 10 and 15 are shown on the left for different subseries lengths (starting at random points, averaged for 100 repetitions). NDT for a randomly selected subseries of 1000 samples is shown on the right (horizons 1-30 and regressor sizes 1-10).

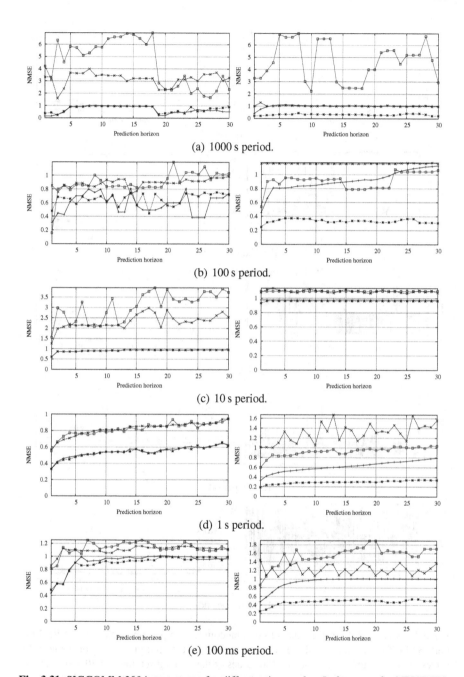

(a) 1000 s period.

(b) 100 s period.

(c) 10 s period.

(d) 1 s period.

(e) 100 ms period.

Fig. 3.31 SIGCOMM 2004: test errors for different time scales. Left: errors for NRVE-FIS (+ for training, × for test) and OP-ELM (∗ for training, □ for test) models. Right: errors for ARIMA (+ for training, × for test) and IOWA-NN (∗ for training, □ for test) models.

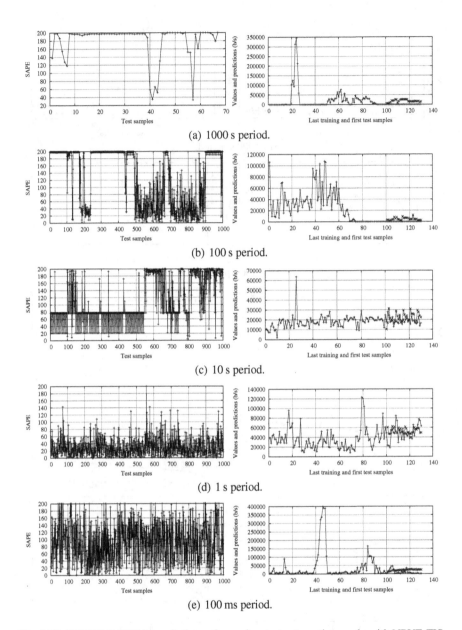

(a) 1000 s period.

(b) 100 s period.

(c) 10 s period.

(d) 1 s period.

(e) 100 ms period.

Fig. 3.32 SIGCOMM 2004: predictions of a random test set per time scale with NRVE-FIS. Left: SAPE for one-step ahead predictions of the test set. Right: predictions of the next 30 values after the training set. The last 100 values of the training set, the first 30 values of the test set (both as +, continuous line) and the 30 corresponding predictions (×, dashed line) are plotted.

for current network measurements [18]. For instance, the degree of self-similarity is remarkably higher in the Bellcore traces, which correspond to Ethernet traffic with a low level of aggregation.

- The predictability analysis with a nonparametric estimator (second figure) leads us to the following conclusions:

  - From the 2D plots a conclusion can be drawn: the DT estimate is stable for series of length ranging from 200 to 3000 samples, although an slightly trend increasing with the number of samples can be observed. This increasing trend can be seen as a logical consequence of the general non-stationarity of the series analyzed. There are some exceptions for some series at certain time scales where the DT estimate can vary up to 100%. This is however the case for series with a high degree of self-similarity typical of traces with a low degree of aggregation, in particular, the Bellcore and DEC series. The values for subseries lengths between 1000 and 2000 samples are in most cases fairly stable tough, with variations within 10% approximately.

    Also, the dependence of the residual variance estimate on subseries lengths is mostly invariant with the prediction horizon, i.e., the increases and decreases of the DT estimates for certain subseries lengths occur consistently for all the prediction horizons shown. For instance, the variance of the noise estimates for the series AMPATH-OC12-200701-0 are significantly irregular with regards to the subseries length for time scales of 10 s, 1 s 100 ms and 10 ms. However, the pattern of variation of the DT estimate is remarkably the same for all the prediction horizons shown in the plots.

    This first part of the predictability plots has an important implication: relatively short subseries, of 1000 or even between 400 and 1000 samples in some cases, can be used for computing approximate estimations of the variance of the noise. It can be concluded that it is feasible to build models for training subseries of 1000 or even less samples.

  - The second (3D) plot of the predictability figure shows that the dependence of the variance of the noise on the maximum regressor size as well as the prediction horizons varies greatly depending on the time series. In general, it is confirmed that the lowest training error possible decreases as the regressor size increases. Regarding the dependence of the variance of the noise on the prediction horizon, the DT estimate is strictly increasing with the prediction horizon in some cases. However, there are a number of series and time scales for which the variance of the noise oscillates, which indicates cyclic behavior to a certain extent. In these cases, the series can be expected to be more predictable for some long- or medium-term horizons than for short-term horizons.

- The performance of four predictive modeling methods has been analyzed in the third and fourth figures for each traffic trace. From the four modeling methods, it is clear that fuzzy inference models and OP-ELM models are the two best options in terms of accuracy. In general, the first kind of models is slightly more accurate, although it can be significantly more accurate for long-term prediction. There are however a few exceptions where OP-ELM models outperform fuzzy models in terms of accuracy for short-term predictions, with slight but consistent better accuracy for a range of prediction horizons. Thus, we can conclude that fuzzy models are usually (but with some exceptions) slightly more accurate for short-term prediction and more accurate for long-term prediction in general. However, both methods are close enough so that we would expect grossly the same capability to predict traffic load from both methods. This can be seen in the left column of figure 3.33. Note however that the plots in this figure show overall results for all the traces analyzed at all the time scales considered, and hence they must be interpreted with care.

IOWA nearest neighbor models and ARIMA models, as examples of more traditional approaches, are consistently less accurate than fuzzy and OP-ELM models. Although ARIMA models are overall less accurate than IOWA-NN models, they are often more robust for long-term prediction. This can be seen in the right column of figure 3.33. There are also a number of exceptions where the ARIMA technique provides more accurate short-term predictions such as the WIDE-F-DITL-200701 series, specially at the 10 s time scale.

Results show that the computational intelligence based techniques are able to take advantage of predictability, i.e., are able to capture the information about the traffic dynamics contained in the training datasets. For example, consider the 10 s period series for the OC48-20030424-0 trace. For a maximum regressor size of 10, the DT estimate quickly increases as the prediction horizon increases up to approximately 20 yet decreases for greater horizons. This fact reveals some kind of periodic behavior that renders the series easier to predict on the mid-term than on the long-term.

The SAPE for the one step ahead prediction at 10 s scale is in general below 10%, with the exception of a maximum at 12%. By looking at the prediction example shown, it can be seen that while the predictions quickly loose track of the real series for the lower prediction horizons, the model is able to provide surprisingly good predictions for the last 15-20 horizons. It follows also that for these cases mid-term predictability should be improved possibly by indirectly increasing the regressor size using projection or other complementary techniques.

For the traffic load series analyzed at different time scales, with a difference of up to five orders of magnitude, the DT estimate of the residual variance, or variance of the noise, has been found to be remarkably stable for all the lengths of the subseries considered. This is the case even at time scales of the order of the millisecond, thus suggesting that traffic is approximately invariant as for predictability for considerably long time periods. For instance, the residual variance for 1 step ahead prediction ranges from 0.45 to 0.55 for the trace Equinix-Chicago-DITL-2008 at 1 ms intervals. Considering that this trace spans more than 1 hour, i.e., more than $3.6 \cdot 10^6$ intervals of 1 ms, it can be concluded that predictability at scales of the order of the millisecond is remarkably constant over scales 4 orders of magnitude higher.

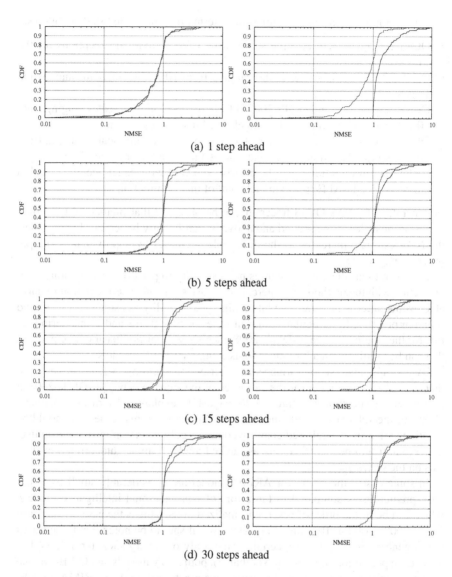

Fig. 3.33 Cumulative distribution function plot of the 1, 5, 15 and 30 steps ahead test errors of NRVE-FIS, OP-ELM, ARIMA and IOWA-NN models for the whole set of traces and all the time scales analyzed. Note the log-normal scale. Left: NRVE-FIS (continuous line) and OP-ELM (dashed line). Right: ARIMA (continuous line) and IOWA-NN (dashed line).

Some remarkable cases of accurate short-term predictions are listed as follows:

- OC48-20020814-1, at all time scales with a few individual exceptions for the 1 ms scale.
- OC48-20020814-0, at time scales ranging from 10 s down to 10 ms and also at 1 ms to a lesser extent.
- OC48-20030115-1 and OC48-20030115-0, specially at time scales ranging from 10 s, down to 10 ms time scales and also at 1 ms to a lesser extent.
- OC48-20030424-0 and OC48-20030424-1, specially at time scales ranging from 10 s down to 10 ms time scales and also at 1 ms to a lesser extent. Interestingly, traffic load in direction 1 can be clearly predicted with better accuracy (with a one-step ahead SAPE approximately half of that of direction 0).

In general, in CAIDA OC48 traces, it can be observed that predictability at time scales of the order of the second and above is significantly better. For lower scales, the NDT estimate increases as the time scale decreases by approximately 0.2 per order of magnitude.

However, in all these cases there is no possibility to perform accurate long-term prediction with the methods used here. In contrast, some other series exhibit remarkable cyclic patterns that translate into a high predictability as estimated by DT and found in practice. In particular, some subseries of the 100 ms and 10 ms period series for the Equinix-Chicago-DITL-2008 trace show clear nonlinear cyclic patterns that can be predicted on a long-term basis using NRVE-FIS models.

By a simple analysis of variance, as described in section 1.5.7, it can be found that the NRVE-FIS model for the subseries of the Equinix-Chicago-DITL-2008 at 100 ms time scale shown in figure 3.20, page 119, explains 84% of the total variability of the subseries despite the clear non-linearity. In this respect, it should be mentioned that in general predictability comes largely from nonlinear determinism rather than from linear correlations, as can be found by simple autocorrelation analysis on the series analyzed [17].

In some cases, the errors for ARIMA model are considerably small for long-term prediction as compared to the errors of computational intelligence models, which can be of the order of several normalized units. This is the case for the trace OC48-20020814-1 at 10 s time scale. However, it should be noted that these traces are in general predictable to a certain extent only on a very short-term basis. For short-term prediction (horizons lower than 5), both fuzzy models and OP-ELM take advantage of the predictability of he series to a greater extent than ARIMA models. ARIMA models are nonetheless more robust in some cases for long-term prediction. In any case, these long-term predictions have a very low accuracy and thus cannot be expected to be useful in general.

Interestingly, for time series which are essentially unpredictable at long-term horizons ARIMA models are more robust against significant changes due to non-stationarities. In these cases, computational intelligence techniques seem to mistakenly (and partially) predict changes identified in the training set that do not actually take place in the test set, which translates into very high prediction errors. This fact can be clearly observed in the case of the series OC48-20020814-1 at 10 s scale.

The MSE has been used as the error metric driving the learning process for all the modeling methods applied. It should be noted that a number of alternative error metrics have been proposed for different purposes and the relations among them are hard to establish. For instance, an NMSE around 1 (between 0.9 and 1.1) can correspond to an SMAPE that varies greatly (approximately between 40-150%). Also, an NMSE around 1 corresponds to a MxAE that is always above 0.50, mostly between 0.60 and 0.70. It is well known that the modeling process may yield very different results for alternative error metrics. Here we have however limited our analysis to MSE driven learning methods. Thus, an important area of future work is the identification of error metrics that better serve the purpose of predicting extreme events.

It is important to note that, in some cases, it can be observed a certain degree of symmetry in predictability for traffic traversing some links in opposite directions. In particular, the Abilene-I and Abilene-O exhibit a high degree of symmetry at all the time scales examined. Also, the CAIDA-OC48-20030424 and CAIDA-OC48-20030115 traces show a a clear symmetry at 10 ms and 1 ms time scales, while there are significant differences in predictability and performance of predictive models for scales of 100 ms and above, being traffic in direction 1 more predictable than traffic in direction 0. This fact indicates some degree of persistence of traffic load characteristics in a same link as for predictability for periods of months. By comparing the results for these two traces in both directions each it can be observed as well that a better predictability at certain time scales does not necessarily imply a better predictability for different time scales.

In previous sections, all series have been analyzed as univariate processes, i.e., no exogenous inputs have been taken into consideration. As a first step towards extending the methodology proposed here, correlations between traffic load with different directions in a same link should be explored. This analysis could possibly be extended to a more general analysis of the correlations among the traffic load series at topologically related links, which would make it possible to find relations between traffic patterns and network topology, one of the areas where little results have been reported to date, as introduced in chapter 1. This would entail the analysis of multi-point, synchronized traces, such as the Abilene aggregate, the two directions of the CAIDA OC48 traces and the traces from the NLANR PMA Special Traces Archive, including Abilene-I, Auckland-VI and Leipzig-II.

In this same area, another venue of future research is the use of exogenous inputs or explanatory variables in general. In particular, for traffic series at scales of days and above, a significant performance improvement can be expected if additional features such as holiday periods and related predictable events are considered as inputs to the models. These and other factors can have a significant impact that has to be analyzed on a per trace basis. In the context of fuzzy inference models, this is specially relevant, as many of these features are usually available in the form of linguistic knowledge that can be seamlessly integrated into rulebases.

The major aim of the methodology for time series prediction by means of fuzzy inference systems developed in chapter 2 was to provide a method for the automatic identification of fuzzy models that can be interpreted linguistically. However, it has been found that with the proper configuration these systems can perform as well or even better that other techniques commonly applied in the time series prediction field. We however by no means imply that the NRVE-FIS methodology is in general more accurate than the other methods applied. It is worth to mention that the comparison performed here can be seen only as an study that shows that for the time series analyzed, *with the selected configurations*, fuzzy inference systems can be compared to or even outperform LS-SVMs, OP-ELM and other methods in terms generalization capability.

Nonetheless, an essential advantage of fuzzy inference models that motivated our work in chapter 2 is the possibility to interpret them in a linguistic manner. Fuzzy inference systems are inherently comprehensible, specially when the rules are defined by human experts. However, when rules are automatically identified from data and optimization methods are applied, interpretability cannot be guaranteed in general [5].

The methodology for times series prediction by means of fuzzy inference systems developed in this monograph has been designed in order to address this issue. First, the input selection stage reduces the number of inputs to a significant extent. Second, the consistency of the rulebases obtained is guaranteed by the identification method applied. Third, as a result of the dimensionality reduction and the choice of a proper optimization method, a reduced number of linguistic terms per input and compact rulebases are obtained.

Similarly to the series studied in chapter 2, in general, rulebases are considerably compact. Also, as a general rule, longer term models have a significant lower number of rules, which denotes these systems are rougher. This way, the NRVE-FIS models are highly interpretable in general. From our viewpoint it should not be expected that NRVE-FIS models will provide a satisfactory, complete and comprehensible explanation of the behavior of a traffic load series in every case. However, this is possible in some particular cases, and hints on the dynamics of time series are often obtained. Of course, the procedure required to provide a physical interpretation of the models is case dependent to a great extent.

Let us show an example of NRVE-FIS model. The one step ahead model for the AMPATH-OC12-200701-0 trace at 1 s time scale has 4 inputs and a rulebase consisting of 16 rules. However, only 15 output values are different, with two rules sharing the same output. The four inputs selected correspond to the traffic load at the last interval, y_{t-1}, two seconds before, y_{t-3}, four seconds before, y_{t-4}, and eight seconds before, y_{t-8}. The selection of these inputs already indicates that these are the relevant variables for predicting the traffic load in this link at 1 s time scale. In addition, the relations between the inputs and the output can be interpreted linguistically. Two linguistic terms are defined for each input, as shown in figure 3.34, that can be thought of as LOW (or L) and HIGH (or H) values in general, represented by Gaussian membership functions in the range of observed values, $[7.6 \cdot 10^6, 3.0 \cdot 10^7]$. The rulebase is shown in table 3.1, where the centers of the output singleton values are specified.

In the table, it can be observed, among other aspects, that a significant increase of traffic load can be expected for the next second when the conditions specified by the

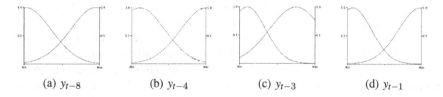

(a) y_{t-8} (b) y_{t-4} (c) y_{t-3} (d) y_{t-1}

Fig. 3.34 Linguistic terms for the example one step ahead predictive model of the AMPATH-OC12-200701-0 series at 1 s time scale

Table 3.1 Rulebase for the NRVE-FIS model of the AMPATH-OC12-200701-0 series at 1 s time scale

y_{t-8}	y_{t-4}	y_{t-3}	y_{t-1}	y_t
L	L	L	L	$8.15 \cdot 10^6$
L	L	L	H	$3.00 \cdot 10^7$
L	L	H	L	$1.12 \cdot 10^7$
L	L	H	H	$3.00 \cdot 10^7$
L	H	L	L	$1.48 \cdot 10^7$
L	H	L	H	$1.95 \cdot 10^7$
L	H	H	L	$1.22 \cdot 10^7$
L	H	H	H	$1.29 \cdot 10^7$
H	L	L	L	$1.52 \cdot 10^7$
H	L	L	H	$2.43 \cdot 10^7$
H	L	L	H	$2.44 \cdot 10^7$
H	L	H	L	$9.72 \cdot 10^6$
H	H	L	L	$1.63 \cdot 10^7$
H	H	L	H	$2.29 \cdot 10^7$
H	H	H	L	$1.77 \cdot 10^7$
H	H	H	H	$2.12 \cdot 10^7$

second and fourth rules hold. In the interpretation process, additional techniques for simplification of fuzzy inference systems can be of considerable help. For this, the simplification methods included in the Xfuzzy environment, where the xftsp tool is integrated, can be used. For instance, if the system is pruned in order to keep the six best rules, a system with an MSE only 2.03% higher than that of the original system is obtained. In particular, the first, second, third, eight, 11th and 15th rules are selected.

The results shown above have been obtained using the fuzzy systems identification and tuning methods described in previous sections. Nonetheless, the same methodology can be applied using alternative identification methods, such as the subtractive clustering based method proposed in [26]. Obviously, different system structures are obtained depending on the identification method applied. Besides differences in prediction performance, the number of rules identified as well as their form can vary significantly. For instance, in the example discussed above for the AMPATH-OC12-200701-0, the cited subtractive clustering based alternative attains a MSE 4.6% higher than that of the method used above, while 12 linguistic terms are identified for every input and the rulebase consists of 12 rules.

It is sometimes argued that a large number of models have been applied to network traffic yet these models provide little insight into traffic dynamics. In this context, black-box models that are purely descriptive in a numerical sense and provide no insight and physical understanding are of little value. However, we have addressed issues that have not been dealt with to date: long-term prediction at different time scales and predictability from a nonparametric viewpoint. In addition, the fuzzy models developed here emphasize interpretability in a linguistic sense, thus providing a novel technique that can provide readily interpretable insights on how network traffic evolves.

On a related subject, we should note that the analysis performed here has followed an essentially black-box, or blind approach in that no specific exploratory data analysis has preceded the application of prediction algorithms. This has made it possible to analyze a large collection of heterogeneous traffic traces. However, experience shows that an initial exploratory data analysis stage tailored to each time series can improve results significantly [39, 11].

It is worth to mention that traffic time series at scales of days, weeks and above have been shown to be predictable in practice to a considerable extent. This provides evidence of the practical applicability of current prediction techniques to network operation and planning, confirming previous work on non-disclosed measurements [28], as well as the accepted common sense that certain daily, weekly and yearly patterns observed in many networks should be predictable. However, series at lower time scales have been found to be essentially unpredictable with very few exceptions. In fact, only coarse predictions are possible even for short-term. This fact poses a challenge on the development of prediction based traffic control schemes.

As final remark, a future research venue would be an extended analysis of traffic load series for higher dimensional models, which would entail developing accurate nonlinear models for very high dimensional problems. This, in turn, involves novel developments in a number of fields related to time series prediction that may be useful to better exploit the long-range dependencies commonly found in network traffic load series.

3.6   Conclusions

A predictability analysis has been conducted for network traffic traces using DT. Results showed that the degree of predictability is satisfactory for a number of applications and that the proposed fuzzy regressive models can approach the theoretical predictability boundary to a high degree under most conditions.

While for some scenarios and time scales prediction is only possible on a very short-term basis and to a limited extent, in some other scenarios and time scales it is possible to achieve fairly accurate predictions for both short- and medium-term. This results have important practical implications for a number of applications, such as prediction based traffic control strategies as well as design of network protocols. Fuzzy regressive models were shown to outperform LS-SVM, OP-ELM and some statistical autoregressive models in terms of accuracy and generalization.

References

[1] Adas, A.M.: Using Adaptive Linear Prediction to Support Real-Time VBR Video under RCBR Network Service Model. IEEE/ACM Trans. Networking 6(5), 635–644 (1998)

[2] Andersen, A.T., Nielsen, B.F.: A Markovian Approach for Modeling Packet Traffic with Long-Range Dependence. IEEE Journal on Selected Areas in Communications 16(5), 719–732 (1998)

[3] Basu, S., Mukherjee, A.: Time Series Models for Internet Traffic. In: Fifteenth Annual Joint Conference of the IEEE Computer and Communications Societies (Proceedings IEEE INFOCOM 1996). Networking the Next Generation, San Francisco, CA, USA, vol. 2, pp. 611–620 (1996)

[4] Box, G., Jenkins, G.M., Reinsel, G.: Time Series Analysis: Forecasting & Control, 3rd edn. Prentice Hall, Englewood Cliffs (1994) ISBN: 0130607746

[5] Casillas, J., Cordón, O., Herrera, F., Magdalena, L. (eds.): Interpretability Issues in Fuzzy Modeling. Studies in Fuzziness and Soft Computing. Springer, Heidelberg (2003) ISBN: 978-3-540-02932-8

[6] Chatfield, C.: The Analysis of Time Series. An Introduction, 6th edn. CRC Press, Boca Raton (2003) ISBN: 1-58488-317-0

[7] Cho, K., Mitsuya, K., Kato, A.: Traffic Data Repository at the WIDE Project. In: 2000 USENIX Annual Technical Conference, FREENIX Track, San Diego, CA, USA, pp. 263–270 (2000)

[8] Cortez, P., Rio, M., Sousa, P., Rocha, M.: Topology Aware Internet Traffic Forecasting using Neural Networks. In: de Sá, J.M., Alexandre, L.A., Duch, W., Mandic, D.P. (eds.) ICANN 2007. LNCS, vol. 4669, pp. 445–454. Springer, Heidelberg (2007)

[9] DatCat, Datcat: Internet measurement data catalog (2008), http://www.datcat.org

[10] Fan, J., Yao, Q.: Nonlinear Time Series: Nonparametric and Parametric Methods. Springer Series in Statistics. Springer, New York (2003)

[11] Faraday, J., Chatfield, C.: Time Series Forecasting with Neural Networks: A Comparative Study Using the Airline Data. Journal of the Royal Statistical Society: Series C (Applied Statistics) 47(2), 231–250 (1998)

[12] Fowler, H.J., Leland, W.E.: Local Area Network Traffic Characteristics, with Implications for Broadband Network Congestion Management. IEEE Journal on Selected Areas in Communication 9(7), 1139–1149 (1991)

[13] Golding, R.A.: End-to-end Performance Prediction for the Internet (Work in Progress). Tech. Rep. UCSC-CRL-92-26, Concurrent Systems Laboratory, University of California at Santa Cruz (1992)

[14] Groschwitz, N., Polyzos, G.: A Time Series Model of Long-term NSFNET Backbone Traffic. In: IEEE International Conference on Communications (SUPERCOMM/ICC 1994), New Orleans, LA, USA, vol. 3, pp. 1400–1404 (1994)

[15] He, Q., Dovrolis, C., Ammar, M.: On the predictability of large transfer TCP throughput. Computer Networks 51(14), 3959–3977 (2007)

[16] internet2observatory, The Internet2 Observatory (2008), http://www.internet2.edu/observatory/

[17] Kantz, H., Schreiber, T.: Nonlinear Time Series Analysis, 2nd edn. Cambridge University Press, Cambridge (2004)

[18] Karagiannis, T., Molle, M., Faloutsos, M., Broido, A.: A Nonstationary Poisson View of Internet Traffic. In: 23th Annual Joint Conference of the IEEE Computer and Communications Societies (IEEE INFOCOM), Hong Kong, vol. 3, pp. 1558–1569 (2004)

[19] Kotz, D., Henderson, T., Abyzov, I.: CRAWDAD data set dartmouth/campus, v. 2007-02-08 (2007), http://crawdad.cs.dartmouth.edu/dartmouth/campus

[20] Lawrence Berkeley National Laboratory/ACM Special Interest Group on Data Communications, SIGCOMM, The Internet Traffic Archive (2008), http://www.sigcomm.org/ITA/

[21] Leland, W., Taqqu, M.S., Willinger, W., Wilson, D.: On the Self Similar Nature of Ethernet Traffic (Extended Version). IEEE/ACM Transactions on Networking 2(1), 1–15 (1994)

[22] Leland, W.E., Wilson, D.V.: High time-resolution measurement and analysis of LAN traffic: Implications for LAN interconnection. In: Tenth Annual Joint Conference of the IEEE Computer and Communications Societies. IEEE INFOCOM 1991, Bal Harbour, FL, USA, vol. 3, pp. 1360–1366 (1991)

[23] Liang, Y.: Real-time VBR Video Traffic Prediction for Dynamic Bandwidth Allocation. IEEE Transactions on Systems, Man, and Cybernetics, Part C: Applications and Reviews 1(34), 32–47 (2004)

[24] Miche, Y., Sorjamaa, A., Lendasse, A.: OP-ELM: Theory, Experiments and a Toolbox. In: Kůrková, V., Neruda, R., Koutník, J. (eds.) ICANN 2008, Part I. LNCS, vol. 5163, pp. 145–154. Springer, Heidelberg (2008)

[25] Montesino Pouzols, F.: Mining and control of network traffic by computational intelligence. PhD thesis, Microelectronics Institute of Seville, National Microelectronics Center, CSIC – University of Seville (2009)

[26] Montesino-Pouzols, F., Barriga, A.: Regressive fuzzy inference models with clustering identification: Application to the ESTSP 2008 competition. In: 2nd European Symposium on Time Series Prediction, Porvoo, Finland, pp. 205–214 (2008)

[27] NLANR:PMA, NLANR PMA: Special Traces Archive. Cooperative Association for Internet Data Analysis, CAIDA (2008), http://pma.nlanr.net/Special/

[28] Papagiannaki, K., Taft, N., Zhang, Z.L., Diot, C.: Long-Term Forecasting of Internet Backbone Traffic. IEEE Transactions on Neural Networks 16(5), 1110–1124 (2005)

[29] Paxson, V., Floyd, S.: Wide Area Traffic: The Failure of Poisson Modeling. IEEE/ACM Transactions on Networking 3(3), 226–244 (1995)

[30] Rodrig, M., Reis, C., Mahajan, R., Wetherall, D., Zahorjan, J.: Measurement-based Characterization of 802.11 in a Hotspot Setting. In: ACM SIGCOMM Workshop on Experimental Approaches to Wireless Network Design and Analysis, Philadelphia, Pennsylvania, USA, pp. 5–10 (2005)

[31] Rodrig, M., Reis, C., Mahajan, R., Wetherall, D., Zahorjan, J., Lazowska, E.: CRAWDAD: Dataset of wireless network measurement in SIGCOMM, Conference (2006), http://crawdad.cs.dartmouth.edu/uw/sigcomm2004

[32] Sang, A., Li, S.Q.: A Predictability Analysis of Network Traffic. In: Nineteenth Annual Joint Conference of the IEEE Computer and Communications Societies (INFOCOM 2000), Tel Aviv, Israel, vol. 1, pp. 342–351 (2000)

[33] Shannon, C., Moore, D., Keys, K., Fomenkov, M., Huffaker, B., Claffy, K.C.: The internet measurement data catalog. ACM SIGCOMM Computer Communication Review 35(5), 97–100 (2005)

[34] Shannon, C., Aben, E., Claffy, K.C., Andersen, D., Brownlee, N.: CAIDA OC48 Traces 2002-08-14, collection (2006),
http://www.caida.org/data/passive/index.xml

[35] Shannon, C., Aben, E., Claffy, K.C., Andersen, D., Brownlee, N.: CAIDA OC48 Traces 2003-04-24, collection (2006),
http://www.caida.org/data/passive/index.xml

[36] Shu, Y., Jin, Z., Wang, J., Yang, O.W.W.: Prediction-Based Admission Control Using FARIMA Models. In: IEEE International Conference on Communications, ICC 2000, New Orleans, LA, USA, vol. 3, pp. 1325–1329 (2000)

[37] Song, L., Deshpande, U., Kozat, U.C., Kotz, D., Jain, R.: Predictability of WLAN mobility and its effects on bandwidth provisioning. In: 25th Annual Joint Conference of the IEEE Computer and Communications Societies (INFOCOM), Barcelona, Spain, pp. 1–13 (2006)

[38] Sorjamaa, A., Miche, Y., Weiss, R., Lendasse, A.: Long-Term Prediction of Time Series using NNE-based Projection and OP-ELM. In: 2008 International Joint Conference on Neural Networks (IJCNN 2008), IEEE World Congress on Computational Intelligence, Hong Kong, China, pp. 2675–2681 (2008)

[39] Weigend, A., Gershenfeld, N.: Times Series Prediction: Forecasting the Future and Understanding the Past. Addison-Wesley Publishing Company, Reading (1994) ISBN: 0201626020

[40] Widely Integrated Distributed Environment (WIDE) Project, MAWI Working Group, Packet traces from wide backbone (2008),
http://tracer.csl.sony.co.jp/mawi/

[41] Wolski, R.: Dynamically Forecasting Network Performance Using the Network Weather Service. Cluster Computing 1(1), 119–132 (1998)

[42] Yager, R.R.: Using fuzzy methods to model nearest neighbor rules. IEEE Transactions on Systems, And, and Cybernetics–Part B: Cybernetics 32(4), 512–525 (2002)

[43] Yager, R.R.: Induced aggregation operators. Fuzzy Sets and Systems 137(1), 59–69 (2003)

[44] Yager, R.R., Filev, D.P.: Induced Ordered Weighted Averaging Operators. IEEE Transactions on Systems, And, and Cybernetics–Part B: Cybernetics 29(2), 141–150 (1999)

[45] Yi, Q., Skicewicz, J., Dinda, P.: An empirical study of the multiscale predictability of network traffic. In: IEEE International Symposium on High performance Distributed Computing (HPDC), Honolulu, HI, USA, pp. 66–76 (2004)

[46] You, C., Chandra, K.: Time Series Models for Internet Data Traffic. In: 24th Annual IEEE International Conference on Local Computer Networks (LCN 1999), Lowell, MA, USA, pp. 164–171 (1999)

Chapter 4
Summarization and Analysis of Network Traffic Flow Records

Abstract. Current network measurement systems are becoming highly sophisticated, producing huge amounts of convoluted measurement data and statistics. As a very common case, those networks implementing statistics reporting based on the NetFlow [15] technology can generate several GBs of data on a daily basis. In addition, these measurements are often very hard to interpret. In this chapter we describe a method that provides linguistic summaries of network traffic measurements as well as a procedure for finding hidden facts in the form of linguistic association rules. Thus, here we address an association rules mining problem. The method is suitable for summarization and analysis of network measurements at the flow level. As a first step, fuzzy linguistic summaries are applied to analyze and extract concise and human consistent summaries from NetFlow collections. Then, a procedure for mining hidden facts in network flow measurements in the form of fuzzy association rules is developed. The method is applied to a wide set of heterogeneous flow measurements, and is shown to be of practical application to network operation and traffic engineering [6, 5], where it can help solve a number of current issues.

4.1 Network Traffic Measurement Systems

As described in section 1, network performance measurement and monitoring are key for both network users and practitioners. The past years have seen a great deal of development in network measurement infrastructures and systems [19, 7, 12, 44]. Currently, the dominant approach to network measurement and monitoring is based on the passive measurement of traffic flows using the NetFlow technology [15, 28]. With an increasing diversity of technologies, applications and traffic patterns, the analysis of network traffic flows is becoming more and more complex. A full understanding of all the relevant facts is now far beyond the practical possibilities of network operators, managers and planners.

As the tasks of network operation and management become more and more complex, network measurement systems are being further developed. Current network measurement systems are becoming highly sophisticated and produce huge amounts

F.M. Pouzols et al.: Mining & Control of Network Traffic by Computational Intelligence, pp. 147–190.
springerlink.com   © Springer-Verlag Berlin Heidelberg 2011

of measurement data that have to be presented in the form of reports and statistics. These can be very hard to interpret. In particular, those networks implementing the NetFlow [15] technology can generate several GBs of data on a daily basis even though sampling techniques are used in order to reduce the amount of data generated. High-precision network measurement in current backbones implies the generation of tens of GBs of data per hour.

The major objective of network measurement systems is to provide an understanding of how networks perform. However, the gap between network measurement systems and users's comprehension is increasing. There are many visualization tools for network measurements (see [17] for an extensive list) which are mostly based on plots and charts to evaluate statistical properties of time series, scaling properties and protocol behavior [43]. The visualization and reporting tools employed nowadays provide reports made of tens of plots, graphs and tables. Thus, it is not easy for experts to extract simple summaries. Additionally, the complexity of tools for flow reporting and monitoring is holding back the adoption of these by end users.

Although there exist a number of tools that generate short summaries of network statistics, such as the analysis tools provided by router vendors and the popular flow-report tool, included in the flow-tools suite [24], it can be argued that human readability is achieved only at a very basic level. Also, complex relations underlying these statistics are missed and require a great deal of research to be unveiled. Complementary visualization tools are required to understand these reports. Additionally, a significant degree of expertise in statistical techniques is required to interpret the descriptors usually available.

Furthermore, flow based measurement systems are often used for real-time monitoring and operation tasks in general. Because of the large amounts of data that flow based measurement systems have to process in current high speed networks, real-time analysis requires fast, often simplified, methods and optimized implementations. This has motivated the development of specific hardware for packet classification and flow monitoring [51].

Many methods for analyzing Internet measurement data have been developed throughout the years. For example, some techniques common in the data mining field, such as Principal Component Analysis, PCA, have been applied to analyzing traffic flows from a structural viewpoint [34]. However, most of these techniques are quantitative, suited for specific data types, and designed for a particular purpose. There is a lack of general-purpose tools for qualitative exploration and analysis of Internet measurements, which is a first step needed for hypothesis-driven discovery, analysis and validation [45].

Our approach here departs from these ideas, as we seek to obtain a linguistic summarization that extracts all the relevant information that can be expressed in a human-readable manner. In this context, it is becoming more and more necessary to extract *concise* summaries that should be several orders of magnitude smaller than the original measurement dataset and should express how the network performs in ideally no more than a few lines of *human-readable* text [23].

We address the problem of summarizing network measurements into brief reports and making them human-readable by means of fuzzy linguistic summaries [38].

Linguistic summaries via fuzzy logic have been shown to be a simple, efficient and human consistent data mining means. This technique can be used as a natural language based knowledge discovery tool. The idea behind linguistic summaries is to use linguistic, natural language, terms to express information and knowledge that are hidden in a potentially large collection of objects. Linguistic summaries as introduced by Yager [47, 49] and further developed by Kacprzyk and Yager [29] and Kacprzyk and Zadrożny [31], are linguistically quantified propositions (as "Most traffic flows have an average packet size small") with a degree of truth. This type of fuzzy linguistic summaries is specially generic and fast to implement, two key properties for analyzing flow collections.

Some other approaches to the linguistic summarization of data based on fuzzy logic have been proposed. For instance, the method in [41] is based on clustering techniques in order to build a hierarchy of summaries while the proposal in [42] focuses on fuzzy and gradual functional dependencies. An alternative approach not considered here would have been to use Dempster-Shafer theory of evidence in order to extract rules [4]. The belief and plausibility functions can be used to define the degree of support for a proposition [11]. Thus, they can be used to perform an inductive rule extraction process based on the mass allocations for various propositions. In addition, belief and plausibility can be seen as pessimistic and optimistic measures of the strength of a rule [50].

Section 4.2 outlines network statistics based on the NetFlow technology. Section 4.3 defines linguistic summaries as applied in this work. Section 4.4 defines linguistic summaries of network flow collections and describes two complementary ways of implementing them. Section 4.5.3 shows experimental results for a set of benchmark NetFlow collections. Finally, we discuss the results and conclude in sections 4.5.7.

4.2   Flow Measurement and Statistics: NetFlow and IPFIX

Most network operations centers currently collect statistics on the performance of their infrastructure. These statistics are mainly based on the concept of flow, defined as a unidirectional sequence of packets between given source and destination endpoints.

NetFlow [15], introduced by Cisco Systems in 1996 as a technology for route caching, is nowadays a de facto standard for passive measurement and monitoring in the Internet. NetFlow based measurement is used for performance analysis, application and user monitoring, traffic engineering, capacity planning, billing, peering agreement and security applications.

From the viewpoint of a router, a flow is made of a sequence of IP packets sharing the same values for a set of properties within a time interval: source and destination IP address, source and destination transport level port, transport (layer 3) protocol type, type of service and incoming interface (see figure 4.1). Thus, a flow in the sense of NetFlow is unidirectional and is defined by a 7-tuple of values. In Internet routers, these 7 values made up the key field for the cache of flows. NetFlow records include flow properties at the link, network (IP) and transport layers, i.e., no specific information from the application layer is included.

When new flows are detected by a NetFlow capable router, a mapping between the flow and an outgoing interface is saved in memory. This way, next packets belonging to identified flows will not require to check routing tables, thus saving time and processing load.

This capability to identify flows can be applied to measure and characterize traffic traversing a router in real-time. Proper aggregation and summarization techniques allow for analyzing network performance.

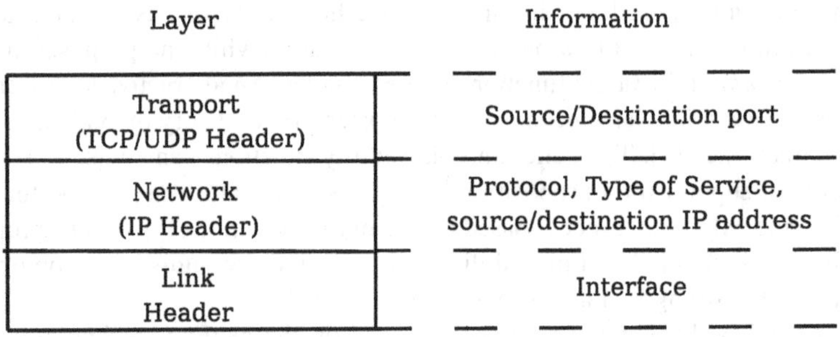

Layer	Information
Tranport (TCP/UDP Header)	Source/Destination port
Network (IP Header)	Protocol, Type of Service, source/destination IP address
Link Header	Interface

Fig. 4.1 Scheme of layers of information contained in NetFlow traces

The flow identifiers and some of the basic attributes considered in the most extended versions of NetFlow (version 5) are listed in Table 4.1. Additional attributes are available as extensions introduced in versions 6, 7, 8 and 9. Derived attributes are also defined from the measured attributes, such as the throughput or transfer rate (bytes/duration). Upon expiry of a flow, its statistics are accumulated and they are reported to a collector using the NetFlow protocol.

In order to standardize the NetFlow technology, the IP Flow Information Export (IPFIX) working group of the Internet Engineering Task Force (IETF) [16] is defining the IPFIX standard for reporting information about established flows in TCP/IP based connections. This standard is based on the NetFlow version 9 implementation and, being supported by a large number of vendors, is expected to be the industry standard for flow monitoring in the near future. Flows are defined as sequences of IP packets that traverse a measurement point in a period of time. IPFIX also defines the procedures required for exporting reports and processing them in devices outside of the network under analysis.

The IPFIX standard only differs from NetFlow in terminology and minor details as well as improvements to the flow transfer protocol but keeps the same principles, architecture, applicability and information model as NetFlow version 9.

Table 4.1 Flow identifiers and some of the attributes defined in NetFlow versions 5 and later

Attribute	Description
Source IP	IP layer address of sender
Destination IP	IP layer address of receiver
Packet count	Amount of packets transmitted
Byte count	Amount of bytes transmitted
Start time	Arrival time of first packet
End time	Arrival time of last packet
Input ifIndex	Index of input/source interface
Output ifIndex	Index of output/destination interface
Type of service	IP header TOS field (Differentiated Services Code Point)
TCP Flags	Logical conjunction of activated flags
Protocol	Protocol code in the IP header
Next hop address	IP address of next router or host
Source AS number	Code of the source Autonomous System
Destination AS number	Code of the destination Autonomous System
Duration	Time the flow was active

Currently, the information and statistics reported by basic collector tools summarize flow collections in the form of aggregation of counters and statistical descriptors such as percentiles. In essence, the reports generated by tools commonly used, such as flow-report [24], softflowd [3], ntop [21] and FlowScan [39], are summaries that specify simple descriptors of NetFlow attributes: minimum, maximum, average and total counters for bytes and packets on an aggregate or per-protocol basis. Analysis and visualization of flow collections is an active area of research [17] and many visualization techniques have been developed, particularly for topology analysis.

Results from analyses of NetFlow records are usually presented through cumulative distribution function plots and various graphs about the distribution of sizes, durations, distribution per transport protocol, subnetworks, protocol numbers, application, etc. However, the general summarization capabilities of available tools do not go beyond basic statistic descriptors, reports of top users and tables and plots of distribution functions, as for example the automatically generated Internet2 weekly reports [44] available online from http://netflow.internet2.edu/weekly/.

These tools are usually based on simple principles of descriptive statistics and provide relevant reports. However, the reports can easily become human unreadable because of the huge amount of tables and graphs generated. Techniques and tools for extracting short yet meaningful reports are sought.

Simple aggregation capabilities have been introduced with recent versions of NetFlow (version 9 onwards) as a simple method for summarization of flow collections. It is possible for instance to request from a router flow records aggregated by autonomous system. This novel capability has been introduced as a response to the need of summarization mechanisms for preprocessing flow collections. This need is motivated by two major factors: reducing the amount of data collected, and improving the understanding of measurement data. Although these methods are powerful and reveal a great deal of useful information about how networks perform, the whole amount of available measurement data and most complex relations underlying them are still difficult to understand.

Recent versions of NetFlow also integrate sampling capabilities [14]. With NetFlow sampling, only a percentage of traffic is accounted for measurement purposes. Most current monitoring systems and infrastructures are based in flow data exported from routers. In order to prevent overloading routers in terms of NPU usage, memory and record look-up time, a sampling of packets is performed typically on the 1-10% of the total traffic, while 90-99% of packets are not accounted for performance measurement purposes [22]. Then, flow statistics are computed from the collected sampled packet statistics. Thus, packet sampling is an inherently lossy process that discards information. As a consequence, some flow statistics are affected by uncertainty. Usually, sampling is performed by a simple deterministic process of choosing 1 packet every certain number of packets on a per-interface basis. For instance, in the Cisco Sampled NetFlow scheme, one packet is randomly chosen within every window of N consecutive arrivals. These sampling capabilities are extensively used in current measurement infrastructures.

4.3 Linguistic Summaries

Linguistic summaries as proposed by Yager [47] are a data mining technique for summarizing data collections using linguistically quantified propositions [53], such as "Most traffic flows are short lived". In this work, we consider the extended definition by Kacprzyk and Zadrożny [31], that leverages on the concept of protoform or prototypical form.

Linguistic summaries have a number of advantages when compared against classical statistical methods of summarization: they can summarize both numeric and non-numeric data, can provide many different summaries for specific purposes and have the ability to provide natural language summaries.

Linguistic summaries are obtained by means of a mining process performed on a usually large set of entities, by which a natural language expression summarizes essential facts about the set. In the sense of Yager [47, 49], a linguistic summary is defined as follows. Given:

- $\mathscr{D} = \{d_1, \ldots, d_N\}$, a set of entities that manifest some attributes, e.g., a set of traffic flows in a NetFlow collection.
- $\mathscr{A} = \{A_1, \ldots, A_M\}$ a set of attributes defined over the entities in the set $\mathscr{D}$, e.g., the set of attributes in a NetFlow collection, such as packet count, destination address, starting time, etc.

A basic linguistic summary is made of:

- A summarizer, $\mathscr{S}$, defined as a linguistic expression (or predicate) semantically represented by a fuzzy set, i.e., "short lived".
- A quantity in agreement or quantifier, $\mathscr{Q}$, defined as a linguistic quantifier that indicates the extent to which the entities satisfy the summary, e.g., "most".
- A measure of validity or quality of the summary. The basic validity criterion is the truth value of the summary, T, defined as a truth value of a linguistically quantified statement. The truth value can be computed using a number of methods, in particular Zadeh's fuzzy-logic-based calculus of linguistically quantified propositions [53, 36] and Yager's OWA operators [48, 36].

Fuzzy subsets are employed to represent the linguistic terms that specify a summarization $\mathscr{S}$ and a quantifier $\mathscr{Q}$ [36, 37]. Thus, the truth value of both can be denoted by their respective membership functions, $\mu_{\mathscr{S}}(x)$ and $\mu_{\mathscr{Q}}(x)$, being its universe of discourse that of one or more of the attributes in the set $\mathscr{A}$.

A summary ($\{\mathscr{S}, \mathscr{Q}\}$) of a data set $\mathscr{D}$ with N elements from a measurement space $\mathscr{X}$ is usually written in generic form as "$\mathscr{Q}$ d's are $\mathscr{S}$", i.e., Q flows are S, as in the statement "most flows are long lived":

$$\{\mathscr{D}, \{\mathscr{Q}, \mathscr{S}\}\}, \text{ read as } \mathscr{Q} \, d_i \text{ are } \mathscr{S} \qquad (4.1)$$

$\mathscr{S}$ is then a fuzzy subset of $\mathscr{D}$ and $\mathscr{Q}$ is a fuzzy set in the range $[0,1]$. For instance, the membership function of the quantifier *most* can be defined as:

$$\mu_Q(x) = \begin{cases} 1, & \text{for } x \geq 0.85 \\ 2x - 0.7, & \text{for } 0.35 < x < 0.85 \\ 0, & \text{for } x \leq 0.35 \end{cases}$$

Then T is a truth value in $[0,1]$ that can be computed from a summary as in equation 4.1 applying Zadeh's calculus:

$$T(\mathscr{D}, \{\mathscr{Q}, \mathscr{S}\}) = \mu_{\mathscr{Q}} \left(\frac{1}{N} \sum_{i=1}^{N} \mu_{\mathscr{S}}(d_i) \right),$$

where the term $\sum_{i=1}^{N} \mu_{\mathscr{S}}(d_i)$ is the precise cardinality of the fuzzy set $\mathscr{S}$ (the summarizer), defined as the sum of the membership degrees of the d_i values [53].

In this formulation the truth value is computed as a linguistic quantifier driven aggregation. The truth value of fuzzy linguistically quantified propositions is just a primary measure of validity or quality of summaries. Additional measures of the goodness of a linguistic summary, in terms of degree of interest, non-triviality or unexpectedness, are usually required in practice in order to select relevant summaries [29].

It is straightforward to generalize the kind of summarizer in equation 4.1 to a compound summarizer made of the conjunction of any number of linguistic expressions about the attributes of the entities in $\mathscr{D}$, as in "Most flows *are* long lived *and* have an average packet size small *and* are high throughput".

Extended linguistic summaries can be defined by adding a qualifier, $\mathscr{R}$, also a subset of $\mathscr{D}$, as "$\mathscr{Q}\mathscr{R}$ d's are $\mathscr{S}$", i.e., $\mathscr{Q}\mathscr{R}$ flows are $\mathscr{S}$, as in the statement "most flows at night are long lived":

$$\{\mathscr{D},\{\mathscr{Q},\mathscr{R},\mathscr{S}\}\}, \text{read as } "\mathscr{Q}\mathscr{R}d's \text{ are } \mathscr{S}" \tag{4.2}$$

Summaries of this kind describe dependencies between specific values of particular attributes. In the case of equation 4.2, the degree of truth of the summary can be determined by Zadeh's calculus as follows:

$$T(\mathscr{D},\{\mathscr{Q},\mathscr{R},\mathscr{S}\}) = \mu_{\mathscr{Q}}\left(\frac{\sum\limits_{i=1}^{N}(\mu_{\mathscr{S}}(d_i) \wedge \mu_{\mathscr{R}}(di))}{\sum\limits_{i=1}^{N}\mu_{\mathscr{R}}(d_i)} \right)$$

Extended linguistic summaries can be interpreted as fuzzy *if-then* rules in which the antecedent is $\mathscr{R}$ and the consequent $\mathscr{S}$, stating that if $\mathscr{Q}$ entities (flows) satisfy $\mathscr{R}$ then they satisfy $\mathscr{S}$. This analogy will be exploited later on in order to define a mining process for linguistic summaries.

Linguistic summaries, whether extended or not, can be compound as well, as in "most high throughput flows are long lived and have a packet size medium". In this case, the universe of discourse of the summarizer is extended to that of a set of attributes.

Thus, linguistic summaries as considered here are essentially linguistically qualified propositions in the sense of Zadeh's calculus [53].

Protoforms of linguistic summaries are defined as abstracted prototypes and may form a hierarchy [54]. A classification of possible protoforms of linguistic summaries is developed in [31]. For instance, replacing $\mathscr{Q}$ with a concrete quantifier *Most* in equation 4.1, we obtain a particular kind of protoform: "Most flows are $\mathscr{S}$.

Another kind of protoform can be specified by fixing the attribute or attributes of interest for $\mathscr{S}$, as "$\mathscr{Q}$ flows are $\mathscr{S}^{A_c}$", where A_c is the attribute of interest. For instance, when one is interested in the duration of flows an appropriate protoform can be defined by restricting the summarizer to the linguistic labels defined for the duration attribute, "$\mathscr{Q}$ flows are $\mathscr{S}^{duration}$", where $\mathscr{S}^{duration}$ may take the form of any of the linguistic variables defined for the attribute duration.

4.4 Definition of Linguistic Summaries of Network Flow Collections

We propose two methods for the linguistic summarization of NetFlow collections. Both are complementary to traditional methods of analysis and visualization of network flow statistics. To this end, a set of linguistic variables for flow attributes as well as a set of fuzzy quantifiers have to be defined.

In this context, the domain of discourse is the set of NetFlow attributes. Some NetFlow attributes can be modeled with crisp values (such as protocol, destination port and interface numbers), while some others are more properly modeled using linguistic variables [52].

In practice, a number of attributes are inherently crisp from our point of view. Their inclusion in summarizers and qualifiers of linguistic summaries can thus be modeled as filters that keep a subset of flows for certain crisp values or ranges of crisp values. For example, if the user is interested in summaries regarding only TCP flows, a first filtering step is carried out in order to account only those flows that correspond to TCP connections. This way, the summary "Most TCP flows are long-lived" differs from "Most flows are long-lived" in the set of flows to which they apply. Both are equivalent as for its evaluation as quantified proposition. Crisp attributes include the IP protocol field (IPv4, IPv6, ICMP, PIM, etc.), the transport layer protocol (TCP, UDP, SCTP, etc.) and transport port (HTTPS, SMTP, SSH, etc.), among others.

Two attributes, the input and output interface numbers, can be analyzed only when a detailed knowledge of the local topology surrounding measurement point is available. In principle, both attributes could be modeled as crisp attributes, though it can make sense to model either or both as fuzzy values provided some type of fuzzy classification of traffic flows according to the input/output interfaces is of interest. In addition, the content of these attributes is usually meaningless in publicly available traces. Thus, we exclude them from our analysis.

Two approaches are followed here in order to define linguistic labels for those attributes that can be more properly modeled by fuzzy sets. First, linguistic labels can be defined on the basis of a priori knowledge drawn from traditional analysis tools and measurement studies found in the literature. As a second and more automatic approach, unsupervised learning techniques such as clustering methods can be applied on previously recorded NetFlow collections in order to automatically define linguistic labels from the clusters identified. In the next section, we will detail linguistic summaries following the first approach. Identification of labels based on clustering techniques will be described later. The procedure for both approaches is depicted in figure 4.2.

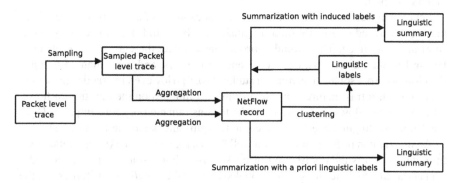

Fig. 4.2 Procedure for extracting linguistic summaries of NetFlow collections. Summaries can be generated using either a priori linguistic labels or labels identified by clustering methods.

A number of schemes have been proposed to define linguistic labels for a given input space [25]. Some schemes partition the domain of the input variables into regions. These include regular grids and more sophisticated methods for hierarchical and local partitioning. Other schemes define labels from clustering procedures.

Our aim is to keep the maximum interpretability for both summarization approaches. Thus, we use a simple regular partitioning scheme. There is no accepted formal definition of interpretability of a fuzzy system. Nonetheless, we will adhere to a few simple principles accepted in the literature [13] in order to maximize interpretability. For each flow attribute a partition of the input domain is defined such that only triangular and trapezoidal membership functions are used for the sake of simplicity and readability. The partitions are defined as standardized grids and are irregular in general. These partitions are defined in such a way that the whole input domain is covered, the degree of overlapping is limited to one, the number of different labels for an attribute ranges between 2 and 4, and the sum of all the membership degrees for any given point of the input domain is 1, i.e., the partition is standardized. This way, we maximize both completeness and distinguishability and thus interpretability.

4.4.1 Defining Linguistic Labels from a Priori Knowledge

For the case of labels defined from a priori knowledge, we propose here a general set of fuzzy sets using domain specific terminology as shown in table 4.2. In the general approach followed, the input space in partitioned with as many labels as relevant classes have been identified in measurement studies.

The dominant values, ranges and classes of elements in attributes of network flows are most often identified by using statistical tools such as common distribution functions (CDF) and histograms [34]. For instance, it is common to analyze the mode and subsequent peaks in the histograms that can be visualized as sudden jumps in the CDF plot.

Let us consider the duration of flows as a first example of definition of labels from previous knowledge. Ideally three linguistic labels should be defined: very short, short and long-lived. This would approximately reflect the classification of flows lasting less than 2 seconds, between 2 and 15 minutes, and more than 15 minutes than follows from some measurement studies [10, 9]. However, in practice most flow measurement infrastructures impose a limit of 5 minutes on the duration of flows. After 5 minutes, flows are expired in order to constrain memory consumption. Thus, we define two linguistic terms for the duration attribute: short- and long-lived.

An invariant in traffic that has been well known for a long time (with implications in router design is the distribution of packet sizes. For instance, one can consider packets around 54 bytes long to be modeled by the label *small*, as this represents the class of TCP ACK packets with little or not payload, which often account for a significant percentage of the total amount of packets. This way, the *small* label accounts for ACK packets, DNS queries, or ARP packets. Following on the same approach, packets around 1500 bytes long can be considered *large*, as this is the approximate value of the currently prevailing MTU in most networks[1]. For packet sizes between the small and large classes, an additional *medium* class can be considered. Finally, packets with sizes higher than 1500 can be classified into the *jumbo* class, accounting for jumbo frames, whose state of deployment currently varies greatly depending on the network.

This is in fact a classification very similar to the one used for generating the Internet2 NetFlow weekly reports [44], where 4 different size ranges are used: small (less than 100 bytes), medium (between 100 and 1400 bytes), large (between 1401 and 1500 bytes) and jumbo (above 1500 bytes).

For the bytes count attribute, *mice*, *bulk* and *elephants* are terms usually employed in the Internet measurement literature to refer to recurrent types of traffic flows with regards to its place in flow size distributions [9, 44, 10]. As for the packets count attribute, three labels are defined in an analogous manner to the flow bytes count (or size) attribute: *pkt-mice*, *pkt-bulk* and *pkt-elephant*.

Derived metrics are also considered, as is the case of throughput, defined from the flow attributes as the ratio bytes/duration. Three labels for the throughput attribute are defined: *low*, *medium* and *high* [44, 10].

The linguistic labels for the average packet size attribute is shown in figure 4.3(b). This is an attribute that can be expected to exhibit a clear increasing evolution with time, as well as show different values for different networks depending on the overall bandwidth capacity. Based on measurement studies, we have defined three labels: *low*, *medium* and *high* [10].

The start and end time attributes have specially to do with daily usage patterns. The definition of linguistic labels that reflect these patterns would in general require a specialized analysis. For the sake of generality and simplicity, we define two labels: *day* and *night* (figure 4.3(f)). Note that more refined terms could be consider in order to integrate additional knowledge, such as distinguishing between morning and afternoon traffic patterns.

Table 4.2 Linguistic Labels for Some Flow Attributes

Attribute	Linguistic Labels
Duration	Short-lived, Long-lived
Average packet size	Small, Medium, Large, Jumbo
Bytes	Mice, Bulk, Elephants
Throughput	Low, Medium, High
Packets	Packet-Mouse, Packet-Bulk, Packet-elephant
Time (start, end)	Day, Night

---

[1] We note the definition of link MTU as the IP MTU over a link uniformly found in IETF documents is considered here. Thus, the IP header is included in the MTU size but link layer headers and other framing which is not part of IP or the IP payload are excluded.

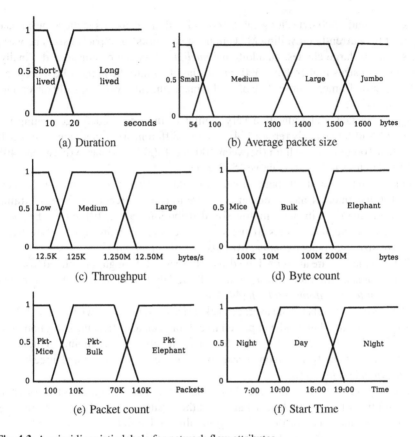

Fig. 4.3 A priori linguistic labels for network flow attributes

4.4.2 Automatic Definition of Linguistic Labels by Unsupervised Learning

We adopt a simple scheme for defining standardized grid partitions that guarantees completeness and provides distinguishability. Values identified as cluster centers are used as vertices of the trapezoidal membership functions. Since a hard constrain of completeness of the partitions of the input domain has been imposed, we have that for each trapezoidal membership function with vertices a_i, b_i and c_i:

$$A_i(x) = \begin{cases} 0, & \text{if } x \leq a_i \\ \frac{x-a_i}{b_i-a_i}, & \text{if } a_i < x < b_i \\ 1, & \text{if } b_i \leq x \leq c_i \ , \\ \frac{x-c_i}{b_i-c_i}, & \text{if } c_i < x < d_i \\ 0, & \text{if } x \geq d_i \end{cases}$$

where $a_i < b_i < c_i < d_i$. Any partition is complete if the following equalities:

$$a_{i+1} = c_i, \ b_{i+1} = d_i$$

hold for all the sets in the partition, where A_i and A_{i+1} are two neighbor fuzzy sets.

4.4.3 Quantifiers

The following sequence of quantifiers has been defined: *very few, few, about 1/3, about 1/2 about 2/3, most* and *almost all*. The quantifiers *very few, few, most* and *almost all* denote different degrees of disparity [9] in the distribution of some property and are specially meant to find disparity conditions. Similar quantifiers have been extensively used in other application domains [32]. Here we emphasize the usefulness of the *Very few* and *Almost all* for discovering cases of disparity.

4.5 Summarization of NetFlow Collections

Once fuzzy quantifiers, qualifiers and summarizers are defined, linguistic summaries for flow collections can be computed. When looking for the summaries that best describe flow collections, two approaches can be considered: 1) the summarizer, the qualifier and the quantifier are given by the user, and 2) the three fuzzy sets corresponding to the quantifier, summarizer and qualifier are not fixed a priori and thus any possible combination must be taken into consideration. On the one hand, a tool that implements case 1) would be of little value for users. On the other hand, an implementation of case 2) is extremely computationally intensive, as it would require the exploration of an unlimited number of possible combinations. This is thus not suitable for fast on-line summarization. However, applying the concept of protoforms [54], intermediate cases can be defined in between. In what follows, we will describe two approaches: fast, on-line summarization for real-time monitoring, and off-line mining of linguistic summaries.

4.5.1 On-Line Summarization of NetFlow Collections

A first way of implementing linguistic summaries of NetFlow collections is applied to on-line monitoring and generation of short reports. Since only one-pass procedures with no significant memory requirements are needed to compute linguistic summaries, a bounded set of summaries can be generated in real-time.

For on-line summarization, a set of protoforms identified as conditions of interest are evaluated. Additionally, specific summaries specified as options to the tool can be evaluated as well.

In order to select a set of relevant protoforms our proposal combines ideas from reports found in traditional flow analysis and visualization tools, in particular from Internet2 NetFlow weekly reports [44]. The basic set of protoforms considered for automatic on-line reports is shown in table 4.3. Additional optional summaries have been defined for control, multicast and routing traffic.

The first six protoforms can be thought of as a linguistic packet classifier. As a case of particular interest, the distribution of packet sizes is a key factor for router performance, as detailed in chapter 6. In the last summary line considered, a generalization of common filtering methods is achieved with the fuzzy bulk TCP constraint

Table 4.3 Basic protoforms for on-line linguistic summaries of NetFlow collections

Concept	Summarizer ($\mathscr{S}$)	Qualifier ($\mathscr{R}$)	Example
Duration distribution	$A^{duration}$	-	"Few flows are long-lived"
Throughput distribution	$A^{throughput}$	-	"Very few flows are high throughput"
Transfer size distribution	A^{bytes}	-	"About 1/3 flows are bulk"
Average packet size distribution	$A^{packetsize}$	-	"Few flows have avgerage packet size medium"
Packets per flow distribution	$A^{packets}$	-	"Almost all flows are pkt-bulk"
Throughput distribution qualified by transfer size	$A^{throughput}$	A^{bytes}	"About 2/3 elephant flows are low throughput"
Bulk TCP transfer duration distribution	$A^{duration}$	TCP crisp filter and fuzzy qualifier *bulk* on A^{bytes}	"Most Bulk TCP flows are short lived"

For instance, in the Internet2 NetFlow weekly reports [27], TCP flows that transfer more than 10 MB of data are considered bulk TCP flows. In the case of the linguistic summaries presented here, TCP flows that transfer a lower volume of data are also partially accounted, and flows transferring more than 200 MB are excluded from the bulk class (see figure 4.3(d)).

In principle, the aggregation of flow collections can be done every 5 minutes, daily, weekly, monthly, etc. However, as pointed out by the authors of the Internet2 weekly reports [27], an aggregation period lower than a week (or a day) tends to show too much statistical volatility. This makes reports generated from flow collections taken on short periods useful for operation purposes but probably not for capturing stable long-term trends.

Regarding the last summary line, it should be recognized that common NetFlow collection mechanisms are always configured so that flows cannot last longer than a certain maximum period of time. Thus, the distribution of transfer sizes is to a certain extent skewed in the upper part.

The set of protoforms shown in table 4.3 has been defined with the aim to provide a reduced set of summaries to be shown. Their interpretation and possible usefulness is left to the user. Thus, a set that is not too large but not too small is provided for fast interpretation. This is usually regarded as an important factor to guarantee user's autonomy [31].

Also, it should be noted that all the summarizers defined in the on-line set consist of a single fuzzy set. Although more complex summarizers would be more informative about particular aspects, simple summarizers have been chosen for two reasons. First, long summarizers are less readable for users. Second, the aim of the on-line summaries is to spot overall properties of flow collections, which would be obscured by more specific summarizers.

Concrete examples for a number of NetFlow collections will be presented below. As mentioned above, the method proposed is complementary to a large number of summarization and visualization tools that are currently used to analyze NetFlow collections. Let us compare the output of the flow-1summary tool against that of widespread summarization tools. Since the first developments on passive measurements, tools for reducing data and providing summaries of traffic traces have been available. The tcp-reduce and tcp-summary (available from http://ita.ee.lbl.gov/html/contrib/tcp-reduce.html) were some of the first to be introduced. At the time, the NetFlow technology was not still deployed in general. The first produces a summary of one liner per TCP connection (or flow) from a packet level trace, whereas the second produces a table with one row per TCP protocol (or port number). The kind of table generated by the second tool is commonly used in current network analysis and monitoring tools.

In current NetFlow analysis tools, besides graphical representations of the distribution of several parameters, it is for instance common to find statistical summaries such as the two examples showed below. For a fragment of the Internet2-LOSA-May08 collection, described below, the softflowd [3] tool can generates a text summary as the following:

```
Expired flow statistics:   minimum       average       maximum
   Flow bytes:                    1       1150861    3570584485
   Flow packets:                  1            35        102735
   Duration:                  0.00s        23.88s       831.46s

Expired flow reasons:
        tcp =          0    tcp.rst =       0   tcp.fin =        0
        udp =          0    icmp =          0   general =        0
    maxlife =          0
   over 2Gb =         68
   maxflows =     302486
    flushed =       8259

Per-protocol statistics:       Octets       Packets    Avg Life     Max Life
        icmp (1):           464764825         31430       3.24s      256.11s
        igmp (2):              326656            34     146.87s      557.75s
     ipencap (4):             2184962           128     320.92s      669.15s
         tcp (6):         221304920583       7059112      76.00s      831.46s
        udp (17):         18858197645        965373      18.59s      300.30s
        ipv6 (41):           11180042           475     188.39s      715.19s
        gre (47):            28080399          1100     231.88s      324.65s
        esp (50):         116930666347       2818043      60.66s      254.71s
    Unknown (54):              87040             5       0.00s        0.00s
      ax.25 (93):              26880             1       0.00s        0.00s
        pim (103):           6730317           447     618.79s      763.73s
    Unknown (169):          95348736          5173     547.54s       796.29
```

A related report, as provided by the flow-report [24] tool reads as follows:

```
Ignores:                  0
Total  Flows:             53588313
Total  Octets:            222703590422
Total  Packets:           273750307
Total  Duration  (ms):    327873242594
Real  Time:               1207094398
Average  Flow  Time:      6118.000000
Average  Packets/Second:  813.000000
Average  Flows/Second:    4155.000000
Average  Packets/Flow:    5.000000
Flows/Second:             23.554512
Flows/Second  (real):     0.044394


Average  IP  packet  size  distribution:

  1-32   64    96   128   160   192   224   256   288   320   352   384   416   448   480
  .004  .482  .071  .034  .024  .012  .012  .009  .009  .012  .009  .006  .007  .004  .005

   512   544   576  1024  1536  2048  2560  3072  3584  4096  4608
  .004  .005  .009  .052  .230  .000  .000  .000  .000  .000  .000

Packets  per  flow  distribution:

    1     2     4     8    12    16    20    24    28    32    36    40    44    48    52
  .738  .098  .065  .042  .017  .009  .006  .004  .003  .002  .002  .001  .001  .001  .001

   60   100   200   300   400   500   600   700   800   900  >900
  .001  .005  .004  .001  .000  .000  .000  .000  .000  .000  .000

Octets  per  flow  distribution:

   32    64   128   256   512  1280  2048  2816  3584  4352  5120  5888  6656  7424  8192
  .004  .383  .132  .081  .069  .077  .132  .011  .032  .009  .011  .006  .006  .003  .004

  8960  9728 10496 11264 12032 12800 13568 14336 15104 15872 >15872
  .002  .003  .002  .002  .002  .001  .002  .001  .001  .001  .024

Flow  Time  Distribution  (ms):

   10    50   100   200   500  1000  2000  3000  4000  5000  6000  7000  8000  9000 10000
  .743  .002  .004  .007  .013  .012  .013  .009  .007  .006  .006  .005  .004  .004  .004

 12000 14000 16000 18000 20000 22000 24000 26000 28000 30000 >30000
  .007  .007  .007  .006  .006  .006  .005  .006  .005  .005  .100
```

The so-called top-10 talkers reports are also frequently used by the networking community, which reflects the importance of disparity in several distributions concerning network traffic. In these reports, global counters are shown for the flows coming from the 10 addresses that generated the most traffic for a given period of time.

It should be noted that some authors argue that an interaction with users has to be assumed for the determination of a class of summaries of interest [31]. This is because it does not seem feasible to perform a fully automatic generation of linguistic summaries. However, in this particular application case, we have defined a consistent and small set of relevant summaries for the on-line summarization procedure by following a number of previous results on the analysis and description of flow collections. With the set in table 4.3 we aim at covering most generic monitoring and concise reporting tasks. Nonetheless, the tool implemented, flow-lsummary, described later on, allows for the inclusion of different protoforms in the set of summaries for on-line summarization. This makes it possible to tailor the on-line procedure to specific requirements.

In previous sections, we have described methods for computing the degree of truth of a fuzzy linguistic summary. Let us now describe additional measures of the quality or informativeness of linguistic summaries in the context of on-line summaries for NetFlow collections. Since linguistic summaries were introduced it has been recognized that the degree of truth is not necessarily a measure of informativeness [47]. That is, the summary with the highest degree of truth may not be significantly informative or even not informative at all.

In a sense, the most informative description of a dataset is the dataset itself. Any summary entails some loss of information. The question is thus to what extent a particular linguistic summary is informative.

In order to overcome this limitation, a number of metrics of quality and informativeness (or the amount of information provided) of linguistic summaries have been proposed [49, 29, 31]. Also, it has been proposed the use of a weighted combination of different metrics.

In what follows we will describe 3 additional metrics: confidence, preciseness and appropriateness The *confidence* of a linguistic summary (called covering in [31]), indicates to which extent the summary holds for the whole set of entities in the data set. That is, the confidence indicates the percentage of entities in the data set that are consistent with the summary, i.e., the the percentage of entities for which the summarizer holds when the qualifier holds. The confidence, T_c, of an extended fuzzy linguistic summary is defined as follows:

$$T_c(\mathcal{D}, \{\mathcal{R}, \mathcal{S}\}) = \frac{\sum_i^N c_i}{\sum_i^N s_i},$$

where

$$c_i = \begin{cases} 1, & \text{if } \mu_f(d_i) > 0 \text{ and } \mu_{\mathcal{R}}(d_i) > 0 \\ 0, & \text{otherwise} \end{cases}$$

$$s_j = \begin{cases} 1, & \text{if } \mu_{\mathcal{R}}(d_i) \\ 0, & \text{otherwise} \end{cases}$$

In the case of non-extended summaries, i.e., summaries without qualifiers, $s_j = 1, i = 1, \ldots, N$ and c_i is defined as follows:

$$c_i = \begin{cases} 1, & \text{if } \mu_{\mathcal{S}}(d_i) > 0 \\ 0, & \text{otherwise} \end{cases}$$

However, the confidence metric may lead to discarding extreme infrequent facts when applied to summaries with no qualifier. In our case, we are particularly interested in some facts of this kind. An example would be the summaries "$\mathcal{Q}$ flows are elephants". One would expect these summaries to hold for a small or even extremely small percentage of flows; however, knowing the presence of these flows is desirable, i.e., when a summary like "Very few flows are elephants" has a high degree of truth it may be the most interesting summary for its protoform in spite of its (likely) low confidence. For this reason, the confidence metric has to be used with care.

In fact, when analyzing network flows, summaries about a small percentage of the flows may be highly informative as these few flows may be responsible for a large amount of traffic load. For instance, in the WIDE-F-DITL-2007 flow collection, described later on, the summary "Very few flows are high throughput" is selected as the most relevant summary on the distribution of throughput. The confidence is $8.6 \cdot 10^{-4}$, whereas the confidence for other summaries corresponding to the same protoform is 0.97. This summary shows the presence of a long-tail in the throughput distribution, and it is the fact about throughput that can be asserted with the highest degree of truth[2].

The *preciseness* (as opposed to the imprecision, non-specificity or fuzziness), T_p, of a fuzzy linguistic summary with a summarizer $\mathscr{S}$ that consists of a set of fuzzy sets $\mathscr{R} = \mathscr{S}_1, \ldots, \mathscr{S}_{N_S}$ is defined as follows:

$$T_p(\mathscr{S}) = 1 - \sqrt[N_S]{\prod_{i=1}^{N_S} \mathrm{in}(\mathscr{S}_i)},$$

with $\mathrm{in}(\mathscr{S}_i)$ defined as:

$$\mathrm{in}(\mathscr{S}_i) = \frac{card\{x \in \mathscr{X}_j : \mu_{\mathscr{S}_j}(x) > 0\}}{card\, X_j},$$

where X_j is the universe of discourse for the fuzzy set $\mathscr{S}_j$. The $\mathrm{in}(\mathscr{S}_i)$ indicate of the degree of fuzziness of a summarizer. That is, the preciseness can be understood as a metric opposed to fuzziness.

It is important to note that the degree of precision depends exclusively on the form of the summary and not on the set of entities. That is, the precision of a summary can be computed from the membership functions of the linguistic terms that define the family of possible fuzzy sets in $\mathscr{S}$.

However, the standard definition of preciseness does not necessarily makes sense in all application cases since it assumes some kind of uniform distribution of semantics over the universe of discourse. Consider for instance the byte count attribute. We have defined three labels with very different values of $\mathrm{in}(\mathscr{S}_i)$. For a summary with a summarizer defined on the byte count attribute, the three linguistic terms that can apply to the attribute are equally precise from a conceptual viewpoint. For instance, the preciseness, T_p, of "$\mathscr{Q}$ flows are elephants" is clearly much lower than that of "$\mathscr{Q}$ flows are mice". In contrast, both seem equally informative, as the range of byte count values for the *elephants* term is naturally much wider.

The *appropriateness* is in a sense the most important metric of informativeness for summaries with complex summarizers. If the summarizer of a summary consists of a set of fuzzy sets, $\mathscr{S} = \mathscr{S}_1, \ldots, \mathscr{S}_{N_S}$, the appropriateness, T_a, of the summary is defined as follows:

---

[2] Or, more precisely, with the highest value for a combination of the degree of truth and the preciseness of the quantifier, as will be detailed later on.

$$T_a(\mathscr{D},\mathscr{S}) = \left\| T_c - \prod_{i=1}^{N_S} s_i \right\|,$$

where

$$s_i = \frac{\sum_{i=1}^{N} a_i}{N},$$

and

$$a_i = \begin{cases} 1, & \text{if } \mu_{\mathscr{S}}(d_i) > 0 \\ 0, & \text{otherwise} \end{cases}$$

That is, the appropriateness is computed with respect to the confidence. However it is a measure that makes sense only when the summarizer consists of the conjunction of two or more linguistic terms. In such cases, the appropriateness is a highly relevant metric. To illustrate the meaning of the appropriateness and its suitability for analyzing flow collections, let us consider a simple example. Let $S = S_1, S_2$ be the summarizer of a summary where S_1 corresponds to $A^{throughput}$ and S_2 corresponds to $A^{duration}$. Let us suppose that 80% of flows are medium throughput and 70% of flows are short lived. In this case, if the distributions of duration and throughput were uniform, one would expect approximately 46% of flows to be short lived and medium throughput. However, it is well known that this is not often the case in real network [10, 9]. The appropriateness aims at measuring this kind of disparity.

Finally, some authors propose the length of a summary [31], measured in terms of the number of sets in the summarizer of a summary, as an additional quality metric. In particular, Kacprzyk et al. [31] propose the following measure of the length of a summary, T_l:

$$T_l(\mathscr{S}) = 2 \left(\frac{1}{2} \right)^{N_S}.$$

As in the case of the precision metric, the length of a summary depends exclusively on the form of the summary and not on the set of entities. Note that this metric does not make any difference for the set protoforms defined for on-line summarization, as all these protoforms include only one fuzzy set in the summarizer.

A straightforward approach to measuring the quality of linguistic summaries is to compute a weighted average of the different metrics described so far. In this case, the final quality metric would be computed as a weighted average as follows:

$$I_5(\mathscr{D},\mathscr{R},\mathscr{S}) = I(T, T_c, T_p, T_a, T_l; w, w_c, w_p, w_a, w_l) =$$
$$= wT + w_c T_c + w_p T_p + w_a T_a + w_l T_l,$$

where the weights assigned to each quality measure take values on the unit interval. Finding the values of the weights, or more generally a good combination of informativeness and quality metrics is a complex problem. In general, this usually requires the cooperation of users in an essentially heuristic process. Some authors propose the use of processes and approaches well-known in the decision theory field [31]. This is however hardly feasible when the summarization process is intended to apply to an essentially dynamic object rather than a particular database, as is the case for on-line summarization of NetFlow collections. Thus, we prefer to define a simple combination of quality measures that is easy to understand for users.

As discussed before, the length and appropriateness measures do not apply to on-line summarization of flow collections. The general applicability of the preciseness measure is arguable as well, we thus prefer not to use it as defined above. Similarly, the confidence metric is not used. In practice, a robust and fully automatic measurement of interest should avoid favoring frequent facts to the detriment of extreme events.

Instead, we define here an informativeness measure for linguistic summaries, I, as a combination of the degree of truth and the preciseness of the quantifier or quantity in agreement. The preciseness of the quantifier, defined as follows:

$$T_{pQ}(\mathcal{Q}) = 1 - \text{in}(\mathcal{Q}).$$

Both factors are considered in order to define a global measure of quality or informativeness about the summaries in the on-line set:

$$I(\mathcal{Q}, \{\mathcal{Q}, \mathcal{R}, \mathcal{S}\}) = 0.9T + 0.1T_{pQ}.$$

That is, a 10% weight is allocated to the preciseness of the quantifier. This way, with the goal of emphasizing extreme events, we favor more precise quantifiers to the detriment of the more fuzzy ones. For instance, $T_{pQ}(most) = 0.5$, while $T_{pQ}(almost\ all) = 0.85$. T_{pQ} plays the role of promoting, for a particular protoform, the most precisely quantified summaries among all the summaries that have a degree of truth close (around 10%) to the maximum.

Finally, it should be noted that alternative proposals of measures of quality of linguistic summaries have been proposed. We have considered here those that have minimum time and space computational complexity, i.e., those that have a time of complexity $O(n)$ and do not require storage of the set of entities. These measures can thus be applied for both on-line and off-line summarization of NetFlow collections.

More sophisticated measures exist that can provide improved interpretability at the expense of higher computational cost. In particular, recently Liétard [35] has proposed a novel measure of validity of a linguistic summary that has a clear meaning in terms of both quantity and quality. This measure is based on a set-oriented function extended to fuzzy sets. The measure, T_{qq}, is a guaranteed minimum of both the quantity and quality of summaries, i.e., T_{qq} can be interpreted as the highest percentage p such that at least $p\%$ of entities satisfy a summarizer at least to a degree p. However, the computational cost of this approach is significantly higher.

Two passes over the whole set of entities (flows) are required to compute this validity measure. In a first pass, the set of α values for the α-cuts to be used in the second pass have to be defined. In order to define this set, all the minimum degrees of membership for the summarizer are computed on a per-entity (per-flow) basis. When the summarizer is made of two or more fuzzy sets, the minimum among any of the respective membership degrees is selected for inclusion in the set of α values. In the second pass, the α-cuts of every set in the summarizer have to be computed for every entity (flow), for every value in the set of α values.

As a consequence, computing T_{qq} requires storing the whole set of entities and can have an algorithmic complexity close to quadratic. Thus, in principle it is not suitable for real-time summarization of NetFlow collections and has a considerable computational cost for off-line summarization of large sets such as common Net-Flow collections.

4.5.2 Data Mining Summaries of NetFlow Collections

Only a part of the potential of linguistic summaries is exploited using a set of fixed protoforms and protoforms specified a priori by the user. If only known facts are considered when looking for informative summaries, more complex, unknown or unexpected relations will are neglected. This issue can be addressed by means of automated data mining techniques. In particular, hidden relations can be found in the form of fuzzy summaries using unsupervised learning algorithms for mining association rules [26].

Association rules are implications of the form $\mathscr{X} \rightarrow \mathscr{Y}$. With association rules mining algorithms, associations between fuzzy itemsets [20], NetFlow records in our case, can be discovered, and, as proposed in [30], equivalent linguistic summaries can be derived from association rules. From these rules, summaries as "$\mathscr{X}$ flows are $\mathscr{Y}$" can be identified, where the qualifier $\mathscr{R}$ is the condition ($\mathscr{X}$) of the rule and the summary $\mathscr{S}$ is the conclusion ($\mathscr{Y}$) of the rule.

Original association rules were defined for transactional data and binary valued attributes. An association rule has the following form: $A_1 \wedge A_2 \wedge \ldots \wedge A_n \rightarrow A_{n+1}$, and states that those items for which attributes $\{A_i\}, i \in \{1 \ldots n\}$ take value 1, will also take value 1 for attribute A_{n+1}. An equivalence between linguistic summaries and association rules can be considered if the summarizer and the qualifier are interpreted as the consequent and the antecedent of an association rule respectively. Then, the confidence of a rule can be interpreted as the combination of the linguistic quantifier and the truth value of the rule.

Two basic measures of the quality of an association rule are usually applied: the *support* and the *confidence*. Intuitively, the *support* is the percentage of itemsets in the collection for which the antecedent of the association rule holds. More formally, the support is defined as the percentage of the total number of itemsets supporting the set of attributes $\{A_i\}, i \in \{1 \ldots n\}$ in the data collection. The *confidence* of a rule can be intuitively thought of as the fraction of itemsets in the support of the association rule for which the consequent holds as well. More formally, the confidence is defined as the percentage of the itemsets supporting $\{A_i\}, i \in \{1 \ldots n+1\}$ among all itemsets supporting $\{A_i\}, i \in \{1 \ldots n\}$, i.e., the support of the set of all the itemsets in the rule divided by the support of the antecedent of the rule.

The confidence can alternatively be thought of as the number of cases where the rule is correct relative to the number of cases where it is applicable. While the support is a measure of the statistical significance of a rule, the confidence is a measure of its strength. The most interesting rules are those with a high confidence and a support higher than a minimal threshold, i.e., strong rules with a substantial statistical significance.

Generalized association rules are redefined for fuzzy linguistic summaries mining as follows:

$$A_1 \text{ is } f_1 \wedge \ldots \wedge A_n \text{ is } f_n \rightarrow A_{n+1} \text{ is } f_{n+1} \wedge \ldots \wedge A_{n+m} \text{ is } f_{n+m},$$

where f_i are fuzzy linguistic variables and m is the number of fuzzy sets that make up the rule consequent. A number of algorithms for association rules mining have been proposed. For the implementation described in the next section, the Apriori algorithm for fast discovery of association rules or frequent itemsets [2] introduced in its initial form by Agrawal et al [1].

The Apriori algorithm follows a bottom-up approach and is applicable databases containing transactions or collections of itemsets. The algorithm looks for subsets that are common to a minimum confidence threshold. Then, frequent subsets are extended one item at a time, generating candidates that are tested against the data collection.

Apriori is a fast and scalable algorithm that has found applications in large scale problems in a variety of fields. We note however that Apriori, while a well established algorithm, suffers from a number of trade-offs and efficiency issues that have to be taken into account when implementing the algorithm. These issues have given rise to a number of modifications to the original proposal. For the work described in the next sections, we used an optimized implementation by Borgelt [8].

4.5.3 Experimental Results

An experimental tool for generating NetFlow linguistic summaries with the two approaches described, flow-lsummary, has been implemented. flow-lsummary is implemented in Perl, based on the Cflow library [40] and thus supporting both cflowd and flow-tools raw flow file formats. The tool allows for the definition of fuzzy linguistic variables and protoforms of interest for the on-line mode in a configuration file. Both the on-line mode and the data mining mode can be executed on NetFlow collections in the format of the widespread flow-tools [24] suite, as well as in Cflowd, argus and ifapd format. IPv4, IPv6 and NetFlow versions 1, 5 and 9 are supported.

We should note that linguistic summaries are not only more concise but also very fast to generate, which is an additional advantage against traditional tools. Thus, when software tools are used, only those that are fast one-pass such as flow-report or flow-lsummary are convenient for real-time monitoring. An alternative option is to apply VLSI technologies to develop hardware implementations of the performance-critical tasks of packet classification and flow monitoring into reconfigurable hardware [51] that can boost the processing capability of a flow measurement and monitoring system by more than 2 orders of magnitude.

Table 4.4 Overall counters of the NetFlow collections analyzed

Name	Duration	Flows	Packets	Bytes
Darmouth-Fall03	15 days	$5.05 \cdot 10^6$	$27.5 \cdot 10^6$	$16.8 \cdot 10^9$
WIDE-F-1-Aug	15 min.	$2.84 \cdot 10^6$	$21.8 \cdot 10^6$	$15.3 \cdot 10^9$
WIDE-F-DITL-2007	50.25 hours	$3.99 \cdot 10^8$	$2.69 \cdot 10^9$	$1.77 \cdot 10^{12}$
WIDE-F-DITL-2008	72.25 hours	$7.41 \cdot 10^8$	$4.15 \cdot 10^9$	$2.44 \cdot 10^{12}$
AMPATH-OC12-0-2007	50 hours	$1.76 \cdot 10^7$	$4.65 \cdot 10^8$	$3.05 \cdot 10^{13}$
AMPATH-OC12-1-2007	50 hours	$1.52 \cdot 10^7$	$3.91 \cdot 10^8$	$2.54 \cdot 10^{13}$
Equinix-Chicago-DITL-2008	63 min.	$7.33 \cdot 10^8$	$1.75 \cdot 10^9$	$1.21 \cdot 10^{12}$
Internet2-LOSA-May08	1 month	$1.38 \cdot 10^9$	$6.48 \cdot 10^9$	$4.99 \cdot 10^{12}$
Internet2-KANS-May08	1 month	$2.44 \cdot 10^9$	$8.96 \cdot 10^9$	$6.95 \cdot 10^{12}$

In order to assess the performance of the method implemented, we have generated linguistic summaries for a number of flow collections. Some of them are generated from packet level captures in pcap format and some other are actual Net-Flow measurements. In the first, case, the process of generating NetFlow collections from packet level traces in pcap format involves the use of an exporting tool and a flow recording tool. The tool softflowd [3] was used for exporting flows in Net-Flow format from the packet level traces. Then, the flow-receive tool included in the flow-tools package [24] was used to record NetFlow collections.

Some overall properties of these collections are shown in table 4.4. The following NetFlow collections were analyzed:

- WIDE-F-1-Aug: network trace taken on August 1, 2007 at samplepoint-F of the WIDE backbone [46], a 155 Mb/s trans-pacific link.
- WIDE-F-DITL-2007 and WIDE-F-DITL-2008: collections extracted from two traces taken during the 2007 and 2008 Day in the Live of the Internet events, at the samplepoint-F measurement point of the WIDE backbone. Time series for these traces were analyzed in chapter 3 (see pages 116 and 116, respectively, for a more detailed description).
- Darmouth-Fall03: Dartmouth/campus data set [33] from the Community Resource for Archiving Wireless Data (CRAWDAD), recorded at a wireless campus network covering 18 buildings.
- AMPATH-OC12-0-2007 and AMPATH-OC12-0-2007: NetFlow collections extracted from the AMPATH (AMericasPATH) traces, belonging to the CAIDA Anonymized 2007 Internet Traces. These traces were also analyzed in chapter 3 (see page 111 for a description).
- Equinix-Chicago-DITL-2008: collection extracted from a trace recorded at an OC192 link during the 2008 Day in the Live of the Internet event (see page 111 for a more detailed description of this trace).
- Internet2-LOSA-May08 and Internet2-KANS-May08: these datasets were collected on the LOSA and KANS backbone nodes, respectively, of the Internet2

Network and made available through the Internet2 Observatory project [28]. This dataset is a remarkable example of status and performance information about an operational network collected on a systematic basis over a long period of time. In the datasets available form the Internet2 Observatory, 1% sampling is used on a per-interface basis, with an additional maximum packet per second rate of 7000 going to the central processor. This way, under most common conditions every 1 in 100 packets are recorded. However, during busy times or under denial-of-service conditions, the sampling rate may be reduced. The flows are anonymized by setting the low-order 11 bits of any non-multicast IPv4 address to zero.

4.5.4   Predefined Set of Summaries

The method presented here has been found to provide insightful and concise summaries of flow collections. Simple on-line summaries for the NetFlow collections analyzed (see table 4.4) are shown in figures 4.4, 4.5, 4.6, 4.7, 4.8, 4.9, 4.10 and 4.11. In simple on-line mode, flow-1summary shows one summary about each of the defined protoforms (see table 4.3 for the basic set), i.e., only the most relevant fact concerning each protoform is shown. In a normal run of the tool, reports are shown periodically. The period can be configured and has a default value of 5 minutes. The figures show the final simple reports after the whole flow collections have been processed.

A more verbose output can be requested by the user. In this case, for each protoform the tool shows the most relevant summaries for each possible summarizer. For instance, the third linguistic summary in the simple report for the Darmouth-Fall03 collection is "very few flows are elephants" (see figure 4.4). In the more verbose report, the three following summaries are shown with regards to the size of flows: "very few flows are elephants" ($I = 0.985$), "almost all flows are mice" ($I = 0.984$), and "very few flows are bulk"($I = 0.973$).

The figures show the degree of truth, T and the informativeness, I, as defined above, for each of the summaries selected. These values are included here for completeness. In the non-verbose output of the flow-1summary tool, these values are not shown for clarity's sake.

Summary	I	T
Most flows are short lived	[0.879	0.921]
Very few flows are high throughput	[0.985	0.999]
Very few flows are elephants	[0.985	0.999]
Very few flows have an average packet size large	[0.961	0.973]
Very few flows are packet elephants	[0.985	0.999]
Very few mice flows are medium throughput	[0.981	0.996]
Few bulk TCP flows are short-lived	[0.553	0.559]

Fig. 4.4 Simple on-line linguistic summary of the Darmouth-Fall03 NetFlow collection (truth values between brackets). A priori labels.

Summary	I	T
Most flows are short lived	[0.908	0.954]
Very few flows are high throughput	[0.983	0.999]
Very few flows are elephants	[0.985	0.999]
Few flows have an average packet size large	[0.895	0.939]
Very few flows are packet elephants	[0.985	0.999]
Almost all elephant flows are medium throughput	[0.985	1]
Most bulk TCP flows are long-lived	[0.593	0.603]

Fig. 4.5 Simple on-line linguistic summary of the WIDE-F-1-Aug NetFlow collection (truth values between brackets). A priori labels.

Summary	I	T
Most flows are short lived	[0.919	0.965]
Very few flows are high throughput	[0.984	0.998]
Almost all flows are mice	[0.982	0.997]
Few flows have an average packet size large	[0.938	0.986]
Very few flows are packet elephants	[0.985	0.999]
Very few mice flows are medium throughput	[0.971	0.984]
About 1/2 bulk TCP flows are long-lived	[0.596	0.588]

Fig. 4.6 Simple on-line linguistic summary of the WIDE-F-DITL-2007 NetFlow collection (truth values between brackets). A priori labels.

Summary	I	T
Few flows are long lived	[0.915	961]
Few flows are high throughput	[0.950	0.999]
Very few flows are elephants	[0.985	0.999]
Few flows have an average packet size large	[0.941	0.991]
Very few flows are packet elephants	[0.985	0.999]
Almost all elephant flows are medium throughput	[0.985	1]
About 1/2 bulk TCP flows are short-lived	[0.691	0.693]

Fig. 4.7 Simple on-line linguistic summary of the WIDE-F-DITL-2008 NetFlow collection (truth values between brackets). A priori labels.

Summary	I	T
Few flows are long lived	[0.875	0.917]
Few flows are high throughput	[0.896	0.940]
Very few flows are elephants	[0.969	0.982]
Most flows have an average packet size jumbo	[0.941	0.991]
Almost all flows are packet mice	[0.984	0.970]
Few elephant flows are low throughput	[0.915	0.961]
Few bulk TCP flows are long-lived	[0.836	0.873]

Fig. 4.8 Simple on-line linguistic summary of the AMPATH-OC12-0-2007 NetFlow collection (truth values between brackets). A priori labels.

Summary	I	T
Few flows are long lived	[0.854	0.893]
Few flows are high throughput	[0.887	0.930]
Very few flows are elephants	[0.972	0.986]
Few flows have an average packet size small	[0.926	0.973]
Most flows are packet mice	[0.948	0.998]
About 1/3 bulk TCP flows are low throughput	[0.948	0.978]
Few bulk TCP flows are long lived	[0.781	0.812]

Fig. 4.9 Simple on-line linguistic summary of the AMPATH-OC12-1-2007 NetFlow collection (truth values between brackets). A priori labels.

Summary	I	T
Almost all flows are short lived	[0.985	1]
Few flows are high throughput	[0.933	0.981]
Almost all flows are mice	[0.985	0.999]
About 1/2 flows have an average packet size medium	[0.825	0.842]
Almost all flows are packet mice	[0.985	0.999]
Very few mice flows are low throughput	[0.984	0.999]
Almost all bulk TCP flows are short lived	[0.985	1]

Fig. 4.10 Simple on-line linguistic summary of the Equinix-Chicago-DITL-2008 NetFlow collection (truth values between brackets). A priori labels.

Summary	I	T
Few flows are long lived	[0.693	0.714]
Very few flows are high throughput	[0.985	0.999]
Very few flows are elephants	[0.984	0.999]
Very few flows have an average packet size jumbo	[0.985	0.999]
Very few flows are packet elephant	[0.950	0.999]
Almost all elephant flows are medium throughput	[0.985	1]
Almost all bulk TCP flows are long-lived	[0.909	0.954]

Fig. 4.11 Simple on-line linguistic summary of the Internet2-LOSA-May08 NetFlow collection (truth values between brackets). A priori labels.

Summary	I	T
Most flows are short lived	[0.770	0.800]
Very few flows are high throughput	[0.985	0.999]
Very few flows are elephants	[0.985	0.999]
Very few flows have an average packet size jumbo	[0.985	0.999]
Very few flows are packet elephant	[0.985	0.999]
Almost all elephant flows are medium throughput	[0.985	1]
Most bulk TCP flows are long-lived	[0.735	0.761]

Fig. 4.12 Simple on-line linguistic summary of the Internet2-KANS-May08 NetFlow collection (truth values between brackets). A priori labels.

In the simple on-line summaries, the last line informs the most relevant fact regarding the distribution of bulk TCP throughputs, which in a regular NetFlow report would usually be shown as a CDF plot, as shown in figure 4.13, or an equivalent table of points selected from the distribution. Note that in the figure, the definition of bulk TCP as given by the linguistic terms defined above is considered, i.e., bulk TCP flows are those with a byte counter higher than 100 KB and lower than 200 MB.

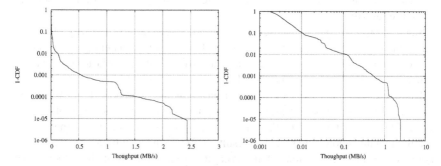

Fig. 4.13 Complementary cumulative distribution function plot of the throughput of bulk TCP flows in the Internet2-LOSA-May08 NetFlow collection, in semi-log (left) and log-log (right) scales.

For a better interpretation of the results presented here, it should be noted that there can be more information in what is not said than in what is said in a simple summary. For instance, for the AMPATH-OC12-1-2007, the most relevant summary about the duration of bulk TCP flows is "Few bulk TCP flows are long lived" (with $I = 0.781$, $T = 0.812$). The fact that this summary is chosen as the most relevant entails that "Very few bulk TCP flows are long lived" holds with less certainty (I=0.211, T=0.140). This means that there is a small but not negligible amount of flows that are long-lived, which in a sense reveals a more interesting aspect of the flow collection.

As an example, the summary for the average packet queue size in the case of the Internet2-LOSA-May08, "Very few flows have an average packet size jumbo", indicates that jumbo frames are present in this segment of the Internet2 network and the degree of deployment of jumbo frames is very low. This is in contrast to all the other flow collections analyzed, where the presence of jumbo frames is at best negligible. It is also worth to mention for this flow collection that the most relevant summary for the duration of bulk TCP flows is "Almost all bulk TCP flows are long-lived". In other words, long durations are dominant within the class of bulk flows. In addition, the most relevant summary for the throughput distribution qualified by byte size count is "Almost all elephant flows are medium throughput". These two summaries concisely express an well known property of duration and throughput distribution in wide are networks, i.e., the presence of the so-called tortoises [10], that seem to be specially clear in the Internet2-LOSA-May08 flow collection.

4.5.5 Identifying Attribute Labels by Clustering

This section presents experimental results from the application of clustering techniques to NetFlow collections. Two clustering methods will be applied: k-means (or hard c-means) and fuzzy c-means. The number of clusters is most often a determining input variable to the process. Our approach is to check a) whether the number of optimal clusters match the number of domain specific terms (current assumptions) and b) how summaries built with clustering derived labels compare with the summaries above.

Since fuzzy sets are being used for defining linguistic terms, fuzzy clustering techniques (i.e., clustering techniques where all data points can belong to all clusters to a certain degree) seem to be more adequate. We will use two clustering methods: hard c-means (HCM, or k-means) and fuzzy c-means (FCM).

The clustering processes could be run on the whole data sets. However, it has been found that it is not necessary to have a large number of flow samples in order to obtain consistent results that give a fairly good approximation to the clusters identified on the whole data sets. The results shown in the next figures correspond to subsets of NetFlow collections of sizes ranging from 10000 through 100000 flows. For each size, 100 repetitions are performed using randomly selected subsets of the flow collection. Figures 4.14, 4.15, 4.16, 4.17, 4.18, 4.19, 4.20 and 4.21 show the clusters identified for the Darmouth-Fall03, WIDE-F-1-Aug, WIDE-F-DITL-200701, WIDE-F-DITL-200803, Equinix-Chicago-DITL-2008, AMPATH-OC12-2007-0, AMPATH-OC12-2007-1 and Internet2-LOSA-May08 NetFlow collections, respectively. Results for the duration, bytes, throughput and packets attributes are shown in the figures.

Four overall conclusions can be drawn from the results shown in the previous figures:

- The centers identified as vertices are consistently distinguishable.
- Although the results obtained with the HCM and FCM clustering methods are very similar, FCM yields cluster centers in an slightly more consistent manner.
- Automatically derived labels match reasonably well (with small numerical differences) a priori labels.
- In order to obtain consistent cluster centers, subsets of no more than a few tens of thousands of flows are required. This implies that a) modest computing resources are sufficient, and b) the labels tuning stage can be performed in a short period of time for on-line monitoring. For instance, in the case of the WIDE-F-1-Aug collection, 50000 data about flows are available after approximately 15.85 seconds of recording. For links with a higher degree of aggregation, the required time is even lower. For instance, in the Internet2-LOSA-May08 collection, the average flows per second rate is around 4200. Thus, a proper tuning of the linguistic labels can be performed after only a few seconds.

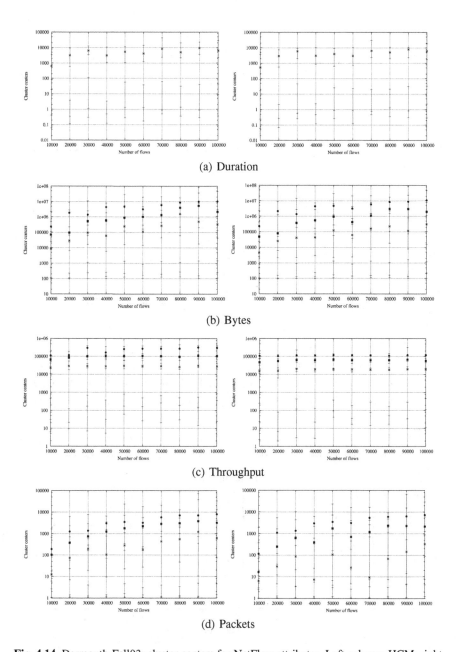

(a) Duration

(b) Bytes

(c) Throughput

(d) Packets

Fig. 4.14 Darmouth-Fall03: cluster centers for NetFlow attributes. Left column: HCM; right column: FCM. Clustering performed on subsets of different sizes, ranging from 10000 through 100000. 100 repetitions per size. The plots show average values and error bars for the 5th and 95th percentiles.

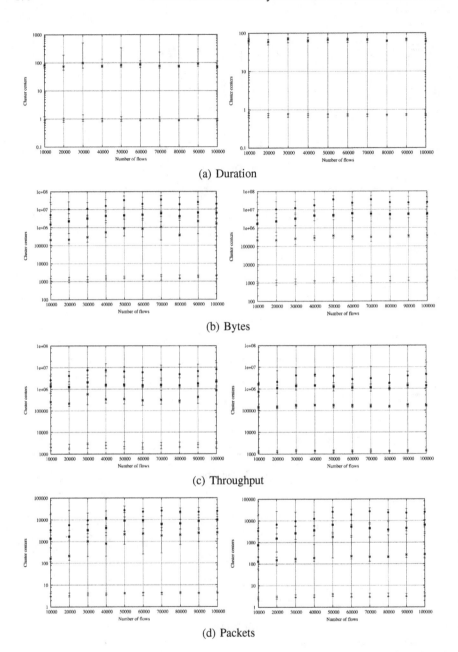

(a) Duration

(b) Bytes

(c) Throughput

(d) Packets

Fig. 4.15 WIDE-F-1-Aug: cluster centers for NetFlow attributes. Left column: HCM; right column: FCM. Clustering performed on subsets of different sizes, ranging from 10000 through 100000. 100 repetitions per size. The plots show average values and error bars for the 5th and 95th percentiles.

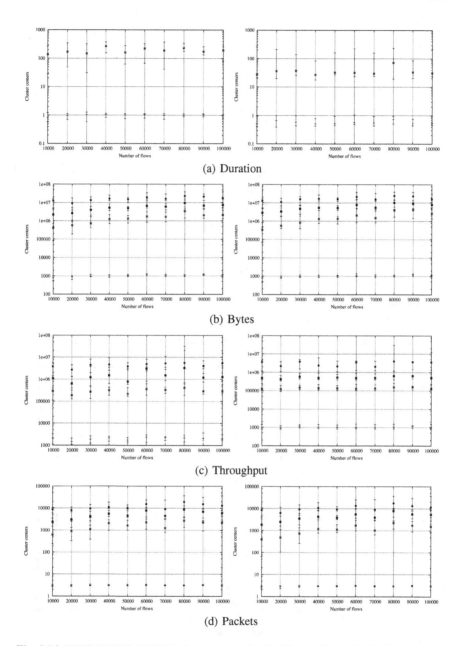

Fig. 4.16 WIDE-F-DITL-200701: cluster centers for NetFlow attributes. Left column: HCM; right column: FCM. Clustering performed on subsets of different sizes, ranging from 10000 through 100000. 100 repetitions per size. The plots show average values and error bars for the 5th and 95th percentiles.

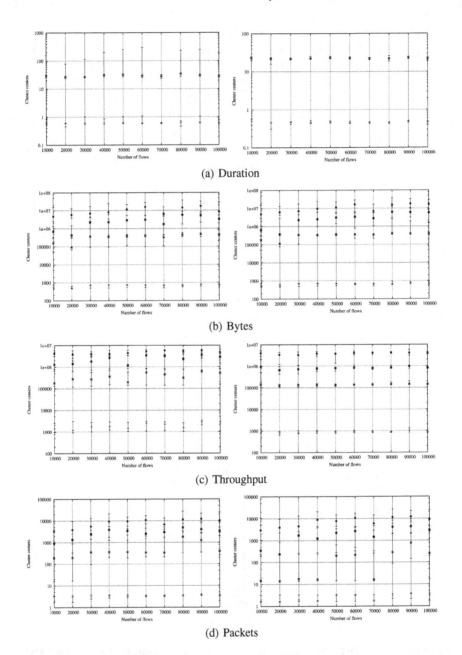

(a) Duration

(b) Bytes

(c) Throughput

(d) Packets

Fig. 4.17 WIDE-F-DITL-200803: cluster centers for NetFlow attributes. Left column: HCM; right column: FCM. Clustering performed on subsets of different sizes, ranging from 10000 through 100000. 100 repetitions per size. The plots show average values and error bars for the 5th and 95th percentiles.

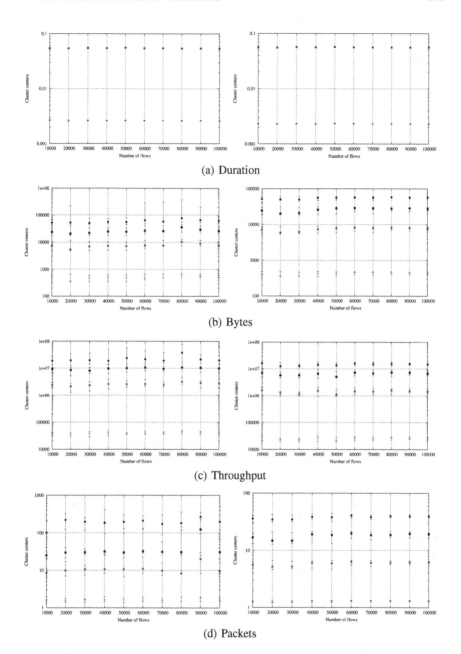

(a) Duration

(b) Bytes

(c) Throughput

(d) Packets

Fig. 4.18 Equinix-Chicago-DITL-2008: cluster centers for NetFlow attributes. Left column: HCM; right column: FCM. Clustering performed on subsets of different sizes, ranging from 10000 through 100000. 100 repetitions per size. The plots show average values and error bars for the 5th and 95th percentiles.

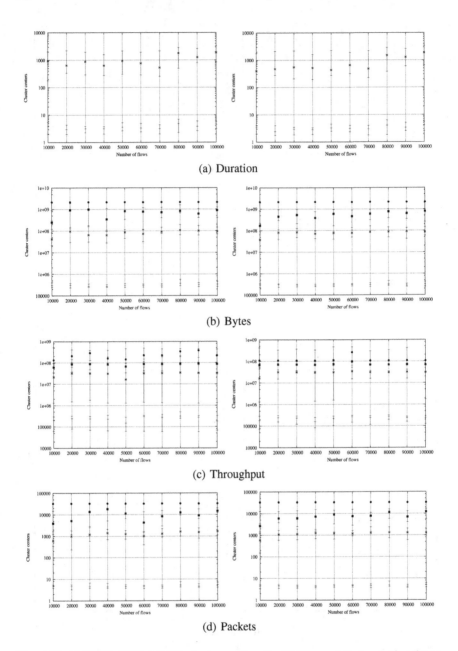

(a) Duration

(b) Bytes

(c) Throughput

(d) Packets

Fig. 4.19 AMPATH-OC12-0-2007: cluster centers for NetFlow attributes. Left column: HCM; right column: FCM. Clustering performed on subsets of different sizes, ranging from 10000 through 100000. 100 repetitions per size. The plots show average values and error bars for the 5th and 95th percentiles.

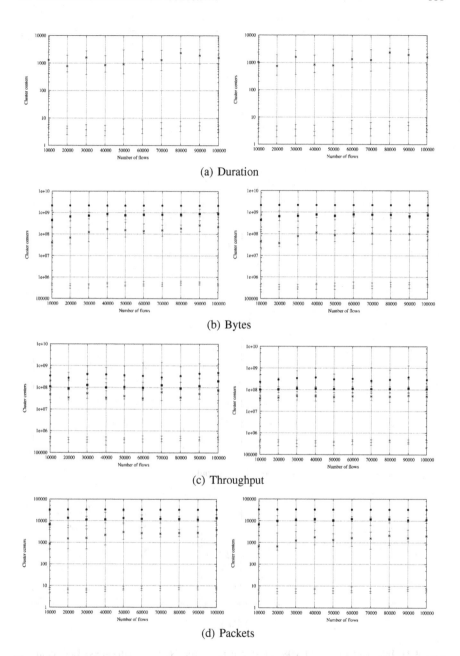

(a) Duration

(b) Bytes

(c) Throughput

(d) Packets

Fig. 4.20 AMPATH-OC12-1-2007: cluster centers for NetFlow attributes. Left column: HCM; right column: FCM. Clustering performed on subsets of different sizes, ranging from 10000 through 100000. 100 repetitions per size. The plots show average values and error bars for the 5th and 95th percentiles.

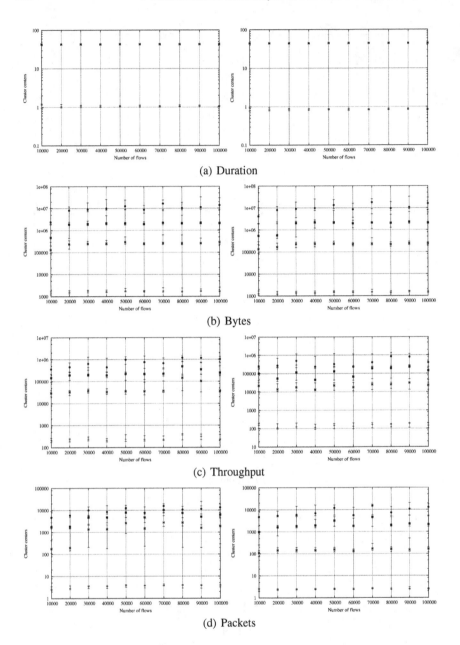

(a) Duration

(b) Bytes

(c) Throughput

(d) Packets

Fig. 4.21 Internet2-LOSA-May08: cluster centers for NetFlow attributes. Left column: HCM; right column: FCM. Clustering performed on subsets of different sizes, ranging from 10000 through 100000. 100 repetitions per size. The plots show average values and error bars for the 5th and 95th percentiles.

4.5.6 Mining Association Rules for Extracting Linguistic Summaries

We discuss a sample set of summaries identified using the Apriori algorithm for association rules mining [2]. This algorithm that has been proven to be fast and effective in a wide variety of applications. Though the amount of association rules found can be overwhelming, a few simple filtering rules can significantly reduce the number of rules to analyze. In particular, we disregarded those rules with a low support or with a low confidence (truth) value. A number of interesting rules were found for the NetFlow collections analyzed.

In order to mine association rules a crisp attribute describing the application was considered. The possible values of this attribute (inferred from the transport port attribute) include DNS, SSH, WWW (HTTP and HTTPS protocols) and mail (SMTP, SSMTP, POP and IMAP protocols). We list as examples a selection of them:

- "Most SSH flows occur during the day, and consists of short lived mice flows", with confidence 0.892 in the Darmouth-Fall03 collection.
- "Most DNS request flows occur both during the day and at night, are mice and short lived", with confidence 0.970, 0.897 and 0.932 in the WIDE-F-1-Aug, WIDE-F-DITL-200701 and WIDE-F-DITL-200803 collections, respectively.
- "Most flows at night are mice", with confidence 0.911, and "Most flows during the day are mice", with confidence 0.988 in the AMPATH-OC12-0-2007 collection.
- "Most flows at night are mice", with confidence 0.898, and "Most flows during the day are mice", with confidence 0.932 in the AMPATH-OC12-1-2007 collection.
- "Almost all WWW flows are mice and short-lived", with confidence 0.940, and "Most SSH flows occur during the day, and consists of short lived mice flows", with confidence 0.932 in the Internet2-LOSA-May08 collection.

The facts pointed out by these summaries can be recognized as simplified expressions of well known facts that have been found in a number of measurement studies [18, 44, 10, 9].

Linguistic summaries provide a novel method to describe qualitative relations in NetFlow records using natural language. Thus, by using association rules mining to find relevant summaries we have a suitable method for addressing a problem related to flow analysis: finding invariants in traffic as well as instances of diversity and disparity, what is known as one the major goals of Internet Science [9].

4.5.7 Discussion

We have addressed network traffic measurement analysis at the flow level from the perspective of linguistic summaries. Two approaches for summarizing NetFlow records have been developed: 1) on-line summarization via a predefined and configurable set of protoforms of interest, and 2) discovery of hidden relevant summaries by means of association rules mining.

The method proposed here is intended as an analysis tool complementary to traditional descriptive statistics. The fact that both techniques are complementary is particularly evident if one considers that fuzzy linguistic labels were defined based on empirical observations that in turn were obtained using traditional statistical techniques such as distribution function plots and histograms of packet size.

Two approaches to generating linguistic summaries have been proposed: on-line and off-line summaries. For on-line summarization, a set of protoforms identified as conditions of interest have been defined. For off-line summarization, a fast association rules mining algorithm is used in order to identify relevant summaries.

A tool that implements both the on-line and off-line approaches, flow-lsummary, has been implemented. Experimental results for a set of benchmark NetFlow collections confirm linguistic summaries as an alternative look into network flow statistics useful for both network users and practitioners. This way, operators are not required to have an in-depth knowledge of many subjects, simple reports for network users can be provided, and the summarized statistics can help planning and dimensioning tasks.

The method presented is a novel technique to generate simple and human-interpretable reports, being useful for both practitioners and users, but also provides a promising technique for finding invariants in network traffic and advancing Internet Science. This can be seen as a first step towards natural language based knowledge discovery tools for Internet Science.

The clustering based approach presented here for the automatic definition of linguistic labels can be extended to additional flow identifiers. For instance, the protocol field, which has been defined as a crisp value, could be mapped into classes of applications.

Figure 4.22 shows the number of flows processed per second for the Internet2-KANS-May08 flow collection. The process was run on a commodity PC based on a Intel© Core™ 2 Duo CPU E6550 at 2.33GHz with 4 MB of L1 cache and 2 GB of RAM memory, running the GNU/Linux operating system with Perl 5.10.0 as build in the standard packages. Memory consumption is constant and below 7 MB, as the summarization process only requires one-pass procedures that do not have significant memory requirements.

In the figure, it can be seen that, after a short initialization stage, the number of flows that can be processed per second is above $20 \cdot 10^3$. This rate is remarkably high and constant due to the purely static and one-pass nature of the algorithms applied. Processing 10^9 flows took approximately 4571 seconds (1 hour, 16 minutes and 11 seconds). That is, a month worth of NetFlow records for the Internet2-KANS-May08 can be summarized in approximately 3 hours and 6 minutes. However, a 1% flow sampling rate was applied for this collection.

If 100% or no sampling were applied, and assuming (as a pessimistic case) a number of flows 100 times greater, a month work of NetFlow records could be processed within 13 days. That is, real-time summarization is feasible with flow-lsummary without particular optimizations on commodity hardware even for 100%

flow sampling rates on very high speed links. In networks with a more moderate overall traffic rate, the summarization process has very low computational requirements. For instance, the simple on-line summary for the whole Darmouth-Fall03 collection, which spans 15 days, can be generated within approximately 3 minutes and 50 seconds on the same commodity PC as before.

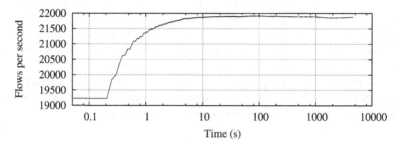

Fig. 4.22 Flows processed per second with flow-lsummary for the Internet2-KANS-May08 NetFlow collection

We should note that this section has studied the analysis and summarization of network measurements at the flow level as opposed to network measurement at the packet level. That is, while in the previous section we analyzed time series directly extracted from packet level measurements, in this section we have described a flow level analysis tool. The latter approach is significantly more widespread among practitioners, finding everyday applications, as flow level measurement facilities are currently available in virtually all medium- and high-end routing equipments. However, performing linguistic summarization at the packet level is also a plausible option we think is worth exploring as a future extension. To this end, the procedures and some of the linguistic terms defined here could be extrapolated from flow level analysis to packet level analysis.

Finally, a promising research direction would be to extend the fuzzy association rules mining method applied in this monograph in order to include topological information. This would make it possible to analyze the interactions between topology and traffic using a soft computing approach. This research line can be expected to lead to interesting discoveries in an area that remains essentially unexplored because of its complexity.

4.6 Conclusions

Methods for extracting linguistic summaries from network measurements and statistics at the flow level were described in this chapter. Both on-line simple summaries and off-line mining of fuzzy association rules were addressed.

We first analyzed how prior knowledge can be used to define linguistic labels for traffic flow attributes and how these match traffic measurements from production networks. By means of unsupervised learning techniques we have shown a remarkable match between the a priori knowledge based linguistic labels proposed and actual network flow measurements. A measure of goodness or informativeness of linguistic summaries tailored to network flow summarization has been proposed.

This chapter described a method for the summarization of network flow records together with a method for inducing association rules from network flow records have been developed and implemented in an experimental tool. Both methods have been applied to a wide set of network flow records. The first has been shown to provide concise and discerning summaries with very low computational cost. The second has been shown to discover relevant facts hidden in network flow measurements. The summarization methods have been implemented in a tool that is valuable for both researchers and practitioners. The tool, flow-lsummary was applied to a diverse set of large flow collections and the techniques described were shown to provide concise, self-explanatory and informative summaries.

References

[1] Agrawal, R., Imielinski, T., Swami, A.N.: Mining Association Rules Between Sets of Items in Large Databases. In: Buneman, P., Jajodia, S. (eds.) ACM SIGMOD International Conference on Management of Data, pp. 207–216. ACM Press, Washington (1993)

[2] Agrawal, R., Mannila, H., Srikant, R., Toivonen, H., Verkamo, A.: Fast Discovery of Association Rules. In: Advances in Knowledge Discovery and Data Mining, pp. 307–328. The MIT Press, Cambridge (1996)

[3] et al DM (2008), Softflowd - fast software NetFlow probe,
    http://www.mindrot.org/projects/softflowd/

[4] Anand, S.S., Bell, D.A., Hughes, J.G.: EDM: A General Framework for Data Mining Based on Evidence Theory. Data & Knowledge Engineering 18(3), 189–223 (1996)

[5] Awduche, D., Malcolm, J., Agogbua, J., O'Dell, M., McManus, J.: Requirements for Traffic Engineering Over MPLS. RFC 2702, Internet Engineering Task Force, Network Working Group (1999)

[6] Awduche, D., Chiu, A., Elwalid, A., Widjaja, I., Xiao, X.: Overview and Principles of Internet Traffic Engineering. RFC 3272, Internet Engineering Task Force, Network Working Group, category: Informational (2002)

[7] Boote, J.W., et al.: Towards Multi-Domain Monitoring for the European Research Networks. In: 21th TERENA Networking Conference, Trans-European Research and Education Networking Association, Poznan, Poland (2005)

[8] Borgelt, C.: Efficient Implementations of Apriori and Eclat. In: ICDM Workshop of Frequent Item Set Mining Implementations (FIMI 2003), Melbourne, FL, USA, pp. 24–32 (2003)

[9] Broido, A., Hyun, Y., Gao, R., Claffy, K.C.: Their Share: Diversity and Disparity in IP Traffic. In: 5th Passive and Active Measurement Workshop (PAM), Antibes Juan-Les-Pins, France, pp. 113–125 (2004)

[10] Brownlee, N., Claffy, K.C.: Understanding Internet Traffic Streams: Dragonflies and Tortoises. IEEE Communications Magazine 40(10), 110–117 (2002)

[11] Cai, D., McTear, M.F., McClean, S.I.: Knowledge discovery in distributed databases using evidence theory. International Journal of Intelligent Systems 15(8), 745–761 (2000)

[12] Calyam, P., Krymskiy, D., Sridharan, M., Schopis, P.: Active and Passive Measurements on Campus, Regional and National Network Backbone Paths. In: 14th IEEE International Conference on Computer Communications and Networks (ICCCN 2005), San Diego, California, USA, pp. 537–542 (2005)

[13] Casillas, J., Cordón, O., Herrera, F., Magdalena, L. (eds.): Interpretability Issues in Fuzzy Modeling. Studies in Fuzziness and Soft Computing. Springer, Berlin (2003) ISBN: 978-3-540-02932-8

[14] Choi, B.Y., Bhattacharyya, S.: Observations on Cisco sampled NetFlow. ACM SIGMETRICS Performance Evaluation Review 33(3), 18–23 (2005)

[15] CISCO IOS NETFLOW: Cisco IOS NetFlow (2007),
http://www.cisco.com/en/US/products/ps6601/
products_ios_protocol_group_home.html

[16] Claise, B., et al.: Specification of the IPFIX Protocol for the Exchange of IP Traffic Flow Information. Revision 26, Internet Engineering Task Force, IPFIX Working Group, Internet Draft (2007)

[17] Cooperative Association for Internet Data Analysis, CAIDA Visualization Tools (2008), http://www.caida.org/tools/visualization/

[18] Crovella, M.E., Bestavros, A.: Self-Similarity in World Wide Web Traffic: Evidence and Possible Causes. IEEE/ACM Transactions on Networking 5(6), 835–846 (1997)

[19] Crovella, M.E., Krishnamurthy, B.: Internet Measurement: Infrastructure, Traffic and Applications. Wiley, Chichester (2006) ISBN: 978-0470014615

[20] Delgado, M., Marín, N., Sánchez, D., Vila, M.A.: Fuzzy Association Rules: General Model and Applications. IEEE Transactions on Fuzzy Systems 11(2), 214–225 (2003)

[21] Deri, L.: ntop (2008), http://www.ntop.org

[22] Duffield, N.: Sampling for Passive Internet Measurement: A Review. Statistical Science 19(3), 472–498 (2004)

[23] Estan, C., Savage, S., Varghese, G.: Automatically Inferring Patterns of Resource Consumption in Network Traffic. In: ACM SIGCOMM 2003, Karlsruhe, Germany, pp. 137–148 (2003)

[24] Fullmer, M., et al.: flow-tools (2007),
http://www.splintered.net/sw/flow-tools/

[25] Guillaume, S.: Designing fuzzy inference systems from data: An interpretability-oriented review. IEEE Transactions on Fuzzy Systems 9(3), 426–443 (2001)

[26] Hastie, T., Tibshirani, R., Friedman, J.: The Elements of Statistical Learning: Data Mining, Inference, and Prediction. Springer, New York (2003) ISBN: 978-0387952840

[27] internet2 netflow weekly reports, Internet2 NetFlow: Weekly Reports (2008),
http://netflow.internet2.edu/weekly/

[28] internet2observatory, The Internet2 Observatory (2008),
http://www.internet2.edu/observatory/

[29] Kacprzyk, J., Yager, R.R.: Linguistic Summaries of Data Using Fuzzy Logic. International Journal of General Systems 30(2), 133–154 (2001)

[30] Kacprzyk, J., Zadrozny, S.: Linguistic Summarization of Data Sets Using Association Rules. In: IEEE International Conference on Fuzzy Systems (FUZZ-IEEE), St. Louis, USA, pp. 702–707 (2003)

[31] Kacprzyk, J., Zadrozny, S.: Linguistic Database Summaries and Their Protoforms: Towards Natural Language Based Knowledge Discovery Tools. Information Sciences 173(4), 281–304 (2005)

[32] Kacprzyk, J., Yager, R.R., Zadrozny, S.: Fuzzy Linguistic Summaries of Databases for an Efficient Business Data Analysis and Decision Support. In: Knowledge Discovery for Business Information Systems. Springer International Series in Engineering and Computer Science, vol. 600, pp. 129–152. Kluwer Academic Publishers, Boston (2001) ISBN: 978079237243

[33] Kotz, D., Henderson, T., Abyzov, I.: CRAWDAD data set dartmouth/campus (v. 2007-02-08) (2007),
http://crawdad.cs.dartmouth.edu/dartmouth/campus

[34] Lakhina, A., Papagiannaki, K., Crovella, M.E., Diot, C., Kolaczyk, E.D., Taft, N.: Structural analysis of network traffic flows. In: Joint International Conference on Measurement and Modeling of Computer Systems (ACM SIGMETRICS), New York, NY, USA, pp. 61–72 (2004)

[35] Liétard, L.: A New Definition of Linguistic Summaries of Data. In: 17th IEEE International Conference on Fuzzy Systems (FUZZ-IEEE 2008), IEEE World Congress on Computational Intelligence, Hong Kong, China, pp. 506–511 (2008)

[36] Liu, Y., Kerre, E.E.: An overview of fuzzy quantifiers (I). Interpretations. Fuzzy Sets and Systems 95(1), 1–21 (1998)

[37] Liu, Y., Kerre, E.E.: An overview of fuzzy quantifiers (II). Reasoning and applications. Fuzzy Sets and Systems 95(2), 135–146 (1998)

[38] Montesino-Pouzols, F., Barriga, A., Lopez, D.R., Sánchez-Solano, S.: Linguistic Summarization of Network Traffic Flows. In: 17th IEEE International Conference on Fuzzy Systems (FUZZ-IEEE 2008), IEEE World Congress on Computational Intelligence, Hong Kong, China, pp. 619–624 (2008)

[39] Plonka, D.: FlowScan: A Network Traffic Flow Reporting and Visualization Tool. In: 14th USENIX conference on System administration, New Orleans, Louisiana, USA, pp. 305–318 (2000)

[40] Plonka, D.: Cflow (2005), http://net.doit.wisc.edu/~plonka/Cflow/

[41] Raschia, G., Mouaddib, N.: SAINTETIQ: a fuzzy set-based approach to database summarization. Fuzzy Sets and Systems 129(2), 137–162 (2002)

[42] Rasmussen, D., Yager, R.R.: Finding fuzzy and gradual functional dependencies with SummarySQL. Fuzzy Sets and Systems 106(2), 31–42 (1999)

[43] Rolls, D., Michailidis, G., Hernández-Campos, F.: Queueing Analysis of Network Traffic: Methodology and Visualization Tools. Computer Networks 48(3), 447–473 (2005)

[44] Shalunov, S., Teitelbaum, B.: TCP Use and Performance on Internet2. In: ACM SIGCOMM Internet Measurement Workshop, San Francisco, CA, USA, pp. 147–160 (2001)

[45] Sommers, J., Barford, P., Willinger, W.: SPLAT: A Visualization Tool for Mining Internet Measurements. In: 7th Passive and Active Network Measurement Workshop, pp. 31–40 (2006)

[46] Widely Integrated Distributed Environment (WIDE) Project, MAWI Working Group, Packet traces from wide backbone (2008),
http://tracer.csl.sony.co.jp/mawi/

[47] Yager, R.R.: A New Approach to the Summarization of Data. Information Sciences 28, 69–86 (1982)

[48] Yager, R.R.: On Ordered Weighted Averaging Operators in Multicriteria Decision Making. IEEE Transactions on Systems, Man and Cybernetics 18(1), 183–190 (1988)

[49] Yager, R.R.: Database Discovery Using Fuzzy Sets. International Journal of Intelligent Systems 11, 691–712 (1996)

[50] Yager, R.R., Engemann, K.J., Filev, D.P.: On the Concept of Immediate Probabilities. International Journal of Intelligent Systems 10(4), 373–397 (1995)

[51] Yusuf, S., Luk, W., Sloman, M., Dulay, N., Lupu, E.C., Brown, G.: Reconfigurable Architecture for Network Flow Analysis. IEEE Transactions on Very Large Scale Integration (VLSI) Systems 16(2), 57–65 (2008)

[52] Zadeh, L.A.: The concept of a linguistic variable and its application to approximate reasoning. Information Sciences 8(3), 199–249 (1975)

[53] Zadeh, L.A.: A Computational Approach to Fuzzy Quantifiers in Natural Languages. Computers and Mathematics with Applications 9, 149–184 (1983)

[54] Zadeh, L.A.: A Prototype-Centered Approach to Adding Deduction Capability to Search Engines-the Concept of Protoform. In: First International IEEE Symposium on Intelligent Systems, Varna, Bulgaria, vol. 1, pp. 2–3 (2002)

Chapter 5
Inference Systems for Network Traffic Control

Abstract. This chapter deals with control of network traffic in routers as well as end-to-end flows. First it is proposed an scheme for implementing end-to-end traffic control mechanisms through fuzzy inference systems. A comparative evaluation of simulation and implementation results from the fuzzy rate controler as compared to that of traditional TCP flow and rate control mechanisms is performed for a wide set of realistic scenarios. Then, fuzzy inference systems for traffic control in routers are designed. A particular proposal has been evaluated in realistic scenarios and is shown to be robust. The proposal is compared against the random early detection (RED) scheme. It is experimentally shown that fuzzy systems can provide better performance and better adaptation to different requirements with mechanisms that are easy to modify using linguistic knowledge.

5.1   Network Traffic Control

End-to-end Internet packet dynamics is a complex problem for which models available to date are at best incomplete. A major research problem in Internet transport layer protocols is the development of rate control mechanisms that can cope with the requirements of a growing diversity of technologies, applications and services.

A better understanding of the end-to-end dynamics of the Internet is key for two reasons. First, it would allow the optimization of network resources and consequently the end-to-end performance experienced by users. Second, it would make it possible to provide improved end-to-end classes of service as well as quality-of-service guarantees. As Keshav remarks [61], *"the Holy Grail of computer networking is to design a network that has the flexibility and low cost of the Internet, yet offers the end-to-end quality-of-service guarantees of the telephone network."*

That is, in the current Internet there is a need for mechanisms that guarantee certain quality of service parameters. This requires the capability to provide improved quality of service, i.e., the capability to avoid performance degradation and failures due to congestion. In extreme cases, such as remote surgery, emergency calling and

F.M. Pouzols et al.: Mining & Control of Network Traffic by Computational Intelligence, pp. 191–262.
springerlink.com   © Springer-Verlag Berlin Heidelberg 2011

other applications in mission critical networks, it may be required to totally prevent any end-to-end unreliability. However, providing quality of service guarantees is not the only open problem. In fact, the development of efficient mechanisms for best effort traffic control poses many challenges. In particular, there is a complex trade-off between aspects such as bandwidth utilization, congestion avoidance, fairness or delay and jitter minimization.

In practice, most current networks, and specially backbone networks, rely on overprovisioning. However, due to the complex and bursty nature of traffic in packet switched networks, users of best effort services may experience degraded performance even when there is sufficient available bandwidth and the average bandwidth utilization is relatively low [17]. In particular, micro-congestion episodes (localized periods of congestion that occur at scales lower than those usually analyzed by network operators) have been found to be frequent even in underutilized links [78].

In this chapter, we address by means of fuzzy inference systems two tightly related problems in traffic control: end-to-end rate and congestion control as well as active queue management in routers. The focus of this chapter is on best effort traffic. We will evaluate software implementations using a packet-level network simulator, experimental software tools and network emulation packages. First, we discuss two simulation scenarios in section 5.2.

In section 5.3 we describe mechanisms for intelligent end-to-end traffic congestion control in Internet by means of fuzzy inference systems. We first outline a fuzzy logic based generalization of TCP (Transport Control Protocol) rate control principles. The design of a fuzzy TCP-like window based congestion controler is then described. A systematic fuzzy systems design methodology is used in order to simulate and implement the system as an experimental tool. A comparative evaluation of simulation and implementation results from the fuzzy rate controler as compared to that of traditional controlers is performed. Besides being a useful modeling approach, the fuzzy rule based rate controler is shown to outperform other approaches with regards to a number of criteria.

A major related research problem in Internet transport and network layers is the development of traffic control mechanisms that can cope with the requirements of a growing diversity of technologies, applications and services. Section 5.4 describes systems for intelligent traffic control in Internet routers by means of fuzzy inference based systems. A systematic design methodology, interpretability principles, evaluation over a broad range of network scenarios as well as practical implementation constraints have been considered. A comparative evaluation of results obtained by means of our fuzzy controlers as compared to that of traditional approaches is outlined.

5.2   Simulation Scenarios

As discussed in [42], simulation of network scenarios is a useful generic tool for building understanding of dynamics, to illustrate a point, or to explore for unexpected behavior. However, care should be taken when performing comparisons among different options and particularly when producing comparative performance measurements. It is well known that small changes (a few lines of code) in popular protocols and applications in the real Internet have a massive potential impact on traffic patterns, users and infrastructures. Consequently, any numeric comparison between two protocols performed by simulation is subject to two fundamental questions: 1) would a small change in the model result in an important change in the results? and 2) what would be the impact on the results of a change in a detail of the software implementation of the underlying model?

Thus, in the next two sections, we will explore by means of simulation the behavior of the proposed schemes for end-to-end flow and congestion control as well as active queue management, respectively. In particular, we have defined a complex scenario in order to analyze the behavior of these schemes under a wide set of realistic network conditions. Here, we describe the two simulation scenarios that will be considered in the next sections.

The aim of this study is to help designing the fuzzy rule based congestion control scheme and the fuzzy active queue management scheme proposed in the next sections. Through simulation, we will check whether these schemes are robust and their more general approach provide a significant performance improvement under a number of conditions.

More specifically, for active queue management, we analyze the robustness of different active queue management algorithms against bursty traffic with different degrees of overall traffic load on the network. For end-to-end control, we consider the robustness, performance and fairness with different degrees of load on a particular end-to-end path of the scenarios with different degrees of competing traffic. We will compare the proposal described here against two alternatives and it will be shown that the fuzzy system performs better or in a comparable way to other alternatives for a number of network scenarios. However, this comparison is limited by two aspects:

- The overwhelming diversity of proposals of end-to-end congestion control algorithms, among which the ones that will be mentioned in section 5.3.1 are only a subset. This makes almost impossible to perform a thorough evaluation of a representative set of proposals under a wide enough set of simulation scenarios.
- Any protocol performance analysis, whether performed through simulation or experimental implementation will always be incomplete and face new challenges as the global Internet evolves.

We will illustrate the behavior of the proposals presented in the next two sections through simulations and compare its performance against that of other alternatives. However, this is sensible as far as the results from any qualitative and -specially-quantitative comparison are regarded with the aforementioned cautions in mind.

The simulations will be complemented with tests performed through emulation as well as full implementations. The emulation and implementation scenarios will be described in the next sections. While emulation and implementation tests provide results that can be expected to match those of deployed implementations more closely, simulation allows us to analyze the dynamics of algorithms and control schemes with much more flexibility in a wider set of conditions.

Simulations will be performed at the packet level. A packet level simulator is required for an effective evaluation of both TCP-like end-to-end congestion control algorithms and active queue management schemes. In the first case, the complex behavior resulting from window based congestion control algorithms, specially when background traffic is not neglected, can be fully modeled only if simulations are performed at the packet level. The same applies to active queue management schemes, for which analytical models are also incomplete at their best. To this end, we use the ns-2 simulator [56], a de facto standard within the Internet research community. The ns-2 simulator supports a wide variety of technologies and protocols.

Two simulation scenarios will be used throughout the simulation tests: dumbbell and GREN. The first plays the role of a simple scenario that illustrates the underlying principles where all but the essential components are abstracted away. In contrast, the second scenario has a large topology where complex traffic patterns are generated [42].

The first scenario is a typical dumbbell topology with 3 nodes in each edge of the network, as depicted in figure 5.1. Dumbbell is an scenario with a fixed topology defined manually. The one-way propagation delays are: 2 and 10 ms for the links of nodes 1-1 and 1-2, respectively, and 1.5, and 12 ms for the links of nodes 2-1 and 2-2, respectively. The one-way propagation delay of the central link is 10 ms. The bandwidth capacities of the edge links are 1 Gb/s, whereas the bandwidth capacity of the central link is 622 Mb/s.

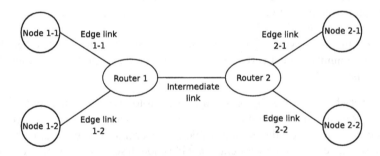

Fig. 5.1 Scheme of a dumbbell topology with 2 nodes in each edge

This simple scenario serves two purposes. First, it allows us to analyze the behavior of different algorithms in a topology with an extensively studied dynamics where it is relatively easy to understand the impact of changes in a particular algorithm. In addition, the results of the simulations in this chapter can be compared with a large number of publications that focus on the performance of end-to-end congestion control algorithms and active queue management schemes in dumbbell-like scenarios [7].

However, simulations and performance studies based on dumbbell-like scenarios suffer from a number of limitations and pitfalls [42]. Simulations performed using the dumbbell topology can be hardly taken as indicative of what could happen in real networks. As described in chapter 1, simulating the behavior of the global Internet or a significant part of it is an immense challenge.

For the purposes of performing a simulation study of the fuzzy active queue management algorithm and the fuzzy end-to-end congestion control scheme presented in this chapter, we define the GREN scenario, an overall simulation scenario that aims to be a realistic simulation setup by considering the following points:

- A set of several thousands of nodes interconnected with a topology that resembles current real networks is first defined.
- Complex traffic patterns are generated as a result of the aggregation of thousands of traffic sources. In order to generate traffic we use distributions that match a number of experimental studies [84, 15, 2, 41].

The GREN scenario has been defined with the aim of generating complex background traffic as a consequence of the aggregation of a large number of individual flows between end nodes distributed over a network with a realistic topology.

Considering the Internet vast and ever increasing scale, the definition of a realistic simulation scenario is a very complex task that involves many aspects of modeling a huge and ever changing target [40]. However, it is possible to describe general models about substantial parts of the global Internet, including the global research and education Network (GREN), depicted in figure 5.2, based on [10], and reports from the Internet2 Observatory, the GÉANT and GÉANT2 projects as well as various other national research and education networks (NRENs).

The topology of the GREN scenario consists of a fixed part and a dynamic, randomly generated part. The fixed part defines the backbone of the network as well as the uppermost nodes. The fixed part is complemented with a dynamic part where end nodes are randomly added following a predefined pattern. In what follows, we will describe the topology and patterns of traffic generation used in the GREN scenario.

In order to generate the GREN scenario, 17 additional subnetworks are added. These subnetworks are intended to play the role of commercial ISPs. 11 of them are connected through two links to each of the NRENs shown in figure 5.2. The other 6 connect bidirectionally APAN to CERNet, TANet2 to APAN, AARNet to Internet2, Internet2 to CANEt3, Internet2 to GÉANT, and GÉANT to RBNet, respectively. The topology of the commercial ISP networks is defined as a backbone mesh of up to 20 nodes connected through OC192 links, with up to 10 additional access points connected to each node of the backbone. The links that made up the backbones of the subnetworks in the GREN scenario (including the NREN networks and the ISPs) as well as the links between these subnetworks are in general OC192 (10 Gb/s) OC48 (2.5 Gb/s), and OC12 (622 Mb/s) links.

This way, different traffic patterns at backbone links [48], national and regional nodes, intercontinental links and exchange points, such as the AMPATH (that were analyzed from a different viewpoint in chapters 3 and 4) are generated. One particular random instantiation of the topology described will be used throughout all the tests described in this section.

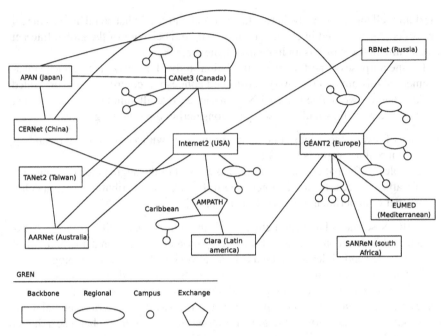

Fig. 5.2 Scheme of the Global Research and Education Network

Figure 5.3 shows a simplified scheme of the part of the GREN model corresponding to the GÉANT2 pan-European R&E network. The topology was defined based on the GÉANT Topology Map, 1st December 2004 (available on-line from http://www.geant.net/server/show/nav.159) and the GÉANT2 Topology (available on-line from http://www.geant2.net/server/show/nav.00d007009). Some links and nodes are not shown for better readability. Also note that the new dark fiber links and additional upgrades that are currently taking place in the GÉANT2 backbone are not taken into consideration in the model.

With the aim of generating a realistic simulation scenario with complex and highly variable background traffic the following aspects are considered for generating tests and analyzing the performance of the proposals:

- Analyses are performed for varying degrees of traffic load, a key aspect for any traffic control algorithm [40]. This way, we consider a wide range of traffic load conditions and we look into how performance degrades for increasing load,
- Most of the metrics discussed in the RFC 5166 on metrics for the evaluation of congestion control [47] are applied for evaluating performance: deployability, throughput, delay, packet loss rates, response to sudden changes or to transient events, minimizing oscillations in throughput or in delay, fairness and convergence times, robustness for challenging environments, robustness to failures and to misbehaving users.

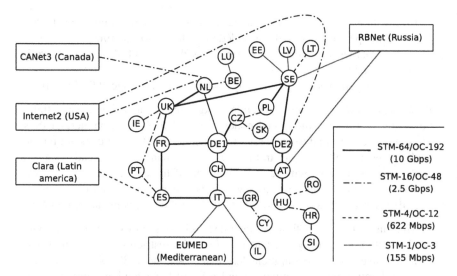

Fig. 5.3 Scheme of the GREN model part corresponding to the GÉANT network and its connections to other NRENs

The procedure followed for generating the GREN scenario can be summarized in the following steps:

1. *End nodes in NRENs.* Up to 50000 end nodes are generated for the NREN sub-networks in the GREN. A 9% of them is allocated to each of the 11 NRENs. End nodes are connected to access points chosen randomly. When adding end nodes to the NRENs, two options are considered:

   - In 1% of the cases, an independent node is directly connected to a node of an NREN backbone.
   - In the remaining 99% of the cases, an access node is added with a varying number of nodes connected to it in a four-level tree structure where the access node is the root. Nodes are added in sets of variable random size between 50 and 800.

2. Link capacities are uniformly distributed within the set of powers of two in the range 1 Mb/s-1 Gb/s.
3. Link propagation delay is uniformly distributed within 0.25 ms-10 ms.
4. *End nodes in commercial networks.* Additionally, for each commercial ISP sub-network, up to 5000 end nodes are generated. End nodes are connected in one of the following two forms of subnetworks:

   - Three-level tree structure with one root node in the first level (interconnection node that is linked to a backbone node) and a number of leaves between 1 and 500 in the second and third level. The number of leaves for each level of each tree is generated following an exponentially decaying distribution.

- Ring with a random number of nodes between 3 and 10. A three-level structure with up to 400 end nodes is added to each node of the ring.
- Link capacities are uniformly distributed within the set of powers of two in the range 64 Kb/s-1 Gb/s.
- Link propagation delay is uniformly distributed in the range of 0.25 ms-10ms.

5. *Background traffic.* In order to generate background traffic in the scenario, traffic flows are established among end nodes of the NRENs and commercial ISPs. These flows correspond to web, remote shell, file transfer, and constant bit rate applications as well as on/off traffic sources. As many flows as end nodes are generated at start time, with random selection of the end nodes for each flow. Upon expiration of each flow, a new flow between two random end points is started. The set of flows is generated so that the following constraints are met:

- The duration of each flow is generated following an exponential decaying distribution with decay constant $1/2 \cdot ln2$, with restrictions such that 60% of flows have very short duration (less than 2 seconds), 38.5% of flows have short duration (up to 15 minutes) and 1.5% of flows are long-running (more than 15 minutes). This distribution of very short, short and long-running flows approximately reflects a number of measurement studies [15, 14].
- For non constant bit rate flows, flow sizes are generated following a Pareto distribution [31, 30], with sizes ranging from 15 KB up to 4 GB. The bit rate of constant bit rate flows is generated following a Zipfian distribution.

6. *Simulation initialization.* The first 30 seconds of simulation time are reserved for initialization. Since the maximum propagation delay in the scenario is lower than 1 second, the initial transitory phase of TCP flows has ended after the initialization time. Measurements are made after the initialization stage during a period that varies depending on the particular test. This guarantees that spurious effects, such as synchronization between flows, that may take place at the beginning of simulations are not included in the results.

7. *Tailoring the scenario for a particular test.* Once the scenario has been generated following the previous steps, a final step is performed in order to test a particular algorithm. The final step depends on the kind of algorithm to be evaluated:

- For end-to-end control schemes:
  - A test flow using fuzzy end-to-end congestion control is added.
  - A set of competing flows is added. These flows are generated in such a way that they share a significant part of their end-to-end path with the test flow with fuzzy control.
- For active queue management schemes:
  - Fuzzy active queue management is implemented in a backbone or access node.
  - Traffic flows that traverse the chosen node are generated.

These last two steps are further detailed in the next sections for each case and each particular test.

The overall scale of the scenario was defined so that simulations could be run with commodity computers and with reasonable memory requirements. As a consequence of the procedure the following properties hold:

- The maximum path length between any two end hosts resulting from the topology defined above is 22, which matches measurement studies of the Internet global topology [29].
- Regarding the definition of the overall traffic matrix, traffic is generated among randomly selected end nodes.
- The minimum RTT (excluding queuing delays) between two end hosts is 0.25 ms.
- On average, 25% of flows are UDP (and thus irresponsive to congestion) while 75% of flows are TCP.

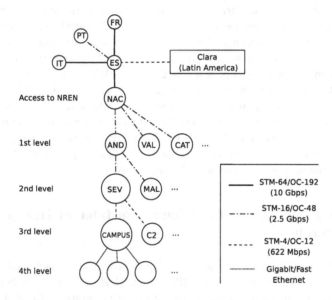

Fig. 5.4 Partial scheme of the RedIRIS regional network, a subnetwork of the GÉANT pan-European network. A 4-level tree structure is connected to the GÉANT network through the NAC access node. In this case, the topology of the first level of the tree was fixed manually in order to loosely resemble the RedIRIS network.

As an example of a set of end nodes in tree structure in an NREN, figure 5.4 shows an scheme of the tree connected to the ES node of the GÉANT backbone, intended to loosely resemble the RedIRIS network[1]. In the scheme, NAC is the access node connected to the GÉANT backbone, the AND router is in the first level of the regional tree, the SEV router is in the second level. Some of the nodes of the third level in this regional network are connected to the CAMPUS node. This

---

[1] Note we consider here that the pan-European GÉANT network is an NREN and RedIRIS is one of the regional networks connected to GÉANT.

regional network will be used for some of the simulation studies in the next sections, where we will analyze the behavior of the link between the SEV and CAMPUS nodes. Note that the set of nodes and the connections in the first level of the regional tree were manually fixed in this case in order to resemble the RedIRIS network. Only a few nodes of each level are shown for readability.

Note that generating a topology with properties that resemble those of the Internet to a significant extent is still an open problem [62]. There is a trade-off between dynamic, measurement based topologies and static topologies. On the one hand, measurement derived topologies are constrained by the measurement method and the observational conditions, being in general difficult to understand. On the other hand, fixed topologies are easier to understand and can be generated in order to analyze specific issues. The former can be used as explanatory models whereas the latter tend to be descriptive. Some proposals have been made for extracting relatively simple graphs that keep the properties of topologies inferred from measurement [63].

In this work, we have generated a static topology with a fixed core network modeled after the GREN in order to analyze a number of issues that arise in current high performance networks. Nonetheless, this particular fixed topology reflects to a considerable degree of detail a set of real networks. In addition, components are added to the topology in order to increase the range and variability of link and path properties in a random manner with a set of restrictions. These restrictions have been defined in order to guarantee some overall topological properties found in measurement studies performed on the Internet [67, 29, 62]. For the simulations performed in both the dumbbell and the GREN scenario, the standard Open Shortest Path First (OSPF) routing scheme as implemented in ns-2 is used.

5.3 Fuzzy End-To-End Rate Control for Internet Transport Protocols

End-to-end Internet packet dynamics is a complex problem for which models available to date are at best incomplete [40]. A great deal of attention is being paid by the research community to two issues in this area: protocols for high performance networking, and solutions for new services and applications for which proper congestion control schemes are sought. These issues have led to an interest in the application of artificial intelligence techniques to end-to-end problems within the End-to-End Research Group of the Internet Research Task Force.

As shown in figure 5.5, end-to-end traffic is subject to control actions not only at the end points but also at every hop in the path, including access nodes, backbone links and peering points, with a wide variety of technologies and conditions that are unknown to the transport layer. Currently deployed schemes for traffic regulation in Internet (as well as proposed alternatives) fit into one of the two following approaches [93] (see also figure 6.10, on page 286):

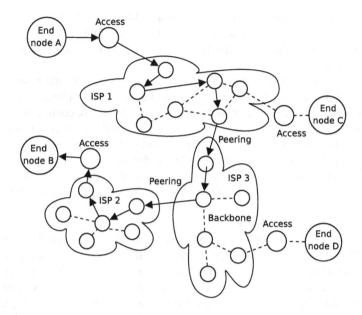

Fig. 5.5 End-to-end path between two nodes and the elements that may made up an end-to-end path

- Distributed control, with functionality distributed among the end nodes in the network and implemented by means of end-to-end transport protocols. End nodes that transmit and receive packet flows cooperate so as to perform flow and congestion control as well as fair distribution of network resources.
- Queue controlers in intermediate nodes (routers). These mechanisms may discriminate packet flows and enforce resource distribution and reservation.

Thus, regulation of packet flows from sender to receivers can involve all the network nodes in the end-to-end path and is performed on both an end-to-end and a per-hop basis. Such a scheme leads to a system that comprises multiple feedback loops with complex interactions.

End-to-end flows traversing routers span a wide range of user requirements and dynamic characteristics, i.e., the number of hops in the path can typically range from a few up to around 20, round-trip times can range from a few milliseconds up to seconds, flows can last from a few milliseconds up to hours and each flow can transfer from a few KBs up to several GBs.

In a network like the one depicted in figure 5.5, the end-to-end paths between nodes A and B are very different to that between A and C, or A and D, or any other two end nodes. Many diverse aspects have an impact on end-to-end performance. These include, among many others, physical distances, methods and agreements of interconnection between ISPs, provisioning and topologies of backbones, characteristics and technologies of access links, the use of traffic shapers, and miscellaneous performance problems due to misconfigurations. Also, other higher scale factors, such as routing protocols, both at an intra-network and an inter-network level, and network upgrades can have an important impact on end-to-end performance that is difficult to model.

Quality of service requirements as well as traffic patterns of emergent services and applications are difficult to characterize and demand deep advances in current rate control schemes. Because of the nature of these problems, complexity, no feasible analytic solution, as well as incomplete and inaccurate information, one of the alternatives for studying them is the employment of intelligent systems based on fuzzy logic and possibly other soft computing techniques.

Both aforementioned approaches can be redefined in terms of fuzzy systems, which does not only provide a deeply backgrounded engineering approach but also a modeling and analysis framework for Internet traffic (which the current Internet research community lacks [40]).

In this section we focus on end-to-end rate control [71]. More specifically, we analyze TCP-like window based rate control. Traffic control in routers will be addressed in section 5.4. With this work we aim at providing a reinterpretation of end-to-end flow and congestion control mechanisms in terms of fuzzy systems. We provide an initial set of results based on simulation and experimental implementation showing a number of advantages as for both performance and model interpretability.

Section 5.3.1 provides an overview of related publications. Section 5.3.2 outlines a fuzzy logic based extension to end-to-end window based rate control schemes. In section 5.3.3 we detail the design of a fuzzy system for end-to-end rate control. Section 5.3.4 gives an overview of the fuzzy systems development methodology and tool chain employed. Simulation and experimental implementation results are then presented in section 5.3.5 and section 5.3.6, respectively.

5.3.1 Related Work

Formal approaches that have been applied to end-to-end rate control include classic control theory and fluid mechanics among others [93]. Fuzzy logic has been applied to related problems, such as active queue management in routers [26, 24] and identification of security incidents [1, 35]. While a number of proposals of fuzzy controlers for packet queues at Internet routers have been reported, see [26, 24] among others, the only reference we are aware of that takes a fuzzy logic based approach to end-to-end rate control is [18], which outlines a preliminary study through an off-line simulation based analysis of a nonlinear Takagi-Sugeno fuzzy controler applied to some of the basic mechanisms of TCP congestion control.

In the area of wireless networks, there have been proposals of fuzzy systems applied to intelligent discrimination of congestion induced packet loss and loss due to physical channel errors and mobility [77, 22].

5.3.2 End-To-End Window Based Rate Control and a Fuzzy Generalization

A number of classes of rate control have been defined and applied in packet switching networks, such as window based and equation based [40]. The prevalent transport protocol in the current Internet is Transmission Control Protocol (TCP), which uses window based rate control mechanisms. In window based congestion control schemes, the window sets a limit on the amount of unacknowledged packets in flight, i.e., the number of packets that are in transit in the end-to-end path for which no acknowledgement has been received.

TCP includes basic mechanisms for flow control since its original specification was published [81]. After a series of congestion collapses in the network [57], mechanisms for congestion control have been added throughout the years. The first proposal of the now widely accepted congestion avoidance algorithm was introduced [57, 75], standardized and further developed [4, 86] within a period of several years. This process lead to the development, standardization and world-wide deployment of congestion control mechanisms that together with the basic flow control mechanisms comprise the core rate control functional block of TCP.

TCP rate control comprises four intertwined algorithms: slow start, congestion avoidance, fast retransmit and fast recovery.

Let us consider the simplified version in algorithm 1 (of historical [57] interest only) for the sake of simplicity in explaining our approach. W_i is the congestion window size and W_{max} is the delay-bandwidth product of the network path. Details of current standard algorithms and how congestion is detected (commonly based on packet loss) propose to extend them are provided in the next section.

Algorithm 1. Basic additive increase multiplicative decrease (AIMD) algorithm.

---

if congestion **then**
  $W_i = d W_{i-1}, \quad (d < 1)$
else
  $W_i = W_{i-1} + u, \quad (u \ll W_{max})$
end if

The algorithm tries to react quickly to congestion conditions. When the network is congested, the amount of traffic from competing flows must be large and the queue lengths will start increasing exponentially. Under the assumption that the system will stabilize if the traffic sources throttle back at least as quickly as the queues are growing and considering that a source controls load in a window based protocol by adjusting the size of the window (W_i), we end up with the sender policy $W_i = d W_{i-1}$, i.e. a multiplicative decrease rate that under persistent congestion leads to an exponential decrease of the sender window.

If there is no congestion, router queues must be near zero and the network load approximately constant. The network announces, via a dropped packet, when demand is excessive but does not notify if a connection is using less than its fair share (since the network is stateless, this information is not available). Thus a connection has to increase its bandwidth utilization to find out the current upper limit. The first thought is to use a multiplicative increase rate possibly with a longer time constant, $W_i = bW_{i-1}, 1 < b < 1/d$. This is however a mistake. The result will show wild oscillations and poor average throughput. The analytic reason for this, well known in queuing systems as the rush-hour effect, is due to the fact that it is easy to drive the net into saturation but hard for the net to recover. An increase policy based on small, constant changes was proposed for TCP [57]. This policy has proven to be effective.

However, the standard additive increase multiplicative decrease (AIMD) scheme can be too cautious under some conditions, which has given rise to a great deal of research towards protocols for high-performance networks. A number of important limitations in TCP rate control have been identified throughout the last years. Among these, two issues have received a great deal of attention recently. First, TCP performs poorly in long-distance high-speed networks where applications require high bandwidth utilization. Second, lossy links also challenge the TCP congestion control scheme, which is specially relevant for wireless networks. Nevertheless, TCP performance can be clearly suboptimal under a wide range of scenarios and traffic conditions, but specially in networks with a considerable amount of background and/or competing traffic.

As a consequence, the Internet research community is developing protocols with alternative rate control schemes, such as the TCP-Friendly Rate Control protocol (TFRC) [52], which uses equation based congestion control mechanisms, and protocols with similar yet extended schemes for rate control, such as the Stream Control Transmission Protocol (SCTP) [87]. A number of extensions and modifications to TCP rate control are being developed as well. These range from minor updates of the standard AIMD scheme of TCP to complete redesigns of the protocol. The aim of these proposals is to define versions of TCP tuned for long-distance high-speed networks, adding robust performance over paths with non-congestive packet loss, intermittent connectivity, significant reordering and related issues.

HighSpeed TCP [38], scalable TCP [60] and H-TCP [64] introduce new response functions integrated in the AIMD scheme. The Westwood [94] and Westwood+ TCP [50] proposals estimate the end-to-end bandwidth in order to set the threshold between the slow start and congestion avoidance stages. Other alternatives detect congestion by measuring the delay of individual packets. These include TCP Vegas [13], Compound TCP [88], and FAST TCP [95]. The CUBIC [51] proposal uses a search method in order to compute the congestion window size.

Nonetheless, to date no proposal has been found to be free of performance or fairness issues. In essence, TCP/IP networks are not deterministic. One of the consequences of this fact is that the available network capacity of an end-to-end path is in principle unpredictable. Thus, TCP-like congestion control schemes have to estimate the available capacity in order to accommodate concurrent or competing traffic. This estimation process is hampered by uncertainty and lack of information.

As stated above, in addition to the complex dynamics that can arise from the AIMD mechanism of TCP, interactions between network layers, end nodes and intermediate nodes lead to a complex nonlinear dynamics that makes it difficult to design, simulate, and test congestion control schemes. A great deal of theoretical analyses are currently being performed and a number of experimental implementations have been proposed. In this work we address the problem by means of fuzzy systems. In what follows we describe a fuzzy model for TCP-like rate control in an incremental manner.

In the simplified algorithm above, we distinguish two window evolution stages. Each stage corresponds to a clearly identified network state and leads to the application of a specific window update policy. However, the actual state can be in between these two crisp conditions. Additionally the knowledge about the current network state is uncertain and delayed. We note here a binary logic problem: when the network state for which the rate control stage has been defined is constant for enough time, the system response is proper. However, when the network state does not exactly match any of the stages but a combination of them, the response of the system may be too aggressive or too conservative.

The starting point of the TCP innovations was to view a TCP connection as a control loop and to ask what the correct behavior of the control loop was under certain impulses. In this same line of development, we apply the fuzzy inference control paradigm to TCP end-to-end window based rate and congestion control. Similarly to the proposal in [18], a first generalization of the algorithm above could be stated by means of a simple reformulation of the basic AIMD principle:

$$w_{i+1} = w_i + \alpha_D f_D(w_i) + \alpha_I f_I(w_i) \tag{5.1}$$

Where f_D and f_I set the decrease and increase policies. α_D and α_I can be thought of as degree of truth values that represent to what extent the system is on the congested (window decrease) or uncongested (window increase) mode; these values can be defined as mutually exclusive, $\alpha_D, \alpha_I \in 0, 1; \alpha_D = \overline{\alpha_I}$. This formulation suggests a fuzzy approach for managing the window update process. Instead of considering the network in one of a set of exclusive states, we will consider the network to be (to a variable degree) in all defined states. The degree to what the network is considered to be in a particular state will be identified by a fuzzy rule based inference system. We however depart from the classical control theoretical approach followed in [18]. Instead, we will define a set of linguistic rules that generalize the behavior of the standard TCP congestion control scheme [71].

5.3.3 Design of a Fuzzy End-To-End Window Based Rate Controler

The aim of this section is to design a generalization of the TCP congestion control algorithm in terms of fuzzy logic. As main design principles we consider simplicity, keeping a certain degree of compatibility with traditional approaches and the overall objective of obtaining a set of rules that can be interpreted with ease from the perspective of protocol design. Thus, we avoid to use automatic learning techniques. On the one hand, analytical approaches and stability analysis in particular is hard to apply in this context. On the other hand, using automatic tuning approaches for neuro-fuzzy systems that can exhibit optimal or near-optimal performance for a particular collection of simulated scenario does not seem a sound approach as the requirements of different scenarios are commonly antagonistic.

A congestion control scheme intended to be applied in the Internet must be flexible and adaptive to a wide range of conditions. This requires testing and thorough analysis of a broad range of complex simulated and real scenarios. This is hard to perform through automated simulation and tuning approaches. In fact, the definition of a standard set of scenarios and tests for evaluating TCP congestion control is a challenging task that is currently work in progress [6].

It is beyond the scope of this section to provide a complete description of the TCP rate control and related algorithms. We will focus on those procedures that perform a direct action on the window, i.e. those procedures that imply the definition of a network state (and a stage in the rate control algorithm) as well as the associated window update policy. For a full specification of TCP refer to the aforementioned documents. Though our approach finds applications in general window based end-to-end transport protocols, we analyze the case of TCP as a case of special interest. In what follows we will describe in an incremental manner those mechanisms defined for rate control in standard TCP implementations and how we have extended them by means of fuzzy inference systems.

The four standard algorithms for controling the congestion window in standard TCP systems (slow start, congestion avoidance, fast retransmit and fast recovery) will be described. In what follows, definitions in table 5.1 are considered.

Table 5.1 TCP rate and congestion control parameters

Parameter	Description
$SMSS$	Sender Maximum Segment Size. Size of the largest segment that the sender can transmit
$rwnd$	The most recently advertised receiver window
$cwnd$	Congestion window. A limit on the amount of data TCP can send. At any given time, TCP does not send data with a sequence number higher than the sum of the highest acknowledged sequence number and the minimum of $cwnd$ and $rwnd$.
$flightSize$	The amount of data that has been sent but not yet acknowledged.
RTT	Round-trip time
RTO	Retransmission timeout (which depends on RTT)
IW	Initial window value for cwnd
LW	Loss window value
$ssthresh$	Threshold between Slow start and Congestion avoidance stages

5.3.3.1   Slow Start

Beginning transmission into a network with initial unknown conditions requires TCP to cautiously probe the network to estimate the available capacity, in order to avoid congesting the network with an inappropriately large burst of data. The slow start algorithm is used for this purpose at the beginning of a transfer. The initial *cwnd* value is usually a few *SMSS* (the most common value is $2 \cdot SMSS$, see [5] for a complete discussion).

The initial value of *ssthresh* may be arbitrarily high (usually the size of the receiver advertised window) though may be reduced in response to congestion. *ssthresh* is used to select which window update policy should be applied. The selection procedure as specified in the standard is shown in algorithm 2:

Algorithm 2.   Standard TCP AIMD.
---
if *cwnd* < *ssthresh* **then**
    Perform slow start
else if *cwnd* > *ssthresh* **then**
    Perform congestion avoidance
else
    Perform either slow start or congestion avoidance
end if

In the standard TCP slow start stage, *cwnd* is incremented each time an acknowledgement packet is received from the sender that acknowledges new data[2]. The increment rate is defined as follows:

$$cwnd_{i+1} = cwnd_i + inc, \quad (inc \leq SMSS) \tag{5.2}$$

It is common for current implementations to choose $inc = SMSS$. The slow start stage (with exponential growth with time in the window) ends when *cwnd* is greater than (or greater or equal to) *ssthresh* or when congestion is observed.

As a fuzzy extension to the slow start algorithm we propose an inference system (*SlowStartConfidence*) that produces as output the extent to which current network conditions should be managed with the slow start algorithm. In other words, the system infers an estimate of certainty about the network being in such a state that should be handled by means of the slow start update policy. The system has the following inputs: *timeout* (to which extent a timeout is expiring), *ssthresh*, *cwnd*, the difference *ssthresh* − *cwnd*, and two inputs that provide information on overall network conditions: the cumulative packet loss fraction, *loss*, and the round-trip time, *rtt*.

Figure 5.6 shows the fuzzy types and linguistic terms for the *timeout* and *cwnd* inputs (types *Ttimeout* and *Tcwnd*). Figure 5.7 shows the types for the input *loss* and the output, *confidence*. In order to simplify definition and easing the use of

---

[2] This point is currently under revision, as better performance can be achieved by counting the number of octets (instead of the number of packets) acknowledged [3].

efficient implementation techniques, only triangular membership functions are used. In general, the fuzzy types are defined using a partition of the crisp input space and triangular and trapezoidal membership functions. 5 linguistic terms are defined for inputs, and 5 linguistic terms are defined for the output (for increasing degrees of certainty from IMPOSSIBLE to CERTAIN).

Fuzzy inference is performed by a simplified Mamdani model were the output membership functions are singletons (or equivalently a zero-order TSK model). and the fuzzy mean defuzzification method is employed to compute the crisp output value. Membership functions were adjusted considering typical performance values considered in recent Internet measurement studies [82].

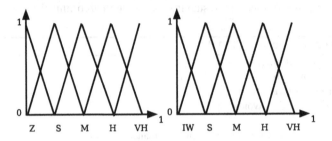

Fig. 5.6 *T timeout* (left) and *T cwnd* (right) membership functions

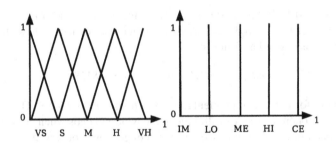

Fig. 5.7 *T loss* (left) and *T confidence* (right) membership functions

Table 5.2 shows the rule base of the *SlowStartConfidence* system. Standard TCP rate control activates slow start update policies when *cwnd* is lower than *ssthresh*. As can be seen in the rule base, the fuzzy system yields a certainty degree that increases with input *ssthresh* − *cwnd* (fourth column). This behavior can be thought of as a generalization of the crisp "lower than" comparison used in standard TCP. The five fuzzy rules triggered by input *ssthresh* − *cwnd* represent the informal expression "the bigger *ssthresh* − *cwnd*, the more possibility of slow start being suitable for current network conditions".

However, additional rules can modify the certainty degree if other network conditions suggest a different update policy. Rules that most directly reproduce the

Table 5.2 SlowStartConfidence rule base

timeout	cwnd	ssthresh	ss − cw	loss	rtt	confidence
Z	x	x	x	x	x	HI
S	x	x	x	x	x	HI
M	x	x	x	x	x	ME
H	x	x	x	x	x	ME
VH	x	x	x	x	x	IM
x	IW	x	x	x	x	CE
x	S	x	x	x	x	LO
x	M	x	x	x	x	LO
x	H	x	x	x	x	LO
x	VH	x	x	x	x	IM
x	x	VS	x	x	x	IM
x	x	S	x	x	x	LO
x	x	M	x	x	x	LO
x	x	H	x	x	x	LO
x	x	VH	x	x	x	ME
x	x	x	VS	x	x	IM
x	x	x	S	x	x	LO
x	x	x	M	x	x	ME
x	x	x	H	x	x	HI
x	x	x	VH	x	x	CE
x	x	x	x	VH	VH	IM
x	x	x	x	VH	H	LO
x	x	x	x	H	VH	LO
x	x	x	x	VS	VS	ME

non-fuzzy standard TCP behavior are those triggered by values of the *ssthresh −
cwnd* input. The degree of certainty given by these five rules is then adjusted by
additional rules that consider further information on current network conditions.
Four rules are at least fired at any given time. Exactly one of the first five rules
(first five rows) is always active (for input *timeout*), as well as exactly one of the
next four rules (for input *cwnd*). The same applies to *ssthresh*, and *ssthresh − cwnd*.
Rules depending on *loss* and *rtt* are triggered only under clear network saturation
conditions.

5.3.3.2 Congestion Avoidance

In the congestion avoidance stage, *cwnd* is incremented by *SMSS* per round-trip
time. In real implementations, it is common to increment *cwnd* for every non duplicated acknowledgement packet received from the receiver as in:

$$cwnd_{i+1} = cwnd_i + SMSS \cdot SMSS / cwnd_i \qquad (5.3)$$

Which is generally considered to be an acceptable approximation and leads to a
growth rate approximately linear with time. Congestion avoidance ends when congestion is detected.

In addition, in any stage *ssthresh* and *cwnd* are modified when a timeout is detected according to:

$$ssthresh = max(flightSize/2, 2 \cdot SMSS) \qquad (5.4)$$

$$cwnd_{i+1} = cwnd_{TO} \leq LW \qquad (5.5)$$

Where $cwnd_{TO}$ is the recommended congestion window value after a timeout, with *LW* usually being set to 1 full-sized segment. As with the slow start algorithm, we propose an extension to congestion avoidance. The extension consists of a fuzzy inference system (*CongestionAvoidanceConfidence*) that produces as output the extent to which current network state should be managed following the congestion avoidance algorithm. The system inputs are the same as those of *SlowStartConfidence*.

Table 5.3 shows the rule base, which has a similar structure to that of *SlowStartConfidence* but employs what can broadly be seen as a complementary rule base. Note however that the rules depending on *loss* and *rtt* are triggered under different conditions. In this case, the five rules triggered by *ssthresh* − *cwnd* represent the following sentence: "the smaller *ssthresh* − *cwnd*, the more possibility of congestion avoidance being a proper algorithm for current network conditions".

5.3.3.3 Fast Retransmit and Fast Recovery

TCP receivers send an immediate duplicate acknowledgement (ACK) packet back to the sender when an out-of-order segment arrives. The purpose of this ACK is to inform the sender that a segment was received out-of-order and which sequence number was expected instead. From the sender's perspective, duplicate ACKs can be caused by a number of network problems. They can be caused by dropped segments. In this case, all segments after the dropped segment will trigger duplicate ACKs. In addition, duplicate ACKs can be caused by the re-ordering of data segments by the network. Finally, duplicate ACKs can be caused by replication of ACK or data segments by the network.

TCP senders use the *fast retransmit* algorithm to detect and repair loss, based on incoming duplicate ACKs. The fast retransmit algorithm uses the arrival of 3 duplicate ACKs. After receiving 3 duplicate ACKs, TCP senders perform a retransmission of what appears to be the missing segment, without waiting for the retransmission timer to expire. After the fast retransmit algorithm sends the apparently lost segment, the *fast recovery* algorithm governs the transmission of new data until a non-duplicate ACK arrives. These two algorithms are usually implemented together as follows.

1. When the third duplicate ACK is received, the lost segment is retransmitted and *ssthresh* and *cwnd* are set to

$$ssthresh = max(flightSize/2, 2 \cdot SMSS) \qquad (5.6)$$

Table 5.3 CongestionAvoidanceConfidence rule base

timeout	cwnd	ssthresh	ss − cw	loss	rtt	confidence
Z	x	x	x	x	x	IM
S	x	x	x	x	x	IM
M	x	x	x	x	x	IM
H	x	x	x	x	x	ME
VH	x	x	x	x	x	CE
x	IW	x	x	x	x	IM
x	S	x	x	x	x	LO
x	M	x	x	x	x	LO
x	H	x	x	x	x	LO
x	VH	x	x	x	x	HI
x	x	VS	x	x	x	CE
x	x	S	x	x	x	LO
x	x	M	x	x	x	LO
x	x	H	x	x	x	IM
x	x	VH	x	x	x	IM
x	x	x	VS	x	x	CE
x	x	x	S	x	x	HI
x	x	x	M	x	x	ME
x	x	x	H	x	x	LO
x	x	x	VH	x	x	IM
x	x	x	x	VH	VH	CE
x	x	x	x	VH	H	HI
x	x	x	x	H	VH	HI
x	x	x	x	H	H	ME
x	x	x	x	M	VH	ME
x	x	x	x	M	H	LO

$$cwnd_{i+1} = cwnd_{DUP3} = ssthresh + 3 \cdot SMSS \qquad (5.7)$$

This artificially inflates the congestion window by the number of segments (three) that have left the network and which the receiver has buffered.

2. For each additional duplicate ACK received, increment the congestion window.

$$cwnd_{i+1} = cwnd_i + SMSS \qquad (5.8)$$

This artificially inflates the congestion window in order to reflect the additional segment that has left the network.

3. Transmit a segment, if allowed by the new value of cwnd and the receiver's advertised window.

4. When the next ACK arrives that acknowledges new data, set cwnd to ssthresh (the value set in step 1).

$$cwnd_{i+1} = ssthresh \qquad (5.9)$$

As with the slow start and congestion avoidance algorithms, we propose an extension that consists of a fuzzy inference system (*FRFRConfidence*) that produces as output the extent to which the fast recovery/fast retransmit strategy is suitable for current network conditions. The inputs to the system are *timeout*, *ssthresh — cwnd*, *loss* and *rtt*. The rule base is shown in table 5.4. Exactly one of the first five rules (first rows) will be triggered at any time for different values of *timeout*. The same applies to the next five rules (for *ssthresh — cwnd* values). The last four rules are triggered when network conditions suggest that the fast recovery, fast retransmit update policies are suitable.

Table 5.4 FRFRConfidence rule base

timeout	ssthresh − cwnd	loss	rtt	confidence
Z	x	x	x	IM
S	x	x	x	IM
M	x	x	x	IM
H	x	x	x	IM
VH	x	x	x	CE
x	VS	x	x	IM
x	S	x	x	LO
x	M	x	x	ME
x	H	x	x	HI
x	VH	x	x	CE
x	x	VH	VH	CE
x	x	VH	H	HI
x	x	H	VH	LO
x	x	M	H	HI

5.3.3.4   Putting All Pieces Together

In our extended approach, we define three inference systems with linguistic rules that generalize the behavior of the standard TCP congestion control scheme. Three fuzzy stages are defined for slow start, congestion avoidance and fast retransmit/fast recovery. Since the actual network state is known with uncertainty, delay, etc.) all network states are considered to occur at the same time with a varying degree of certainty. The extent to which the system is in one of these stages is evaluated by three fuzzy inference systems whose outputs are regarded as a degree of certainty about current network state. This way, we address the uncertainty and lack of information about the actual network congestion state as well as the difficulties in identifying congestion conditions.

Update policies of *ssthresh* and *cwnd* given in equations 5.2 to 5.9 are kept. However, as the network state is in general uncertain, all policies are simultaneously evaluated and applied to a varying degree given by the three fuzzy inference systems described. Thus, the three inference systems identify complex network states on the basis of linguistic rules.

Slow start and congestion avoidance equations are always evaluated as in standard TCP implementations (equations 5.2 to 5.5). Additionally, when duplicated

ACKs are detected, fast retransmit and fast recovery (equations 5.6 to 5.9) are evaluated as well (and considered to the extent given by the *FRFRConfidence* fuzzy inference system).

In the simplest case, the three sets of policies are combined as follows. If we denote by μ_{ss}, μ_{ca}, and μ_{FRFR} the outputs of the three fuzzy inference systems, and by f_{SS}, f_{CA} and f_{FRFR} the values given by the standard update policies (which we will generalize to μ_i and f_i $(1 \leq i \leq n)$ for a variable number n of network states (or stages in the rate control algorithm, or sets of update policies)), we define the certainty degree c_i of a stage as:

$$c_i = \frac{\mu_i}{\sum_{j=1}^{n} \mu_j}$$

Where the rule sets of the fuzzy inference systems should verify $\sum_{j=1}^{n} \mu_j > 0$. The final update policy for *cwnd* is computed as the weighted average of the update policies associated to all possible stages (as in equation 5.1):

$$cwnd_{i+1} = cwnd_i + \sum_{i=1}^{N} c_i f_i (cwnd, ssthresh)$$

This way, in essence we have a rule based method of combining update policies that have been shown to be effective under certain conditions. The rule sets can lead to compromise solutions under complex conditions. For instance, policy 5.9 usually implies a large decrease of the congestion window. When additional inputs indicate network congestion, the policy may be adequate. However, when the network does not appear to be congested, a more aggressive behavior could improve performance in terms of throughput and responsiveness.

5.3.4 Development Methodology and Tool Chain

In order to develop fuzzy inference systems, we adhere to a design methodology [16] for the whole development process that covers from initial high-level description to implementation as software and hardware components. A complete tool chain for the development stages [73, 74] has been employed.

Leveraging on the Xfuzzy [73] CAD suite of tools and a methodology [11, 16] for the development of fuzzy controlers, we have defined a methodology and a tool chain tailored for the development of fuzzy Internet rate controlers.

The design flow and tool chain employed to develop fuzzy inference modules will be described in the next section and is partially depicted in figure 5.8. The whole development process is covered, from initial description to final implementation whether as software or hardware. In this section we use however a simplified flow which is shown in figure 5.8. The first development stage (description) is performed using a high level fuzzy systems specification language, XFL [72], which can be automatically turned into C and VHDL code among other implementation options.

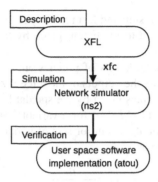

Fig. 5.8 Fuzzy systems design flow and tool chain for end-to-end traffic control

The development stages after specification have been tailored for end-to-end rate control as follows. For network simulation, we have used ns-2 [56]. ns-2 is an object oriented discrete event driven simulator with support for a vast variety of transport protocols, queuing systems, routing schemes and access media, thus enabling us to evaluate the performance of traffic controlers under complex and realistic simulated scenarios. Fuzzy controlers are integrated into ns-2 as components implemented in C.

Verification can be performed over software and hardware implementations of fuzzy controlers. Software verification is performed over a controler implementation within a tool that implements the TCP protocol in user space (atou), which is further described in section 5.3.6.

5.3.5   Simulation Results

Following on the limitations of simulation discussed in section 5.2, there is a need for the definition of benchmarks for end-to-end protocols, which includes common simulation topologies, traffic patterns, reference implementations and a number of additional factors to consider in any simulation study. In this sense, there has been a considerable deal of effort within the research community during the last years towards defining common principles for new congestion control as well as specifying practices for evaluating new end-to-end protocol proposals.

The development and deployment of new end-to-end congestion control algorithms is currently an ill-defined problem and raises controversy. In fact, it has been only very recently that a set of criteria to consider when evaluating new congestion control algorithms has been agreed within the IETF (not representing all the implied parties) [39]. Still, these very general principles do not represent hard and fast requirements for an appropriate congestion control scheme and should rather be considered and weighted in the context of each proposal. As far as deployment, many factors impact the adoption of new mechanisms and the success of some proposals often depends on unexpected conditions and apparent randomness [39].

It is currently work in progress in early stage of development within the IETF the definition of a best current practice request for comments on a common TCP evaluation suite [6]. Nonetheless, this suite does not aim to result in an exhaustive evaluation of proposed TCP modifications or new congestion control schemes. Rather, it is being developed as an evaluation suite for the initial evaluation of protocols. The final goal of this effort is to define a common evaluation suite so that quick and easy evaluations can be performed in simulations and testbeds using a common set of well-defined, standard test cases in order to compare and contrast proposals.

The simulation study that we describe in this section aims to consider the above cautions and requirements as far as possible in order to perform an overall comparative evaluation of the proposal presented here. The simulation study also addressess concrete current issues such as the limitations of standard versions of TCP so as to take full advantage of available bandwidth for paths with high bandwidth-propagation delay products.

Simulations of fuzzy rate controlers have been performed by means of the ns-2 [56] simulator. The methodology applied in order to integrate fuzzy inference systems into ns-2 simulations was described in section 5.3.4. A performance evaluation study was conducted on traditional and fuzzy rate controlers so as to compare both approaches. We will show results for a comparison of TCP using three alternative congestion control algorithms:

- The Reno congestion control algorithm [4] with the NewReno modification [45] and the SACK options [68, 43]. This variant will be called TCP SACK. TCP NewReno has been the most commonly deployed algorithm for a few years on a worldwide scale. The SACK extensions are very common as well [69]. However, at the time of this writing it is very difficult to estimate the real-world deployment of the many TCP variants proposed to date, as some operating systems are integrating new TCP variants in their later versions.
- HighSpeed TCP [38].
- The fuzzy rate controler proposed here.

In the simulation experiments that follow, we will use the standard implementations of TCP SACK and HighSpeed TCP in ns-2 as well as a C implementation of the fuzzy TCP variant implemented using the xfc tool of the Xfuzzy environment [73]. Due to the AIMD nature of TCP-like window based congestion control, the slow-start stage will condition the behavior of short-lived flows, whereas long-lived flows will transition into congestion avoidance state and stay for a long time (if no congestion is detected). Thus, the congestion avoidance stage conditions long-lived flows. Consequently, by studying a long-lived flow, we can gain understanding of the behavior of the proposed scheme for both short-lived and long-lived flows. In particular, we will show the results of test flows established for 30 seconds.

First, we will analyze a set of experiments performed in the dumbbell scenario. A test TCP flow is established between node 1-1 and node 2-1. Different tests are performed using TCP SACK, High Speed TCP and fuzzy TCP, respectively. All the competing flows are generated by either UDP traffic sources or TCP SACK traffic sources. Low, medium and high competing traffic load was defined in the following form:

- Low competing traffic load corresponds to a CBR UDP flow between node 1-2 and node 2-2, with bit rate 5 Mb/s.
- For medium competing traffic load, low traffic load is increased by adding two flows between node 1-1 and node 2-2 and two flows between nodes 1-2 and 2-2.
- High competing traffic load corresponds to medium traffic load and two flows between nodes 2-2 and 2-1 and two flows between 2-1 and 1-1.

Figure 5.9 shows the evolution of the congestion window for TCP SACK, High Speed TCP and Fuzzy TCP with low, medium and high competing traffic load in the dumbbell scenario. Notice that in this and following figures the congestion window size is measured in bytes. It should be noted that traditional variants of TCP change the window size in units of segments (or sender maximum segment size), which has some performance and fairness issues [3]. In contrast, in the fuzzy TCP proposal analyzed here all computations concerning the congestion window size are made in bytes.

As it could be expected, there is a large decrease in the amount of bytes in flight allowed by the congestion window for the medium and high competing traffic load, a consequence of the lower available bandwidth in the end-to-end path. Figure 5.10 shows the evolution of the throughput of the test flow for the fuzzy, SACK and HighSpeed TCP variants in the dumbbell scenario. Low, medium and high competing traffic load are taken into consideration as well.

For the GREN scenario, The test flow is generated between an end node 2 hops away from the LOSA node of the Internet2 backbone (west coast of the USA) and a node 2 hops below the SEV node (South Western Europe) of the RedIRIS backbone, a regional subnetwork of the GÉANT network, shown in figure 5.4 (page 199). These test flows traverse an end-to-end path with approximately 80 ms of minimum propagation delay (or an equivalent 160 ms minimum RTT), a maximum link bandwidth capacity of 10 Gb/s and a bottleneck bandwidth capacity of 100 Mb/s. Unless explicitly stated, all competing flows are generated by either UDP traffic sources or TCP SACK traffic sources.

In this scenario, we define low, medium and high competing traffic load as follows:

- Low competing traffic load corresponds to background traffic in the scenario with no additions. The generation of background traffic in the GREN scenario was explained in section 5.2.
- For medium competing traffic, an additional TCP SACK flow is established between a node connected to one of the backbone nodes of the Internet2 backbone and a node of one of the subnetworks of the GÉANT backbone (excluding RedIRIS). This competing flow has a bottleneck bandwidth of 1 Gb/s and shares the central part of the end-to-end path with the test flow.
- High competing traffic load corresponds to 20 flows between nodes connected to nodes of the Internet2 and GÉANT backbones plus 20 flows in the reverse way.

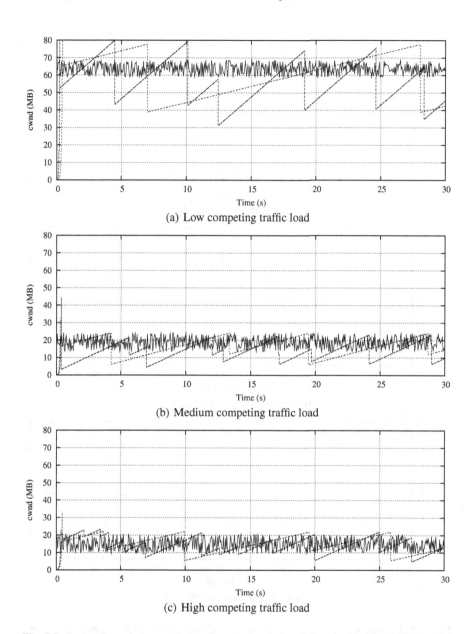

(a) Low competing traffic load

(b) Medium competing traffic load

(c) High competing traffic load

Fig. 5.9 Congestion window evolution for a test end-to-end flow in the dumbbell scenario with low, medium and high traffic load. Continuous line: fuzzy TCP; long dashed line: High-Speed TCP; short dashed line: TCP SACK.

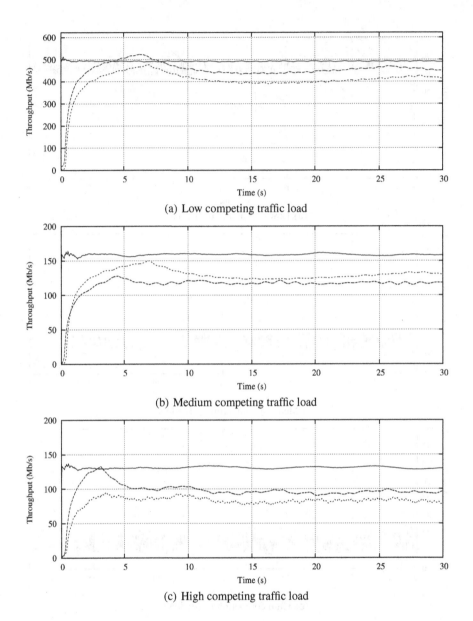

(a) Low competing traffic load

(b) Medium competing traffic load

(c) High competing traffic load

Fig. 5.10 Throughput evolution for a test end-to-end flow in the dumbbell scenario with low, medium and high traffic load. Continuous line: fuzzy TCP; long dashed line: HighSpeed TCP; short dashed line: TCP SACK.

This way, the variance of the RTTs of the competing flows (besides the basic background traffic of the GREN scenario) is sufficiently high and follows a pattern found in measurement studies [84], with approximately 160 ms RTT between the West Coast of the USA and South Western Europe.

Figure 5.11 shows a comparison of the evolution of the congestion window for the three congestion control schemes being analyzed in the GREN scenario. Figure 5.12 compares the evolution of the throughput of the test flow for the three alternatives considered, with low, medium and high competing traffic load in the GREN scenario.

As overall conclusions we can draw that the fuzzy extended version of TCP rate control shows higher robustness against competing, cross and reverse-path traffic and the derived RTT variance and loss events. This translates into higher final throughput (improved by approximately 15% and 11% as compared to TCP SACK and HighSpeed TCP, respectively in the case of low competing traffic load). This fact becomes more evident as the competing traffic load increases. These simulations were run with TCP flows with different RTTs within a wide range. Additional simulations performed on network scenarios under high congestion conditions confirm that the fuzzy rate controler still provides proper and quick reactions to congestion. We note that the higher stability of the throughput of fuzzy TCP flows contributes to reduce RTT variance.

Common fairness principles considered in the design of Internet end-to-end congestion control schemes [93] are satisfied as well. In particular, the decrease in throughput that can be observed for the tests with medium and high competing traffic load as compared to the tests with low competing traffic load is a consequence of the available bottleneck bandwidth being shared among competing flows rather than high congestion. Also, it can be observed that the presence of traffic with complex behavior (in particular, high bandwidth and RTT variability [2]) in the GREN scenario has a considerable impact on the stability of the throughput for the cases of TCP SACK and HighSpeed TCP. The higher robustness of the fuzzy variant proposed here becomes more evident under such conditions.

5.3.6 Implementation Results

In order to make the results of our work available for the Internet research community as a research tool, we have developed an experimental user space (instead of kernel space) TCP implementation with fuzzy extensions. A user space implementation allows maximum experimentation flexibility for further refinement of fuzzy rule bases and identification of new rules.

The tool is based on the "Almost TCP over UDP" (atou) [36] utility, developed as part of the web100 project of the High-Performance Networks research program of the U.S. Department of Energy. We introduced changes to atou in order to integrate the fuzzy inference scheme presented in previous sections. The three fuzzy inference systems were generated as C code using the xfc tool of the Xfuzzy

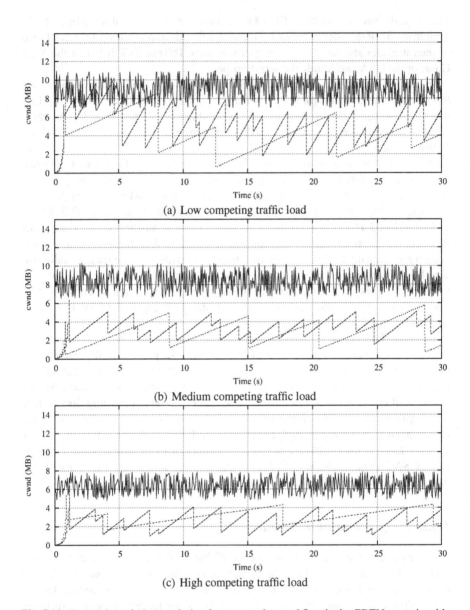

(a) Low competing traffic load

(b) Medium competing traffic load

(c) High competing traffic load

Fig. 5.11 Congestion window evolution for a test end-to-end flow in the GREN scenario with low, medium and high traffic load. Continuous line: fuzzy TCP; long dashed line: HighSpeed TCP; short dashed line: TCP SACK.

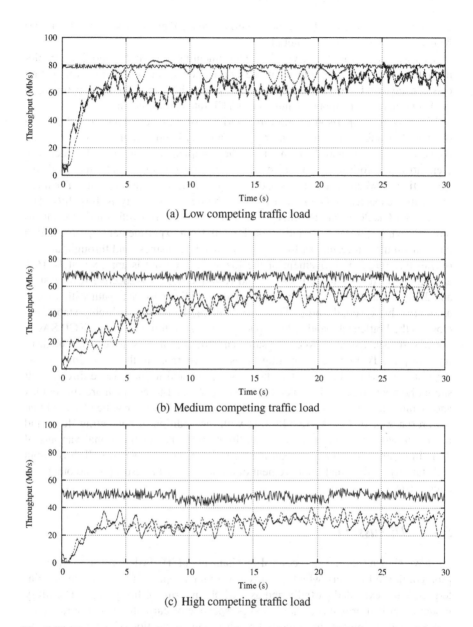

(a) Low competing traffic load

(b) Medium competing traffic load

(c) High competing traffic load

Fig. 5.12 Throughput evolution for a test end-to-end flow in the GREN scenario with low, medium and high competing load. Continuous line: fuzzy TCP; long dashed line: HighSpeed TCP; short dashed line: TCP SACK.

environment. The modified atou tool generates detailed event logs and packet traces with a configurable degree of detail.

This way, a flexible framework for experimenting with highly modified implementations of TCP rate control at the application level is available for testing between any two Internet hosts. Among the options that can be configured, the modified version of atou supports a number of TCP rate control variants.

We show a comparison of the three variants of TCP rate controlers that were compared through simulation. Figures 5.13 and 5.14 compare the evolution of the congestion window, and throughput for three implementation tests. The network scenario is the following: two hosts transfer a file for 30 seconds; the TCP-like connection between the two hosts is established along a 5 hops long path. The maximum link capacity is 1 Gb/s and the path bottleneck capacity is 100 Mb/s. The average RTT is 20 ms approximately. Several tests were run with similar results to those shown in figures 5.13 and 5.14. Results from experimental implementation confirm simulation results as for improvements in robustness and throughput. The overall throughput improvement of the fuzzy TCP variant is approximately 16% and 12% with respect to TCP SACK and HighSpeed TCP, respectively, for tests 1 and 2. Tests 1 and 2 were performed in an almost idle end-to-end path with a small amount of background traffic. Thus, the available bandwidth is highly stable and close to the bottleneck bandwidth capacity. For test 3, four competing TCP SACK flows were established between the same end nodes as the test flow using the thrulay tool [85]. Two of the competing flows are generated in the same direction as the test flow, while the other two flows are generated in the reverse direction. It should be noted that all the links that make up the end-to-end path are full-duplex and symmetrical. In this case (figures 5.13(c) and 5.14(c)), it can be observed that the presence of real background traffic with the addition of competing flows and reverse flows can significantly degrade the performance of traditional variants of TCP. In contrast, the fuzzy variant is able to keep the a higher overall throughput and stability, with a final improvement of arount 50% with respect to the other two alternatives.

5.3.7   Discussion

We have shown that the rate control mechanisms of TCP (the prevalent transport protocol in the Internet), which is currently a major topic of research, can be reinterpreted and extended partially and as a whole in terms of fuzzy logic. The fuzzy model described provides a rule based perspective of currently evolving rate control schemes, being the first reported result in the application of fuzzy systems to intelligent network state inference at end nodes.

In summary, simulation experiments confirm that the fuzzy congestion control scheme proposed here provides significant improvements by better handling the trade-off between different network states, being more adaptive to variations in network conditions. These results were confirmed through an experimental implementation.

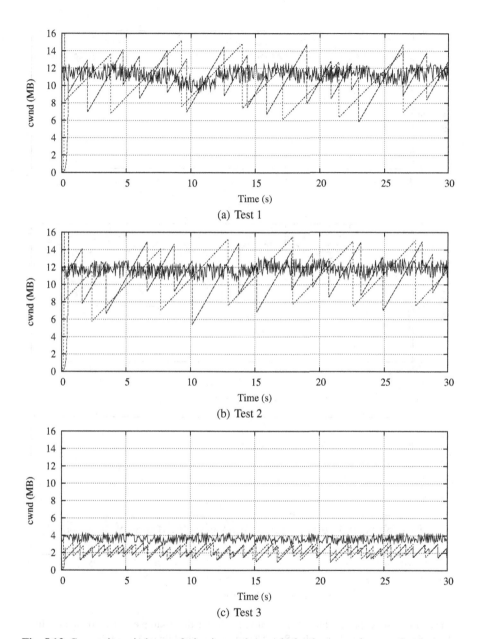

Fig. 5.13 Congestion window evolution in a real scenario for three sample tests. Continuous line: fuzzy TCP; long dashed line: HighSpeed TCP; short dashed: TCP SACK.

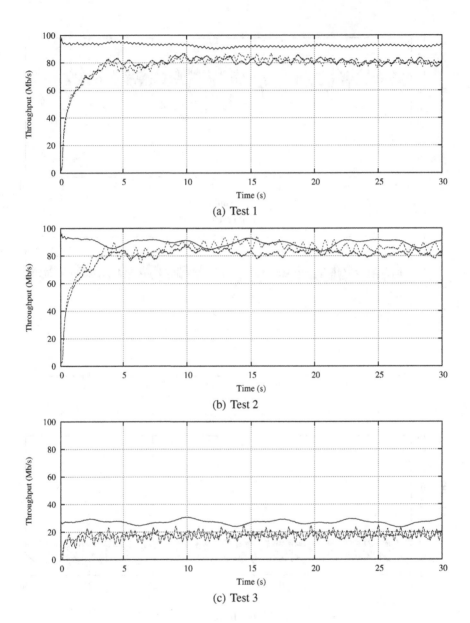

(a) Test 1

(b) Test 2

(c) Test 3

Fig. 5.14 Throughput evolution in a real scenario for three sample tests. Continuous line: fuzzy TCP; long dashed line: HighSpeed TCP; short dashed line: TCP SACK.

Both simulation and implementation results show that the proposed fuzzy extension to TCP rate control can improve performance with regards to a number of criteria, namely, faster convergence to achievable transference rate, higher throughput and reduced oscillations around the stabilized rate for long transfers. The proposed fuzzy system also eases finding compromise solutions for different user requirements. By performing simulation tests, we have analyzed the behavior of the proposal presented here in a realistic scenario in the presence of background traffic with high RTT and bandwidth variability under controled conditions.

We note however that there is still a lot of work to do as extensions to the system described in this section. In particular, regarding the identification of new rules, the exploration of the whole set of possible rules has not been explored. The following subjects are of special interest:

- Application of adjustment and learning techniques on particular scenarios to gain further insight on the system dynamics in these cases.
- Experimentation with a number of extensions that are current topics of research, such as the initial *cwnd* value, fuzzy extensions to RTT and retransmission time-out computation[3], and extensions for intelligent loss-congestion differentiation [77, 22]. This work paves the way and provides tools for experimentation with a number of these mechanisms.

In this work, we have performed tests using both simple and complex scenarios, i.e., in a dumbbell topology as well as in a complex topology modeled after some real networks with thousands of nodes and simultaneous flows. With the complex scenario, we have attempted at generating background traffic with high variability and realistic overall statistical properties.

Altough a fairly complex scenario has been defined, many fundamental aspects in network simulation are still in early stages of development from our viewpoint. In particular, there is the fundamental question of how large an scenario has to be in order to be large enough. This requires further exploration of complex simulation scenarios such as the one described here.

As discussed above, the definition of topologies, traffic patterns, metrics, and evaluation criteria is currently an ongoing effort within the research community. Generation of background traffic is an area of current research where a number of important developments are sought. In particular, whether relative simple models can account for the impact on traffic dynamics that emerges from complex topologies is still an open question [90]. Thus, a great deal of extensions can be considered for further work, while these are mostly related to recent and future developments in the areas of network simulation and emulation, such as measurement based traffic generation methods that can reproduce statistical and application specific patterns from traffic traces [89].

We should note that the end-to-end traffic control scheme proposed here has been designed and studied in standard scenarios. Thus, an area worth to explore as future

---

[3] Current standard retransmission timer computation employs interpolation techniques to smooth variations in measurements through a simple low-pass filter and take into account RTT variance.

work is the application of the scheme proposal in two particular environments of interest: a) high-speed long-distance networks with large available bandwidths, and b) lossy links networks and in particular those with wireless connectivity.

A future research area worth to explore as a natural extension to the approach presented here is the description of end-to-end window based rate control schemes using fuzzy state machines and Markov chains.

Finally, we should mention that regardless of the performance of the proposal for end-to-end traffic control presented here, the main contribution of this work lies in the formulation of a linguistic rule based scheme for window based end-to-end congestion control.

5.4 Active Queue Management by Means of Fuzzy Inference Systems

A major research problem in Internet transport and network layers is the development of traffic regulation mechanisms that can cope with the requirements of a growing diversity of technologies, applications and services. More generally, Internet traffic dynamics is an increasingly complex topic of research [40, 82].

Quality of service requirements as well as traffic patterns of emergent services and applications are difficult to characterize and demand deep advances in current flow and congestion control schemes. Because of the nature of these problems (complexity, no feasible analytic solution as well as incomplete and inaccurate information) the employment of intelligent systems based on fuzzy logic and other soft computing techniques is an appealing alternative.

As discussed in the previous section, currently deployed schemes for traffic regulation in the Internet (as well as proposed alternatives) fit into one of the two following approaches [12]: distributed control and queue controllers in routers.

The complex dynamics that arises from convoluted interactions among end-to-end mechanisms and router queue controllers is difficult to understand and poses many challenges. New algorithms aimed at general deployment in the Internet require to careful consider a number of practical arguments [46], congestion control principles [37] and metrics for the evaluation of congestion control schemes [47].

5.4.1 Approach and Related Work

Around the first half of the 1990s the surge of ATM technologies and proposals for quality of service architectures triggered the interest of the research community in intelligent scheduling architectures and mechanisms for network traffic control. Short after the first applications of computational intelligence and fuzzy inference systems in particular to traffic control appeared [79, 21, 25]. Many proposals have followed on during the 1990s and the present decade [83, 49, 23] mainly focused on different ATM classes of service and related aspects. The core idea behind these proposals is the use of a fuzzy inference system as a controler for packet queues. Among these, we cite [55, 92, 34, 91, 28, 27, 32].

A number of proposals of fuzzy systems for specific areas of network traffic control have been made. Among these, [96] describes fuzzy systems for balancing priorities when considering multiple classes of service within the DiffServ architecture [12]. Results in optimizing flow control in asynchronous transfer mode (ATM) based B-ISDN networks by means of Takagi-Sugeno fuzzy controllers are reported in [24]. In [26] limited simulation results for balancing service rates among classified queues have been reported. Genetic algorithms have been successfully employed for optimizing queue controlers in specific cases [33].

Additional works that deal with the application of fuzzy control to the area of active queue management as well as traffic control can be found in the literature (see for example [28]). Type-2 fuzzy logic systems have been proposed for admission control in ATM networks [66]. Type-2 fuzzy systems have been also proposed for modeling and classification of VBR video traffic [65]. In addition, a general theory of fuzzy systems for queuing control [97, 80] has been developed.

Here we further elaborate on the same idea. First, we consider simplicity as the main design principle. Second, the focus of our work is however on the evaluation over a broad range of scenarios that resemble network nodes with highly aggregated traffic and other key characteristics as closely as possible. Our work differs from previous reported results regarding the following points:

- Simulation is performed through the ns-2 [56], a de facto standard within the Internet research community. Realistic state-of-the-art models of Internet aggregate traffic are taken into consideration.
- We use a design methodology [16] and tool chain [73] for the whole development process that cover from initial high-level description to implementation as software and hardware components. The methodology and tool chain are overviewed in the next section.
- Practical implementation constraints are considered, (current protocols, implementations and technological constraints). In particular, the next chapter will deal with efficient hardware implementations that can achieve the high inference rates required by current and future high performance Internet links [53].

The dominant queue control scheme in the current Internet is the passive FIFO queue without classes of service (known as drop-tail), that discards packets when the storage space is full. Active schemes (known as Active Queue Management, AQM) are however being developed and promoted [12, 96] since AQM mechanisms are required to provide quality of service, differentiate services or penalize misbehaving flows, among other demanded functionalities.

Current Internet routers at core networks process aggregate traffic [58] which typically comprises millions of packets per second as well as millions of active end points and simultaneous flows established by services and applications with an increasing diversity of traffic patterns. Analytically modeling these aggregates at core routers is a challenging task. Furthermore, those scheduling architectures that require state information for all active flows are eventually not deployable at global scale because of its complexity. It is thus assumed that Internet routers schedulers and traffic controlers must cope with a high degree of uncertainty.

Although a number of AQM schemes have been proposed [54], properties of aggregate traffic [58] (such as self-similarity and burstiness at multiple time scales) make it difficult to stabilize packet queues. There are a great deal of challenges in tuning AQM schemes in real environments and no generally accepted solution has been found.

Some works can be found in the literature that use advanced techniques in order to optimize fuzzy inference systems for queue control in routers, including computational intelligence techniques such as genetic algorithms and swarm optimization [34, 76]. These proposals shed some light on to what extent the performance of an active queue management system can be optimized by computational intelligence methods. However, we argue that these optimization methods are applied for a specific network scenario and thus the results cannot be extrapolated to active queue management in general. An optimized controler that shows better performance than its non-optimized counterpart in a particular scenario may however perform poorly in another scenario. In addition, most simulation scenarios widely used throughout the literature are far from resembling real networks in terms of topology and traffic patterns.

In order to define an optimization procedure that guarantees good results in general, it would be necessary to: a) define a set of simulation scenarios representative of the highest possible range of conditions that can be found in the real world and b) define a set of optimization goals that includes all the relevant performance metrics, both link specific and end-to-end. Both a) and b) are complex issues for which some proposals have been developed throughout the last years [42, 47, 39]. The proposals, oriented towards the definition of more generic performance evaluation principles, are however in early stages of development and adoption. In this context, defining a complete and representative set of simulation scenarios and performance metrics for proper optimization of active queue management mechanisms is an ill-posed problem. Rather than off-line optimization methods, the evolving intelligent systems approach [8, 59, 70] seems a better alternative for dynamically adapting neuro-fuzzy systems for AQM in particular scenarios.

To cope with the aforementioned problems, we propose here the notion of FAQM (fuzzy active queue management) and show the usefulness of a simple fuzzy inference based controler in a complex simulation scenario in terms of a set of performance metrics. By means of FAQM we aim at defining a general class of traffic controlers for aggregate traffic that can perform in a flexible and adaptive manner. Given the current stage of development of generic performance evaluation principles for AQM, we leave the optimization of this class of controlers as future work.

Figure 5.15 shows an scheme of a fuzzy inference based controler in an output queue of a typical Internet router. A fuzzy inference system regulates a variable number of packet queues. The fuzzy inference system implements a real-time task with hard deadline, whereas the queue controller periodically retrieves data from the inference system within the hard deadline imposed on the latter. In the most basic

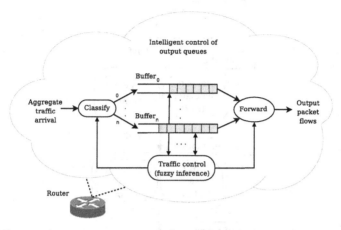

Fig. 5.15 Fuzzy active queue management in Internet routers

scheme, inputs (linguistic variables of the rule base antecedents) are queue sizes as well as their variation whereas the output variable is defined as a probability value or reference to determine which packet should be sent next. This scheme eases the integration of intelligent traffic analysis systems as inputs to the controler.

5.4.2 Development Methodology and Tool Chain

Leveraging on the Xfuzzy [73] CAD suite of tools and a methodology [11, 16] for the development of fuzzy controllers, in this section we describe a methodology and a tool chain tailored for the development of active queue management mechanisms based on fuzzy inference.

The design flow and tool chain employed to develop fuzzy inference modules is depicted in figure 5.16. The whole development process is covered, from initial description to final implementation whether as software or hardware. The first development stage (description) is performed using a high level fuzzy systems specification language, XFL [72], which can be automatically turned into a C software implementation and synthesizable VHDL code, among other implementation options.

The development stages after specification have been tailored for Internet traffic control as follows. For network simulation, we have used ns-2 [56]. ns-2 is an object oriented discrete event driven simulator with support for a vast variety of transport protocols, queuing systems, routing schemes and access media, thus enabling us to evaluate the performance of traffic controlers under complex and realistic simulated scenarios. Fuzzy controlers are integrated into ns-2 as components implemented in C.

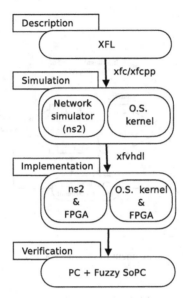

Fig. 5.16 Fuzzy systems design flow and tool chain for active queue control

Verification can be performed over software and hardware implementations of fuzzy controlers. Verification of software implementations is performed over a controler implemented within the networking stack of the kernel of the general purpose operating system of a commodity PC. The steps required for the realization and validation of hardware implementations are discussed in detail in chapter 6. In this section we will analyze the performance of software implementations.

5.4.3 Fuzzy Internet Traffic Control of Aggregate Traffic

Fuzzy Internet traffic controlers can be developed as replacements for traditional traffic controlers proposed for Internet routers. These employ three non-exclusive mechanisms in order to regulate traffic [54, 12]:

- Basic AQM, which attempt to prevent congestion by discarding packets when queues grow.
- Explicit congestion notification (ECN), whereby the controler sends notification packets back to senders and/or intermediate routers in case of congestion.
- Admission control, i.e., filtering of packets that match an admission criteria (such as flow rate and source/destination subnetwork).

Any of these mechanisms imply that some packets are selected to trigger a certain action. Those packets which are selected are said to be marked by the controler. Depending on the protocols and architectures involved, marking can correlate to one or more of the following actions: modify packet headers, discard packet (do not forward to next node in the network), send ECN notifications, and activate filtering rules.

Among the different traffic controllers proposed for Internet routers, the most widely accepted is RED [54], an AQM controller which discards packets so as to enforce end-to-end traffic regulation. Though a number of variants and specializations of RED have been defined, they are generally based on the definition of a queue length threshold and a discard probability value that is proportional to the packet queue length. According to figure 5.15, we propose a fuzzy controller that marks packets the same way as RED controllers, i.e., discarding them. We distinguish buffer size and queue length values following recent developments on the subject [9].

This kind of fuzzy controllers show some similarities to classic real time regulators, such as PD controllers. In a basic setup, the inputs to the fuzzy system are two: the current size of the packet queue and its variation. The fuzzy inference system must produce as output the forwarding decision to apply to the next packet in the queue.

5.4.4 Fuzzy Controler of Best-Effort Aggregate Traffic

This section describes the design of the FAQMBestEffort fuzzy system, which has been developed as a controller for best-effort traffic according to the AQM paradigm. FAQMBestEffort implements a traffic controller for congestion control on routers with no support for classes of service.

In order to keep the system simple a number of restrictions will be applied. These have a twofold implication. First, the system is easy to interpret. Second, it can be realized using optimized hardware implementation techniques, as will be discussed in chapter 6.

Two inputs and one output are defined. An scheme of membership functions for both inputs is shown in figure 5.17, which depicts the fuzzy variable types Te_i and Te_{i-1}. Input e_i is the deviation between the number of currently queued packets and a desired value reference, normalized between 0 and 1. Input e_{i-1} is the deviation at the last time interval, normalized betwen 0 and 1. For both variables, seven linguistic terms are defined ranging from Z to H for increasing queue sizes. The linguistic terms are arranged so that a uniform partition of the input space is performed. Triangular membership functions were chosen for simplicity.

The output of the system, p_i, is defined as a probability value for marking the next packet to be forwarded. In this case, as in AQM schemes currently most accepted within the Internet research community, marking is defined as dropping the packet. An scheme of membership functions for the fuzzy type Tp_i is shown in figure 5.18. 7 linguistic terms are defined ranging from P1 to P7 for increasing levels of probability, uniformly distributed through the output space. For simplicity, the output linguistic terms are defined as singleton fuzzy sets. The rule base is presented in table 5.5 whereas the resulting control surface is depicted in figure 5.19.

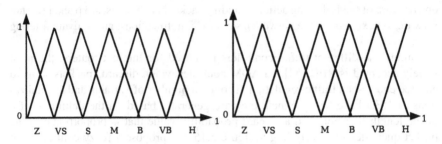

Fig. 5.17 Te_i (left) and Te_{i-1} (right) membership functions

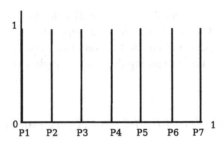

Fig. 5.18 Tp_i membership functions

Table 5.5 FAQMBestEffort rule base

p_i		e_{i-1}						
		Z	VS	S	M	B	VB	H
	Z	P1	P1	P1	P1	P1	P1	P1
	VS	P2	P2	P1	P1	P1	P1	P1
	S	P3	P3	P3	P2	P1	P1	P1
e_i	M	P5	P4	P4	P4	P3	P2	P1
	B	P6	P5	P5	P5	P5	P4	P3
	VB	P7	P6	P6	P6	P6	P6	P5
	H	P7	P7	P7	P7	P7	P7	P7

In order to simplify the definition task and easing the employment of efficient implementation techniques, only triangular and singleton membership functions are used. Fuzzy inference follows a TSK order 0 model, and the fuzzy mean defuzzification method is employed to compute crisp output values.

The choice of the number of linguistic terms for the inputs and the output was performed as follows. Equivalent FAQMBestEffort systems were designed for an

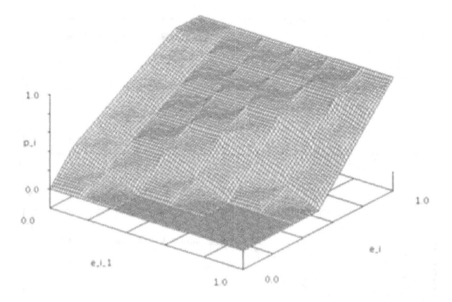

Fig. 5.19 FAQMBestEffort control surface

increasing number of linguistic terms starting from 3. Simulations were run for these systems and the overall application level throughput, was considered as criterion for selecting the best option. We define the application level throughput as the transport level goodput (amount of data transferred minus the amount of data in packets that never reach their destination). Simulations were run in both the dumbbell and GREN scenario for low, medium and high competing load. The definition of these load levels is further described in the next section. Figure 5.20 shows the performance of the FAQMBestEffort as a function of the number of linguistic terms for the inputs and output. The performance is normalized against that of the system with 7 linguistic terms. The system with 7 linguistic terms was chosen as the simplest system that provides almost the best performance observed.

5.4.5 Simulation Results

The objective of this section is to explore by means of simulation the behavior of the proposed active queue management scheme under a wide set of realistic network scenarios and conditions. Our proposal will be compared against the RED algorithm and in particular its Adaptive RED variant as specified in [44] and implemented in the ns-2 simulator.

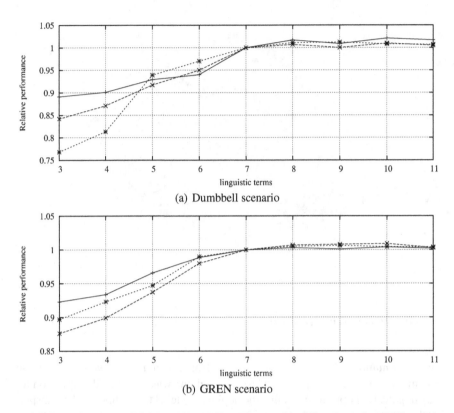

(a) Dumbbell scenario

(b) GREN scenario

Fig. 5.20 Performance of the FAQMBestEffort scheme with different numbers of linguistic terms for the two inputs and the output. The performance index is the overall application level throughput. Three different degrees of competing traffic load are considered. Continuous line: low load, long dashed line: medium load, short dashed line: high load. For each traffic load level, performance is normalized against that of the system with 7 linguistic terms.

A performance evaluation study was conducted on the Adaptive RED and FAQMBestEffort schemes so as to assess the latter and compare both approaches. What follows is a summary of results from the FAQMBestEffort controler as compared to the results from an Adaptive RED system in the two simulation scenarios described in section 5.2: dumbbell and GREN. In the case of the dumbbell scenario, we analyze the queue controler in router 1 for the intermediate link. In the case of the GREN scenario, we analyze a queue controler in the SEV node of the RedIRIS regional network, belonging to the GÉANT pan-European network.

Three different levels of traffic load will be distinguished in the simulations below. Traffic load is generated using UDP traffic sources or TCP SACK traffic sources. All TCP flows implement the TCP Reno congestion control scheme, with the NewReno extension and TCP SACK options. As in the simulation study performed for end-to-end congestion control in the previous section, this variant will be called TCP SACK. For the dumbbell scenario, the three levels of traffic load are defined as follows:

- Low traffic load corresponds to

  - two CBR UDP flows at 10 Mb/s, established between node 1-1 and 2-2 and 1-1 and 2-1, respectively.
  - 4 TCP flows established between nodes 1-1 and 2-2, 1-2 and 2-1, 2-1 and 1-1, and 2-2 and 2-2, respectively.

- Medium competing traffic is generated by duplicating each flow considered in the low traffic load case. In addition, two CBR UDP flows at 10 Mb/s are established between node 2-1 and 1-2, and 2-2 and 1-1, respectively.
- High traffic load is generated following the same scheme as in the medium load case. However, in this case, the load is multiplied by 3. Thus, 12 UDP flows and 24 TCP flows are generated.

An scheme of the interconnections of the SEV node was shown in figure 5.4 (page 199). We analyze the queue for the link between the SEV node and the CAMPUS node, with 622 Mb/s of bandwidth capacity. This particular part of the GREN scenario resembles a campus access node, a typical network configuration where traffic controllers can have a direct impact on overall network performance as experienced by end users [17].

The procedure followed to tailor the generic GREN scenario to evaluating the FAQMBestEffort scheme is as follows:

- TCP flows are generated for low, medium and high traffic load, between end nodes of the Internet2 network of the GREN scenario and nodes in the 4th level of the network being analyzed (these nodes are directly connected to the CAMPUS node).
- For low competing traffic, 10 flows are generated between end nodes belonging to Internet2 and end nodes connected to the CAMPUS node. These flows traverse the SEV-CAMPUS link in addition to the background traffic, i.e., traffic resulting from the random initialization of the GREN scenario. For medium competing traffic load, 50 competing flows are generated. Finally, 100 competing flows are generated for high competing traffic load.

A performance comparison of both Adaptive RED and FAQMBestEffort controlers is presented in what follows in terms of a number of metrics. In particular we look at the packet queue sizes and the application level throughput, or goodput.

Table 5.6 Queue length statistics for the dumbbell scenario

Load	Controler	Mean	Stdv	Max.	Min.	5 pc	95 pc
Low	RED	23.73	9.26	50	0	8.93	39.56
	FAQM	34.36	4.39	46	21	26.81	41.39
Medium	RED	23.72	9.26	50	0	8.92	39.56
	FAQM	34.49	4.67	48	20	26.89	42.06
High	RED	20.75	11.99	50	0	1.00	41.68
	FAQM	34.27	5.78	50	16	24.84	43.50

Table 5.7 Queue length statistics for the GREN scenario

Load	Controler	Mean	Stdv	Max.	Min.	5 pc	95 pc
Low	RED	23.83	10.51	50	0	6.96	41.60
	FAQM	34.27	5.26	50	17	25.69	42.67
Medium	RED	23.95	12.66	50	0	1.01	45.55
	FAQM	34.43	7.03	50	11	22.66	46.16
High	RED	24.15	15.03	50	0	0.25	47.84
	FAQM	34.06	8.61	50	0	19.84	47.79

Figures 5.21 and 5.22 show the evolution of queue length for a period of 30 seconds for the Adaptive RED and the FAQMBestEffort controlers, respectively, in the dumbbell scenario. For all the traffic load conditions considered, queue length oscillations are significantly lower for the FAQMBestEffort controller which also manages to keep a higher mean queue length. Overall statistical properties of the oscillations of both cases are summarized in tables 5.6 and 5.7, which shows mean queue length, standard deviation, maximum peak value, minimum peak value, and 5% and 95% percentiles. It can be seen that the performance improvement provided by the FAQMBestEffort scheme increases with the traffic load.

Figures 5.23 and 5.24 show the evolution of the queue length of the Adaptive RED and FAQMBestEffort schemes for 30 seconds in the GREN scenario. In this case, the differences between Adaptive RED and FAQMBestEffort are higher than in the dumbbell scenario.

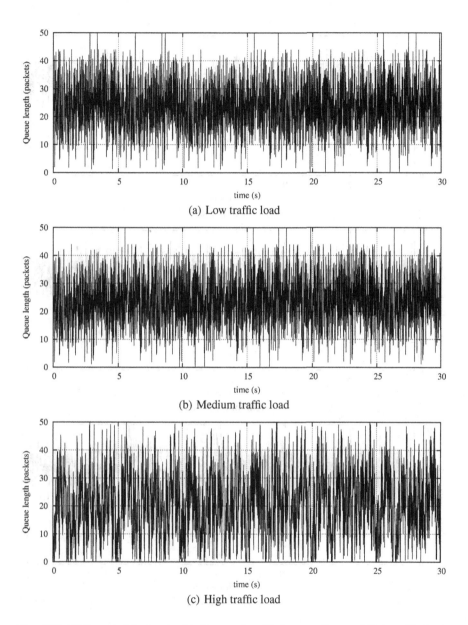

(a) Low traffic load

(b) Medium traffic load

(c) High traffic load

Fig. 5.21 RED control in the dumbbell scenario with low, medium and high traffic load. Queue length evolution for 30 seconds.

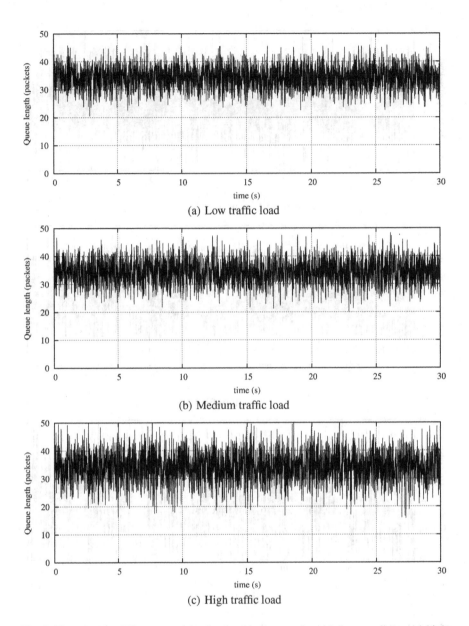

(a) Low traffic load

(b) Medium traffic load

(c) High traffic load

Fig. 5.22 FAQMBestEffort control in the dumbbell scenario with low, medium and high traffic load. Queue length evolution for 30 seconds.

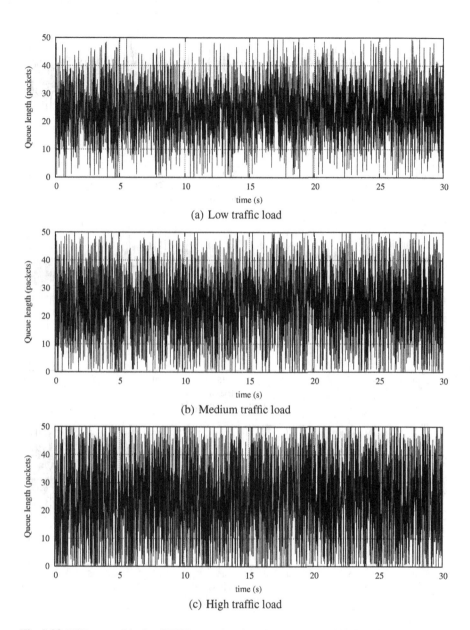

(a) Low traffic load

(b) Medium traffic load

(c) High traffic load

Fig. 5.23 RED control in the GREN scenario with low, medium and high traffic load. Queue length evolution for 30 seconds.

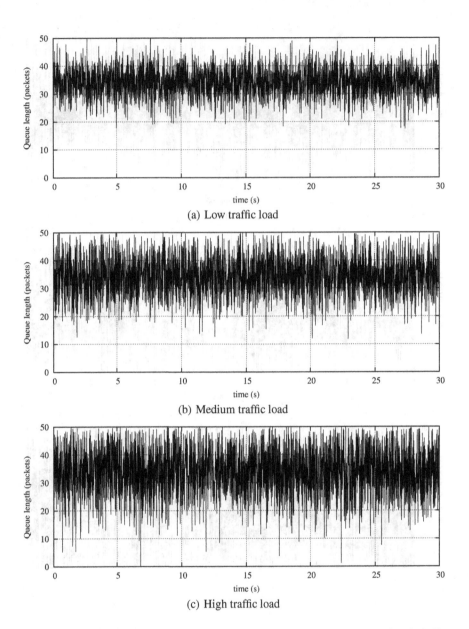

Fig. 5.24 FAQMBestEffort control in the GREN scenario with low, medium and high traffic load. Queue length evolution for 30 seconds.

The higher stability of FAQMBestEffort can also be seen in figures 5.25 and 5.26, which show the evolution of the output (marking value) for both AQM schemes in the dumbbell scenarios. The same applies to the simulations performed in the GREN scenario (figures 5.27 and 5.28).

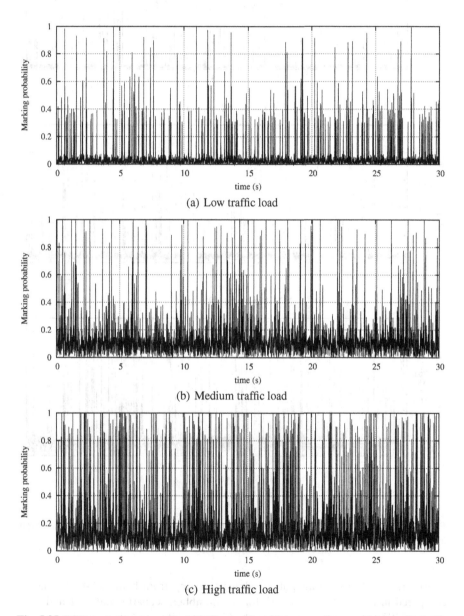

(a) Low traffic load

(b) Medium traffic load

(c) High traffic load

Fig. 5.25 RED controler output. Dumbbell scenario with low, medium and high traffic load. Marking probability evolution with time for 30 seconds.

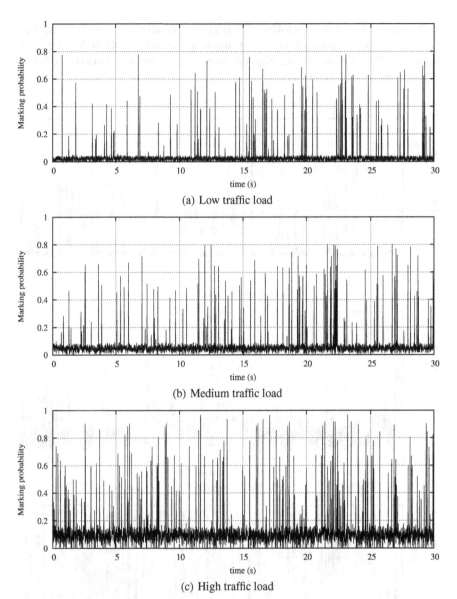

Fig. 5.26 FAQMBestEffort controler output. Dumbbell scenario with low, medium and high traffic load. Marking probability evolution with time for 30 seconds.

Application level throughput (goodput) resulting from both AQM schemes is compared in figures 5.29 and 5.30 for the dumbbell scenario, and 5.31 and 5.32 for the GREN scenario. As expected, the higher stability and mean queue length in the FAQMBestEffort case lead to higher goodput.

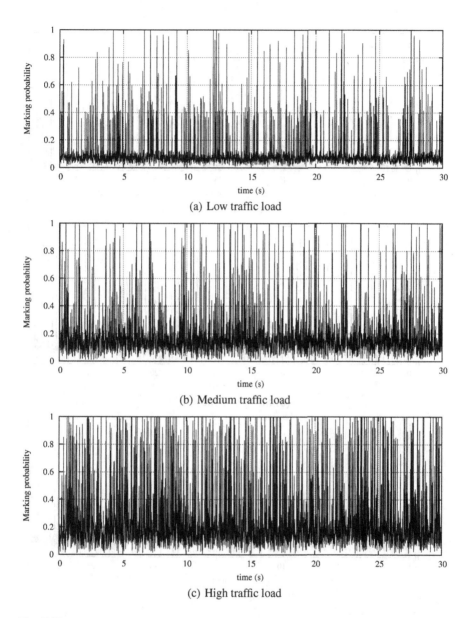

(a) Low traffic load

(b) Medium traffic load

(c) High traffic load

Fig. 5.27 RED controler output. GREN scenario with low, medium and high traffic load. Marking probability evolution with time for 30 seconds.

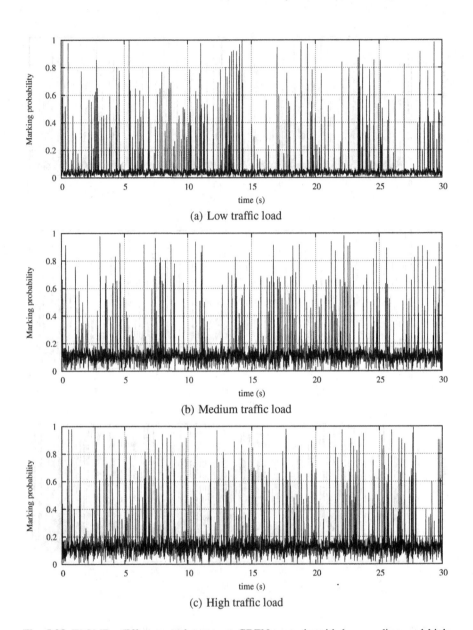

(a) Low traffic load

(b) Medium traffic load

(c) High traffic load

Fig. 5.28 FAQMBestEffort controler output. GREN scenario with low, medium and high traffic load. Marking probability evolution with time for 30 seconds.

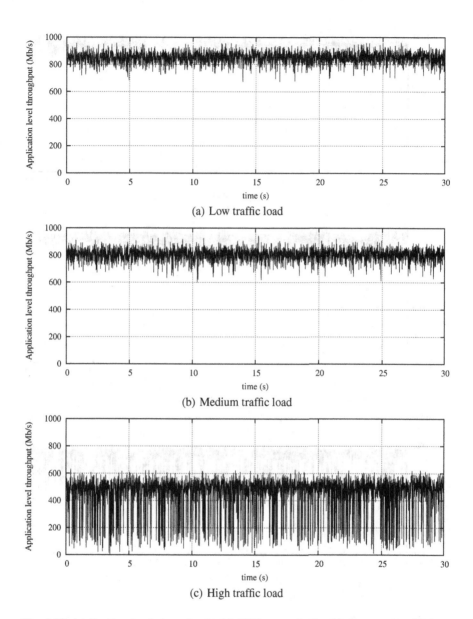

(a) Low traffic load

(b) Medium traffic load

(c) High traffic load

Fig. 5.29 Application level throughput with RED control. Dumbbell scenario with low, medium and high traffic load.

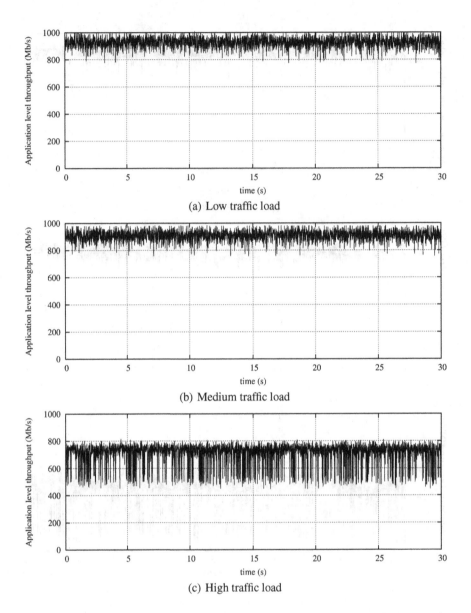

Fig. 5.30 Application level throughput with FAQMBestEffort control. Dumbbell scenario with low, medium and high traffic load.

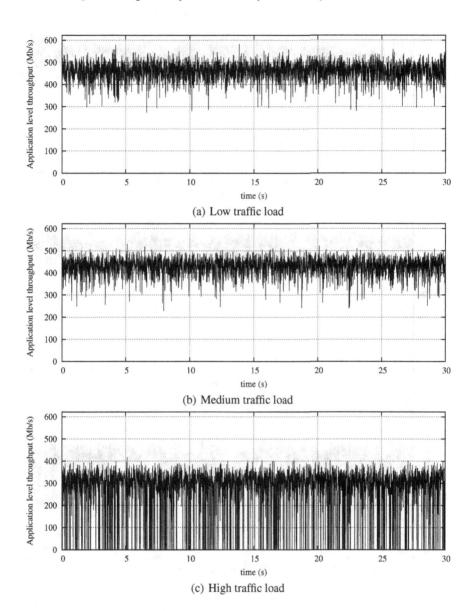

(a) Low traffic load

(b) Medium traffic load

(c) High traffic load

Fig. 5.31 Application level throughput with RED control. GREN scenario with low, medium and high traffic load.

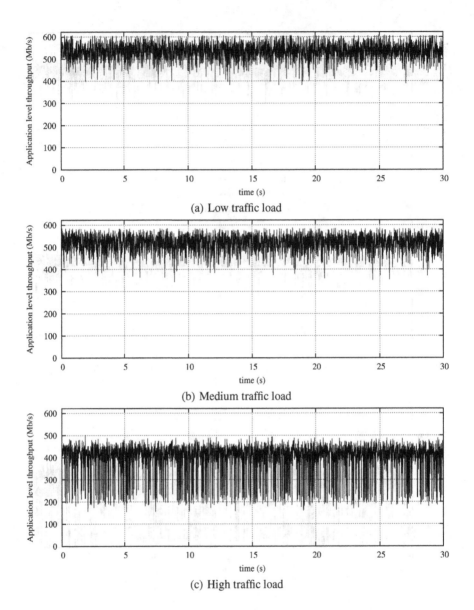

(a) Low traffic load

(b) Medium traffic load

(c) High traffic load

Fig. 5.32 Application level throughput with FAQMBestEffort control. GREN scenario with low, medium and high traffic load.

Overall statistics of the simulations are summarized in tables 5.8 and 5.9 for the dumbbell and GREN scenarios, respectively. The tables show the following statistics for the application level throughput: average, standard deviation, maximum peak value, minimum peak value, and 5% and 95% percentiles.

Table 5.8 Application level throughput statistics for the dumbbell scenario

Load	Controler	Mean	Stdv	Max.	Min.	5 pc	95 pc
Low	RED	850.4	43.16	960.9	670.6	775.0	916.5
	FAQM	926.1	38.59	999.9	774.5	859.5	985.1
Medium	RED	803.3	46.10	942.9	600.9	724.3	870.0
	FAQM	907.8	39.26	982.9	757.5	838.4	966.9
High	RED	454.0	139.1	631.7	2.109	99.65	578.1
	FAQM	709.8	83.82	815.2	446.2	498.0	785.5

Table 5.9 Application level throughput statistics for the GREN scenario

Load	Controler	Mean	Stdv	Max.	Min.	5 pc	95 pc
Low	RED	459.6	42.31	582.5	274.3	383.6	521.7
	FAQM	534.0	39.0	607.9	382.5	463.3	592.3
Medium	RED	428.9	41.21	531.2	228.1	353.3	485.3
	FAQM	518.3	40.43	584.8	342.7	441.8	573.9
High	RED	273.7	109.7	417.6	0.0	23.1	370.7
	FAQM	397.0	77.26	501.4	154.6	203.7	467.6

5.4.6 Implementation Results

The objective of this section is to analyze the behavior of implementations of RED and FAQMBestEffort in an emulated scenario. In particular, we will analyze the dumbbell scenario. The tests were performed using the NIST Net router [20] in order to emulate the dumbbell topology described before. Traffic load was generated using the thrulay tool [85]. Three levels of traffic load, low, medium and high, were defined in the same manner as in the simulation tests.

Here we show tests performed with software implementations of the FAQMBest-Effort and Adaptive RED integrated in a network emulator, namely the NIST Net software router and network emulation package [19, 20] running on the GNU/Linux operating system in a commodity PC. The problem of performing emulation tests with hardware implementations of fuzzy controlers will be addressed in chapter 6.

FAQMBestEffort was integrated into the NIST Net routing system by applying a table look-up procedure on a table of previously generated values of the output of the FAQMBestEffort inference system. This way, at the expense of higher memory usage, it is possible to emulate queues for bandwidth capacities of the order of the Gb/s, which would not be feasible with a functional software implementation as provided by the xfc tool of the Xfuzzy environment.

Figures 5.33 and 5.34 show the evolution of the packet queue length for a period of 30 seconds for the Adaptive RED and the FAQMBestEffort active queue management schemes, respectively, in the emulated dumbbell network. Despite the differences as compared to the simulation tests, the overall same behavior can be observed for both schemes. Again, the packet queue length oscillates less with the FAQMBestEffort scheme and the advantages of FAQMBestEffort as compared to Adaptive RED become more evident as the global traffic load increases.

Application level throughput (goodput) resulting from both schemes is compared in figures 5.35 and 5.36. As in the simulation tests, the higher stability and average queue length in the FAQMBestEffort case lead to a performance improvement that increases with overall load.

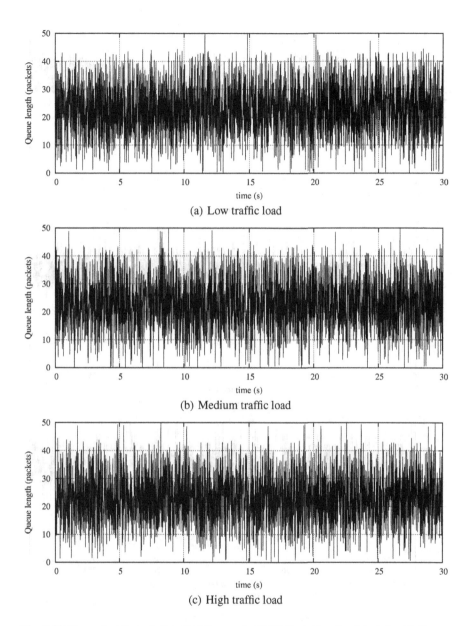

(a) Low traffic load

(b) Medium traffic load

(c) High traffic load

Fig. 5.33 Queue length evolution for 30 seconds. RED Controler. Emulated dumbbell scenario with low, medium and high traffic load.

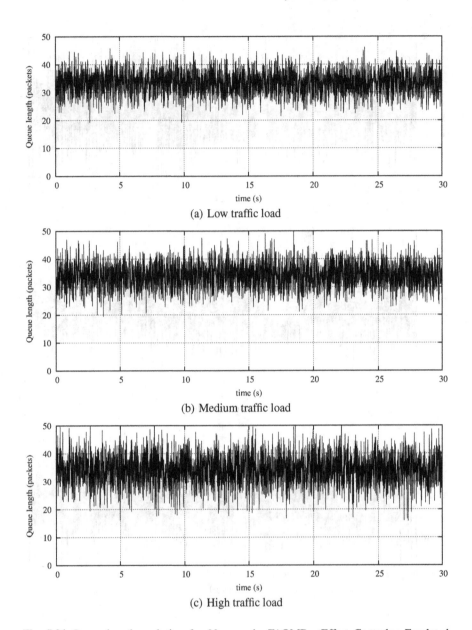

(a) Low traffic load

(b) Medium traffic load

(c) High traffic load

Fig. 5.34 Queue length evolution for 30 seconds. FAQMBestEffort Controler. Emulated dumbbell scenario with low, medium and high traffic load.

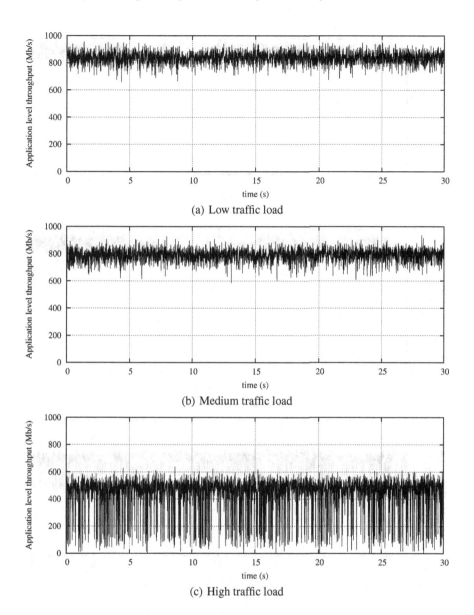

(a) Low traffic load

(b) Medium traffic load

(c) High traffic load

Fig. 5.35 Application level throughput with RED Controler. Emulated dumbbell scenario with low, medium and high traffic load.

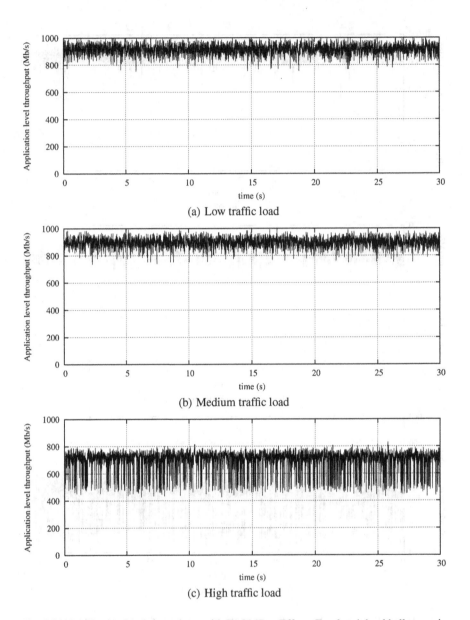

(a) Low traffic load

(b) Medium traffic load

(c) High traffic load

Fig. 5.36 Application level throughput with FAQMBestEffort . Emulated dumbbell scenario with low, medium and high traffic load.

5.4.7 Discussion

We have shown the results of a comparative evaluation of the proposed FAQMBest-Effort controler considering a number of characteristics of real traffic patterns as well as a complex topology. It has been argued that defining a complete and representative set of simulation scenarios and performance evaluation metrics for optimizing an AQM scheme is currently an ill-posed problem. Thus, the optimization of the FAQMBestEffort controler for specific conditions has been left as subject of future research.

The focus of this section has been on the evaluation of a simple controler that can be regarded as a basic and general from of a class of fuzzy inference systems for active queue management. It is suggested that rather than off-line optimization methods, on-line methods such as evolving neuro-fuzzy systems should be explored. However, this would require an in depth analysis of how to use available information in routers to evolve the active queue control system in such a way that the performance in terms of measurements not directly known is optimized.

We explored the performance of FAQMBestEffort and RED through simulation experiments and then confirmed the correct behavior of the fuzzy inference based proposal on a real emulated scenario. Results from emulation tests matched those obtained from simulation tests except for small numeric variations due to implementation details. As overall conclusion, FAQMBestEffort shows a higher robustness in the presence of bursty traffic and outperforms RED for both bulk transfer and real-time traffic, showing better performance in terms of queue length and stability, link utilization, as well as impact on end-to-end delay and jitter.

The results summarized above imply that end users will experience a higher mean delay in the FAQMBestEffort. The delay increase is nevertheless negligible for current network technologies and indeed generally lower than 5% of the overall end-to-end delay in our simulation scenario. On the other hand, FAQMBestEffort improves peak delay values as compared to those of the RED controler by approximately 50%, which implies a significant improvement in end-to-end jitter. Thus, FAQMBestEffort performance is better for best effort traffic while the benefits it introduces for real-time traffic clearly outperforms the hardly noticeable mean delay increase.

We note that, although developed for bulk transfer traffic with no time constraints, the higher degree of robustness and responsiveness to packet bursts shown by FAQMBestEffort leads to an improvement in end-to-end performance as experienced by time constrained services and applications. FAQMBestEffort is thus a practical compromise solution for currently deployed routers.

We also note that small variations of the fuzzy controler (mostly through membership functions shifts) can provide results suited for specialized traffic. It is thus feasible to develop controlers tailored for particular applications and traffic patterns. These controlers can be combined in an intelligent manner within class of service enabled infrastructures.

The FAQM scheme for controling aggregate traffic in Internet routers has been described. Within the FAQM scheme, results from FAQMBestEffort show a higher robustness than traditional traffic controlers in the presence of self-similar bursty traffic and outperforms RED results for both bulk transfer and real-time traffic, showing better performance in terms of queue length, stability, utilization, and delay variability, among other parameters.

5.5 Conclusions

Here we have presented two proposals for performing end-to-end flow and congestion control as well as active queue management in routers by means of fuzzy inference systems.

We first defined complex network scenarios in order to simulate network control mechanisms under realistic conditions. These scenarios have been shown to be substantially different to those commonly used throughout the literature to design and optimize computational intelligence based traffic control methods.

Then, we designed and tested a new paradigm for performing end-to-end flow and congestion control in a TCP-compatible manner by means of fuzzy rule based systems. This new approach was tested and compared against TCP Reno and High-Speed TCP. Simulation and implementation results showed that our proposal can improve overall performance under a wide set of conditions.

In addition, an scheme for active queue management in routers using fuzzy logic control was presented. As opposed to previously proposed controlers, the focus here was on the testability and performance on realistic and varied scenarios. This control scheme was tested and compared against RED and drop-tail schemes. Simulations showed that that the proposed controler has better performance in common conditions and can achieve satisfactory performance under a wider set of conditions than the alternatives compared. An implementation on emulated scenarios confirmed the simulation results.

The design of the rule based schemes developed for both end-to-end congestion control and active queue management in this chapter enables tailoring these mechanisms to meet specific application requirements in a linguistic manner. This makes it possible, for instance, to gradually adapt congestion control functions for services with varying degrees of end-to-end elasticity.

References

[1] Abraham, A., Smith, K., Jain, R., Jain, L.: Network and information security: A computational intelligence approach (editorial). Journal of Network and Computer Applications 30(1), 1–3 (2007)
[2] Aikat, J., Kaur, J., Smith, F.D., Jeffay, K.: Variability in TCP Round-trip Times. In: 3rd ACM SIGCOMM Conference on Internet Measurement, Miami Beach, FL, USA, pp. 279–284 (2003)
[3] Allman, M.: TCP Congestion Control with Appropriate Byte Counting (ABC). RFC 3465, Internet Engineering Task Force, Network Working Group, category: Experimental (2003)

[4] Allman, M., Paxson, V., Stevens, W.R.: TCP Congestion Control. RFC 2581, Internet Engineering Task Force, Network Working Group, status: Proposed Standard (1999)

[5] Allman, M., Floyd, S., Partridge, C.: Increasing TCP's Initial Window. RFC 3990, Internet Engineering Task Force, Network Working Group, status: Proposed Standard (2002)

[6] Andrew, L., Floyd, S.: Common TCP Evaluation Suite. Tech. rep., Internet Engineering Task Force, Network Working Group, Internet Draft. Intended status: Best Current Practice (2008)

[7] Andrew, L., Marcondes, C., Floyd, S., Dunn, L., Guillier, R., Gang, W., Eggert, L., Ha, S., Rhee, I.: Towards a Common TCP Evauation Suite. In: 6th International Workshop on Protocols for Fast Long-Distance Networks, Manchester, UK (2008)

[8] Angelov, P., Filev, D., Kasabov, N. (eds.): Evolving Intelligent Systems. Methodoloy and Applications. IEEE Series on Computational Intelligence. John Wiley & Sons/IEEE Press (2010)

[9] Appenzeller, G., Keslassy, I., McKeown, N.: Sizing Router Buffers. In: ACM Special Interest Group on Data Communications (SIGCOMM) Conference, Portland, OR, USA, pp. 281–292 (2004)

[10] Banerjee, S., Griffin, T.G., Pias, M.: The Interdomain Connectivity of PlanetLab Nodes. In: Barakat, C., Pratt, I. (eds.) PAM 2004. LNCS, vol. 3015, pp. 73–82. Springer, Heidelberg (2004)

[11] Barriga, A., Sánchez-Solano, S., Brox, P., Cabrera, A., Baturone, I.: Modelling and implementation of fuzzy systems based on VHDL. International Journal of Approximate Reasoning 41(2), 164–178 (2006)

[12] Blake, S., Black, D.L., Carlson, M.A., Davies, E., Wang, Z., Weiss, W.: An Architecture for Differentiated Services. RFC 2475, Internet Engineering Task Force, Network Working Group, category: Informational (1998)

[13] Brakmo, L., Peterson, L.: TCP Vegas: End to End Congestion Avoidance on a Global Internet. IEEE Journal on Selected Areas in Communications 13(8), 1465–1480 (1995)

[14] Broido, A., Hyun, Y., Gao, R., Claffy, K.C.: Their Share: Diversity and Disparity in IP Traffic. In: 5th Passive and Active Measurement Workshop (PAM), Antibes Juan-Les-Pins, France, pp. 113–125 (2004)

[15] Brownlee, N., Claffy, K.C.: Understanding Internet Traffic Streams: Dragonflies and Tortoises. IEEE Communications Magazine 40(10), 110–117 (2002)

[16] Cabrera, A., Sánchez-Solano, S., Brox, P., Barriga, A., Senhadji, R.: Hardware/ Software Codesign of Configurable Fuzzy Control Systems. Applied Soft Computing 4(3), 271–285 (2004)

[17] Calyam, P., Krymskiy, D., Sridharan, M., Schopis, P.: Active and Passive Measurements on Campus, Regional and National Network Backbone Paths. In: 14th IEEE International Conference on Computer Communications and Networks (ICCCN 2005), San Diego, California, USA, pp. 537–542 (2005)

[18] Carbonell, P., Zhong-Ping, J., Panwar, S.: Fuzzy TCP: A Preliminary Study. In: 15th IFAC World Congress, Barcelona, Spain, pp. 21–26 (2002)

[19] Carson, M., Santay, D.: NIST Net: a Linux-based network emulation tool. ACM SIGCOMM Computer Communication Review 33(3), 111–126 (2003)

[20] Carson, M., Santay, D.: NIST Net Network Emulation Package. NIST Internetworking Technology Group. National Institute of Standards and Technology (2008), http://www-x.antd.nist.gov/nistnet/

[21] Catania, V., Ficili, G., Palazzo, S., Panno, D.: Using Fuzzy Logic in ATM Source Traffic Control: Lessons and Perspectives. IEEE Communications Magazine, 70–81 (1996)

[22] Chang, L., Marsic, I.: Fuzzy Reasoning for Wireless Awareness. International Journal of Wireless Information Networks 8(1), 15–26 (2001)

[23] Chen, B.S., Peng, S.C.: Traffic Modeling, Prediction, and Congestion Control for High-Speed Networks: A Fuzzy AR Approach. IEEE Transactions on Fuzzy Systems 8(5), 491–508 (2000)

[24] Chen, B.S., Yang, Y.S., Lee, B.K., Lee, T.H.: Fuzzy Adaptive Predictive Flow Control of ATM Network Traffic. IEEE Transactions on Fuzzy Systems 11(4), 568–581 (2003)

[25] Cheng, R.G., Chang, C.J.: Design of a fuzzy traffic controller for ATM network. IEEE/ACM Transactions on Networking 4(3), 460–469 (1996)

[26] Cho, H.C., Fadali, M.S., Lee, H.: Dynamic Queue Scheduling Using Fuzzy Systems for Internet Routers. In: IEEE International Conference on Fuzzy Systems (FUZZ-IEEE 2005), Reno, USA, pp. 471–476 (2005)

[27] Chrysostomou, C., Pitsillides, A., Hadjipollas, G., Sekercioglu, A., Polycarpou, M.: Fuzzy Explicit Marking for Congestion Control in Differentiated Services Networks. In: 8th IEEE Symposium on Computers and Communications, Antalaya, Turkey, vol. 1, pp. 312–319 (2003)

[28] Chrysostomou, C., Pitsillides, A., Rossides, L., Sekercioglu, A.: Fuzzy Logic Controlled RED: Congestion Control in TCP/IP Differentiated Services Networks. Soft Computing Journal 8(20), 79–92 (2003)

[29] Claffy, K.C., Monk, T.E., McRobb, D.: Internet Tomography. Nature Web Matters (1999), http://www.nature.com/nature/webmatters/tomog/tomog.html

[30] Crovella, M.E., Bestavros, A.: Self-Similarity in World Wide Web Traffic: Evidence and Possible Causes. IEEE/ACM Transactions on Networking 5(6), 835–846 (1997)

[31] Crovella, M.E., Taqqu, M.S., Bestabros, A.: Heavy-tailed Probability Distributions in the Worl Wide Web. In: Adler, R., et al. (eds.) A Practical Guide to Heavy Tails: Statistical Techniques and Applications, pp. 3–25. Birkhauser, Boston (1998)

[32] Tsang, D.H.K., Bensaou, B., Lam, S.T.C.: Fuzzy-Based Rate Control for Real-Time MPEG Video. IEEE Transactions on Fuzzy Systems 4(4), 504–516 (1998)

[33] Di Fatta, G., Lo Re, G., Urso, A.: A Fuzzy Approach for the Network Congestion Problem. In: Sloot, P.M.A., Tan, C.J.K., Dongarra, J., Hoekstra, A.G. (eds.) ICCS-ComputSci 2002. LNCS, vol. 2329, pp. 286–295. Springer, Heidelberg (2002)

[34] Di Fatta, G., Hoffmann, F., Re, G.L., Urso, A.: A Genetic Algorithm for the Design of a Fuzzy Controller for Active Queue Management. IEEE Transactions on Systems, Man and Cybernetics, Part C: Applications and Reviews 33(3), 313–334 (2003)

[35] Dickerson, J.E., Juslin, J., Koukousoula, O., Dickerson, J.: Fuzzy intrusion detection. In: IFSA World Congress and 20th North American Fuzzy Information Processing Society (NAFIPS) International Conference, Vancouver, British Columbia, Canada, vol. 3, pp. 1506–1510 (2001)

[36] Dunigan, T., Fowler, F., et al.: Almost TCP over UDP (atou). Oak Ridge National Laboratory, U.S Department of Energy (2006)

[37] Floyd, S.: Congestion Control Principles. RFC 2914, Internet Engineering Task Force, Network Working Group, category: Best Current Practice (2000)

[38] Floyd, S.: HighSpeed TCP for Large Congestion Windows. RFC 3649, Internet Engineering Task Force, Network Working Group, category: Experimental (2003)

[39] Floyd, S., Allman, M.: Specifying New Congestion Control Algorithms. RFC 5033, Internet Engineering Task Force, Network Working Group, category: Best Current Practice (2007)

[40] Floyd, S., Kohler, E.: Internet Research Needs Better Models. ACM SIGCOMM Computer Communication Review 33(1), 29–34 (2003)

[41] Floyd, S., Kohler, E.: Tools for the Evaluation of Simulation and Testbed Scenarios. Tech. rep., Internet Engineering Task Force, Network Working Group, Internet Draft. Intended status: Best Current Practice (2008)

[42] Floyd, S., Paxson, V.: Difficulties in Simulating the Internet. IEEE/ACM Transactions on Networking 9(4), 392–403 (2001)

[43] Floyd, S., Mathis, M., Mahdavi, J., Podolsky, M.: TCP Selective Acknowledgment Options. Tech. Rep, Internet Engineering Task Force, Network Working Group, status: Proposed Standard (2000)

[44] Floyd, S., Gummadi, R., Shenker, S.: Adaptive RED: An Algorithm for Increasing the Robustness of RED. Tech. rep., ICIR, ICSI Center for Internet Research (2001), http://www.icir.org/floyd/adaptivered/

[45] Floyd, S., Henderson, T., Gurtov, A.: The NewReno Modification to TCP's Fast Recovery Algorithm. RFC 3782, Internet Engineering Task Force, Network Working Group, status: Proposed Standard (2004)

[46] Floyd, S., Kempf, J., et al.: Concerns Regarding Congestion Control for Voice Traffic in the Internet. RFC 3714, Internet Engineering Task Force, Network Working Group, category: Informational (2004)

[47] Floyd, S., et al.: Metrics for the Evaluation of Congestion Control Mechanisms. RFC 5166, Internet Engineering Task Force, Network Working Group, category: Informational (2008)

[48] Fraleigh, C., Moon, S., Lyles, B., Cotton, C., Khan, M., Moll, D., Rockell, R., Seely, T., Diot, C.: Packet-level traffic measurements from the sprint IP backbone. IEEE Network 17(6), 6–16 (2003)

[49] Ghosh, S., Razouqi, Q., Schumacher, H.J., Celmins, A.: A Survey of Recent Advances in Fuzzy Logic in Telecommunications Networks and New Challenges. IEEE Transactions on Fuzzy Systems 6(39), 443–447 (1998)

[50] Grieco, L.A., Mascolo, S.: Performance evaluation and comparison of Westwood+, New Reno and Vegas TCP congestion control. ACM SIGCOMM Computer Communication Review 34(2), 25–38 (2004)

[51] Ha, S., Rhee, I., Xu, L.: CUBIC: a new TCP-friendly high-speed TCP variant. ACM SIGOPS Operating Systems Review 42(5), 64–74 (2008)

[52] Handley, M., Floyd, S., Padhye, J., Widmer, J.: TCP Friendly Rate Control (TFRC): Protocol Specification. RFC 3448, Internet Engineering Task Force, Network Working Group, status: Proposed Standard (2003)

[53] Hidell, M., Sjödin, P., Hagsand, O.: Control and Forwarding Plane Interaction in Distributed Routers. Tech. Rep. TRITA-S3-LCN-0501, Laboratory for Communication Networks, Department of Signals, Sensors, and Systems. KTH Royal Institute of Technology, Stockholm, Sweden (2005)

[54] Hollot, C., Misra, V., Towsley, D., Gong, W.: Analysis and Design of Controllers for RED Routers Supporting TCP Flows. IEEE Transactions on Automatic Control 47, 945–959 (2002)

[55] Hu, R.Q., Petr, D.W.: A Predictive Self-Tuning Fuzzy-Logic Feedback Rate Controller. IEEE/ACM Transactions on Networking 8(6), 697–709 (2000)

[56] Information Sciences Institute University of Southern California, Viterbi School of Engineering, The Network Simulator – ns-2 (2008), http://www.isi.edu/nsnam/ns/

[57] Jacobson, V., Karels, M.J.: Congestion Avoidance and Control. In: ACM Computer Communication Review SIGCOMM 1988 Symposium: Communications Architectures and Protocols, vol. 18(4), pp. 314–329 (1988)

[58] Karagiannis, T., Molle, M., Faloutsos, M., Broido, A.: A Nonstationary Poisson View of Internet Traffic. In: 23th Annual Joint Conference of the IEEE Computer and Communications Societies (IEEE INFOCOM), Hong Kong, vol. 3, pp. 1558–1569 (2004)

[59] Kasabov, N.: Evolving Connectionist Systems: The Knowledge Engineering Approach, 2nd edn. Springer, Heidelberg (2007)

[60] Kelly, T.: Scalable TCP: Improving Performance in Highspeed Wide Area Networks. ACM Communication Review 32(2), 83–91 (2003)

[61] Keshav, S.: An Engineering Approach to Computer Networking: ATM Networks, the Internet, and the Telephone Network. Computing Series. Addison-Wesley Longman Publishing Co., Inc., Amsterdam (1997) ISBN: 978-0201634426

[62] Krioukov, D., Chung, F., Claffy, K.C., Fomenkov, M., Vespignani, A.: Willinger The Workshop on Internet Topology (WIT) Report. ACM SIGCOMM Computer Communication Review 37(1), 69–73 (2007)

[63] Krishnamurthy, V., Faloutsos, M., Chrobak, M., Cui, J.H., Lao, L., Percus, A.G.: Sampling large internet topologies for simulation purposes. Computer Networks 51(15), 4284–4302 (2007)

[64] Leith, D., Shorten, R.: H-TCP: TCP for High-Speed and Long-Distance Networks. In: Second International Workshop on Protocols for Fast Long-Distance Networks, PFLDNet (2004)

[65] Liang, Q., Mendel, J.M.: MPEG VBR Video Traffic Modeling and Clasification Using Fuzzy Techniques. IEEE Transactions on Fuzzy Systems 9(1), 183–193 (2001)

[66] Liang, Q., Karnik, N., Mendel, J.M.: Connection Admission Control in ATM Networks Using Survey-Based Type-2 Fuzzy Logic Systems. IEEE Transactions on Systems, Man and Cybernetics Part C: Applications and Reviews 30(3), 329–339 (2000)

[67] Mahadevan, P., Krioukov, D., Fomenkov, M., Huffaker, B., Claffy, X.D.K.C., Vahdat, A.: The internet as-level topology: Three data sources and one definitive metric. ACM SIGCOMM Computer Communication Review 26(1), 17–26 (2006)

[68] Mathis, M., Mahdavi, J., Floyd, S., Romanow, A.: Tcp selective acknowledgment options. Tech. Rep, Internet Engineering Task Force, Network Working Group, status: Proposed Standard (1996)

[69] Medina, A., Allman, M., Floyd, S.: Measuring the Evolution of Transport Protocols in the Internet. ACM SIGCOMM Computer Communication Review 35(2), 37–52 (2005)

[70] Montesino Pouzols, F., Lendasse, A.: Evolving fuzzy optimally pruned extreme learning machine for regression problems. Evolving Systems 1(1), 43–58 (2010)

[71] Montesino-Pouzols, F., Lopez, D.R., Barriga, A., Sánchez-Solano, S.: Fuzzy End-to-End Rate Control for Internet Transport Protocols. In: 15th IEEE International Conference on Fuzzy Systems (FUZZ-IEEE 2006), Vancouver, Canada, pp. 1347–1354 (2006)

[72] Moreno-Velo, F., Sánchez-Solano, S., Barriga, A., Baturone, I., López, D.: XFL3: a New Fuzzy System Specification Language. In: 5th WSEAS/IEEE Multiconference on Circuits, Systems, Communications and Computers (CSCC 2001), Rethymon, pp. 361–366 (2001)

[73] Moreno-Velo, F.J., Baturone, I., Sánchez-Solano, S., Barriga, A.: Rapid Design of Fuzzy Systems With Xfuzzy. In: 12th IEEE International Conference on Fuzzy Systems (FUZZ-IEEE 2003), St. Louis, MO, USA, pp. 342–347 (2003)

[74] Moreno-Velo, F.J., Baturone, I., Barriga, A., Sánchez-Solano, S.: Automatic Tuning of Complex Fuzzy Systems with Xfuzzy. Fuzzy Sets and Systems 158(18), 2026–2038 (2007)

[75] Nagle, J.: Congestion Control in IP/TCP Internetworks. RFC 896, Internet Engineering Task Force, Network Working Group (1984)

[76] Nyirenda, C.N., Dawoud, D.S.: Multi-objective particle swarm optimization for fuzzy logic based active queue management. In: 15th IEEE International Conference on Fuzzy Systems (FUZZ-IEEE 2006), Vancouver, Canada, pp. 2231–2238 (2006)

[77] de Oliveira, R., Braun, T.: A Delay-based Approach Using Fuzzy Logic to Improve TCP Error Detection in Ad Hoc Networks. In: IEEE Wireless Communications and Networking Conference, Atlanta, USA (2004)

[78] Papagiannaki, K., Veitch, D., Hohn, N.: Origins of Microcongestion in an Access Router. In: Barakat, C., Pratt, I. (eds.) PAM 2004. LNCS, vol. 3015, pp. 175–184. Springer, Heidelberg (2004)

[79] Pedrycz, W., Vasilakos, A.V. (eds.): Computational Intelligence in Telecommunications Networks. CRC Press, Boca Raton (2001) ISBN: 1420040952

[80] Phillis, Y.A., Zhang, R.: Fuzzy Service Control of Queueing Systems. IEEE Trans Systems, Man, and Cybernetics 29(4), 503–517 (1999)

[81] Postel, J.: Transmission Control Protocol. RFC 793, Information Sciences Institute. University of Southern California, status: Proposed Standard (1981)

[82] Rolls, D., Michailidis, G., Hernández-Campos, F.: Queueing Analysis of Network Traffic: Methodology and Visualization Tools. Computer Networks 48(3), 447–473 (2005)

[83] Sekercioglu, A., Pitsillides, A., Vasilakos, A.: Computational Intelligence in Management of ATM Networks: A Survey of Current State of Research. Soft Computing Journal 4(5), 257–263 (2001)

[84] Shakkottai, S., Srikant, R., Brownlee, N., Broido, A., Claffy, K.C.: The RTT distribution of TCP flows in the internet and its impact on TCP-based flow control. Tech. rep., Cooperative Association for Internet Data Analysis, CAIDA (2004), http://www.caida.org/publications/papers/2004/tr-2004-02/

[85] Shalunov, S., Lutzmann, B., Montesino-Pouzols, F.: thrulay, network capacity tester. Internet2 End-to-End Performance Initiative (2008), http://e2epi.internet2.edu/thrulay/

[86] Stevens, W.R.: TCP Slow Start, Congestion Avoidance Fast Retransmit, and Fast Recovery Algorithms. RFC 2001, Internet Engineering Task Force, Network Working Group, status: Proposed Standard (1997)

[87] Stewart, R.R., et al.: Stream Control Transmission Protocol. RFC 4960, Internet Engineering Task Force, Network Working Group, status: Proposed Standard (2007)

[88] Tan, K., Song, J., Zhang, Q., Sridharan, M.: A Compound TCP Approach for High-Speed and Long Distance Networks. In: 25th IEEE International Conference on Computer Communications (INFOCOM), Barcelona, Spain, pp. 1–12 (2006)

[89] Vishwanath, K.V., Vahdat, A.: Realistic and responsive network traffic generation. In: ACM Special Interest Group on Data Communication (SIGCOMM) Conference, Pisa, Italy, pp. 111–122 (2006)

[90] Vishwanath, K.V., Vahdat, A.: Evaluating distributed systems: does background traffic matter? In: ACM/USENIX 2008 Annual Technical Conference, Boston, MA, USA, pp. 227–240 (2008)

[91] Wang, C., Li, B., Sohraby, K., Peng, Y.: AFRED: An adaptive fuzzy-based control algorithm for active queue management. In: 28th Annual IEEE International Conference on Local Computer Networks, Bonn/Königswinter, Germany, pp. 12–20 (2003)

[92] Wang, C., Li, B., Hou, Y.T., Sohraby, K., Lin, Y.: LRED: A Robust Active Queue Management Scheme Based on Packet Loss Ratio. In: Annual Joint Conference of the IEEE Computer and Communications Societies (IEEE INFOCOM 2004), Hong Kong, China, vol. 1, pp. 1–12 (2004)

[93] Wang, J., Wei, D.X., Low, S.H.: Modelling and Stability of FAST TCP. In: 24th Annual Joint Conference of the IEEE Computer and Communications Societies (INFOCOM), Miami, FL, USA, pp. 938–948 (2005)

[94] Wang, R., Yamada, K., Sanadidi, M.Y., Gerla, M.: TCP with sender-side intelligence to handle dynamic, large, leaky pipes. IEEE Journal on Selected Areas in Communication 23(2), 235–248 (2005)

[95] Wei, D.X., Jin, C., Low, S.H., Hegde, S.: FAST TCP: motivation, architecture, algorithms, performance. IEEE/ACM Transactions on Networking 14(6), 1246–1259 (2006)

[96] Zhang, R., Phillis, Y.A., Ma, J.: A Fuzzy Approach to the Balance of Drop and Delay Priorities in Differentiated Services Networks. IEEE Transactions on Fuzzy Systems 11(6), 840–846 (2003)

[97] Zhang, R., Phillis, Y.A., Kouikoglou, V.: Fuzzy Systems for Queuing Control. Springer, London (2005)

Chapter 6
Open FPGA-Based Development Platform for Fuzzy Inference Systems

Abstract. This chapter looks into the practical implementation of some of the fuzzy inference systems proposed in previous chapters. Both architectural and operational constraints are considered. The focus is on an open FPGA-based hardware platform for the implementation of efficient fuzzy inference systems for solving problems in high-performance packet switched networks. A feasibility study is conducted in order to show that the techniques developed can be deployed in current and future network scenarios with satisfactory performance.

6.1 Fuzzy Inference Systems for High-Performance Networks

Computational intelligence techniques are gaining momentum as tools for network traffic modeling, analysis and control. Efficient hardware implementations of these techniques that can achieve real-time operation in high-speed communications equipment as well as many other demanding application fields is however an open problem. Current routing architectures pose two major challenges in the design of new mechanisms: *scalability* and *flexibility* of implementations. Here we introduce a platform and a companion development methodology for developing fuzzy systems that does not only fulfill operational requirements but also addresses the challenges posed by current routing architectures.

An FPGA development board with PCI/PCI-E interface is employed to support an open platform that comprises open CAD tools as well as IP cores. For the development process, we set up a methodology and a set of tools that cover from initial specification in a high-level language to implementation on FPGA devices. PCI compatible fuzzy inference modules are implemented as SoPC based on the open WISHBONE interconnection architecture. Results from the design and implementation of fuzzy analyzers and controlers for network traffic are analyzed. These systems are shown to satisfy operational and architectural requirements of current and future high-performance routing equipment.

F.M. Pouzols et al.: Mining & Control of Network Traffic by Computational Intelligence, pp. 263–304.
springerlink.com   © Springer-Verlag Berlin Heidelberg 2011

In section 6.2 we overview current practices and trends in routing architectures and design. We discuss how these practices and trends constrain the integration of fuzzy mechanisms in routers. Architectures and platforms for research in the networking field are overviewed as well. This serves for us to introduce the challenges and constraints in this field and motivate the development of the platform that is the main subject of this chapter. Section 6.4 outlines the architectures and technologies proposed for efficient hardware implementations of fuzzy systems, with the focus on the digital architecture used here. Section 6.5 describes an FPGA-based platform for fast development of prototypes of fuzzy inference systems with applications to networking, namely network traffic analysis and control. The platform is applied to implement some of the fuzzy inference systems described in previous chapters.

6.2 Routing Architectures

A major research problem in Internet transport and network layers is the development of traffic regulation mechanisms that can cope with the requirements of a growing diversity of technologies, applications and services. More generally, Internet traffic dynamics is an increasingly complex topic of research [67]. In the previous chapter we dealt with fuzzy inference systems for traffic control tasks that are to be implemented in Internet routers. In what follows we analyze how these fuzzy inference blocks would fit in current and next generation router architectures.

Technological trends in Internet core routers and high-end communications hardware in general (see figure 6.1) lead to hard constraints specially regarding packet processing rates. During the last years total Internet traffic has grown at over 80 percent per year, which directly translates into a similar or even higher increase of traffic volume in backbone routers. Overall, network traffic volume increases at a rate that outpaces advances in VLSI technology. Within this context, two main constraints arise: scalability (processing units must be able to process up to millions of packets per second (Mpps or Mp/s)), and flexibility and reconfigurability of implementations (a requirement imposed by the fast increasing diversity of protocols and technologies involved) [38, 72].

Both academia and major vendors are currently pushing for distributed and modular router designs, where routers are composed of modules that can be mapped onto different processing elements and communicate through open well-defined interfaces over an internal network [38]. Hardware for high-end communications systems has been traditionally developed in a custom and unstructured manner. Nevertheless, reconfigurable architectures are employed in practice by most vendors and design methodologies for easing the development process are sought.

Routers can be classified into three classes depending on the level they are deployed within the Internet. These classes correspond to access routers, campus or enterprise routers and core routers. We will focus on core or high-end routers, those designed for network backbones. In many aspects, today high-performance routers resemble supercomputers and have the hardest operational constraints.

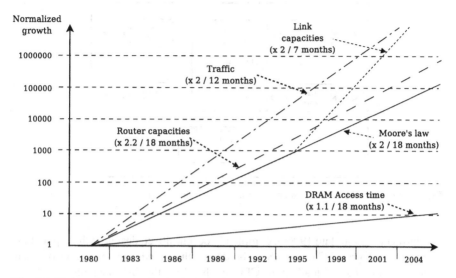

Fig. 6.1 High-end routers technological trends [38, 39, 48]

By tying together networks of the global Internet, routers made up a unified whole. While the main function of a router is to forward packets from a set of input links to a set of output links, they must also implement complex distributed routing algorithms, deal with diverse link technologies and provide support for traffic engineering tasks, differentiated services as well as quickly evolving quality of service schemes.

The architecture of Internet routers has evolved at a fast rate since the first implementations appeared [34, 69, 18, 47]. This evolution has been driven by a number of technological trends and functional requirements. On the one hand, the divergence in performance increase seen by the diverse components of a router (such as memory elements, interconnection links, programmable devices and processors) challenges router design. On the other hand, new functional requirements have arisen as new applications, services and technologies are being deployed. As a consequence, a great deal of research efforts are ongoing to address these challenges in router design.

The architecture of network systems in general has dramatically changed during the past two decades [21, 23, 50, 5]. From a historical perspective, architectures can be broadly classified into three generations depending on the degree of centralization:

- First generation (late 1980s and early 1990s): software running on a standard processor (for example, an IP router built by adding software to a standard minicomputer).
- Second generation (mid 1990s): classification and a few other functions offloaded from the CPU with special-purpose hardware, and a higher-speed switching fabric replacing a shared bus.

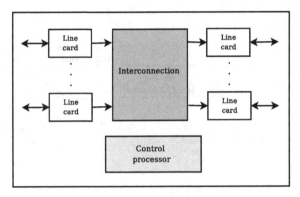

Fig. 6.2 Overall scheme of a router: control and switching plane

• Third generation (late 1990s): completely decentralized design with Application Specific Integrated Circuit (ASIC) hardware plus a dedicated processor on each network interface offloading the CPU and handling the fast data path.

In addition, since more than a decade high performance routers are designed following a conceptual partition that separates two major functional blocks: the control and the switching or interconnection planes [39] (see figure 6.2). A detailed description of switching fabrics architectures is outside the scope of this monograph, for an extensive review see [38, 15, 48]. These partitions do not necessarily match the physical structure of hardware implementations.

Figure 6.3 shows the overall physical structure and components of the control and switching planes for routers of the second and third generation [5, 22, 38]. First generation equipment uses a shared bus as interconnection element and packets are transmitted twice over the bus (from the input line card to the central processor and from the processor to the output line card). In second generation equipment, line cards incorporate the necessary buffers so that packets have to be transmitted just once from the input to the output line card.

Third generation routers introduced switched backplanes as interconnection elements, which allow for simultaneous packet transfers thus increasing the global bandwidth capacity. In general, line cards are not mere medium access control systems to the physical connection means. With the introduction of second and third generation architectures, the main functions of the line cards of a core router include route lookup, packet classification and traffic management for quality of service control among many others. As a consequence, line cards typically include the following elements as well:

• Packet processing devices that perform functions belonging to the data plane, such as classification (see chapter 1) and routing decisions [73]. Consequently, these devices integrate the routing database in whole or part.
• Memory elements that perform as buffers for different connection rates.

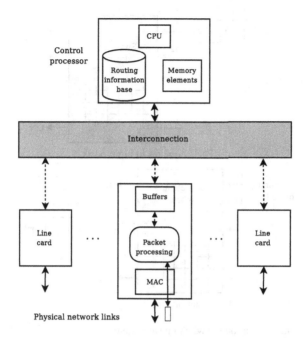

Fig. 6.3 Overall architecture and components of a second/third generation router

On the other hand, the control plane broadly consists of a specialized CPU, memory elements and a system for managing the routing database. Although we do not get into details about the routing database, usually referred to as the routing information base (RIB), it is one of the most complex and critical components of a router as a whole [22, 38].

The interconnection block usually takes the form of a switching matrix [22]. It is currently implemented as a switching system with complex estructures that aim at providing a balance among its switching speed, its scalability, the combinatorial direct interconnection possibilities among the maximum number of boards as well as its cost in terms of area and consumption [45].

The performance as for bandwidth that the switching system can attain are limited not only by its design but also by the organization of the memories that are used for buffering packet queues. The function of these queues is to adapt different bandwidths available in input and output ports, with the aim to reduce losses due to incoming packet bursts. These buffers can be implemented in the input or output boards.

In the first case (where packet queues are implemented in the input boards, see figure 6.4(a)) outgoing traffic may be unnecessary blocked at some output ports due to the lower bandwidth of other output ports. It has been shown that in the case of uniform traffic patterns this issue can reduce the overall performance of the

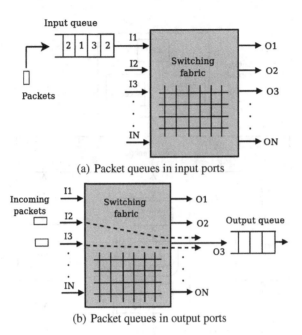

(a) Packet queues in input ports

(b) Packet queues in output ports

Fig. 6.4 Packet queues in a router switching subsystem

interconnection system down to 59%. It should be noted also that these configurations require the bandwidth of the memory elements be equal to that of the physical links.

An alternative configuration can be defined by including packet queues in output ports (see figure 6.4(b)). In this case, since incoming packets in different input ports can be simultaneously transmitted to the same output port, the bandwidth of the memory elements is required to be the same as that of the switching system in order to avoid a performance degradation.

Given that the bandwidth of memory elements is the hardest technological constraint on routers performance, many current architectures use a hybrid scheme for input queues known as VOQ (Virtual Output Queueing), depicted in figure 6.5. Under this scheme, each and every input port integrates a complex packet queue implemented as a set of parallel queues, as many as output ports are enabled. These virtual output queues prevent that packets routed to available output ports be blocked. This way, the bandwidth required on memory elements is the same as that of the input queue case while overall performance can attain 100% of the switching system nominal capacity.

For some conditions and configurations of the switching matrix, packets arriving at input ports can be directly routed to the proper output port (through a direct connection in the switching matrix between both ports) without any participation of the control processor. The routing time in these cases is remarkably lower than in those cases where the control processor participates in the routing process [70].

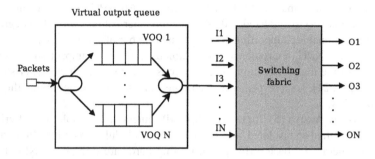

Fig. 6.5 Virtual Output Queuing (VOQ) scheme

Thus, there is a partition between two paths: the fast and slow paths, conditioning the design of current routers where the aim is to keep most of the traffic in the fast path.

The slow path provides poorer performance and is generally implemented as software running on a microprocessor. On the contrary, the fast path is generally implemented as a set of ASIC, is not flexible and hard to extend and update besides being costly to implement. The partition between the fast and slow paths is of a great importance in current architectures due to the technological trends shown in figure 6.1 [54], as the gap between links and memory elements keeps increasing. However, taking full advantage of the fast path is increasingly difficult as new protocols and services are being supported which require a great deal of flexibility in the packet processing task. Thus, in conclusion three major design issues are currently posed in this context:

• New architectural developments are sought in order to keep the capacity of routers increasing.
• Transmission rates increase faster than packet processing rates.
• More flexibility is needed for packet processing, including those packets in the fast path.

6.2.1 High-End Routing Hardware

A number of approaches and architectures have been proposed throughout the last years for implementing high-end Internet routers. In this section we analyze the compatibility of fuzzy modules with some of these architectures as well as their feasibility considering speed, area and power consumption constraints. Our analysis focuses on the applicability to architectures used for high-end routers from major vendors [34]: Cisco series 7600, 12000 and CRS [18, 70] and Juniper T-series [69].

Currently, the major problems in the design of high-end routers derive from the difficulties in designing processors for the control plane. Processors being implemented, Network Processing Units (NPU), aim at providing the speed of an ASIC and the programmability of a CPU, i.e., providing high performance and also the necessary flexibility that an ASIC cannot provide. An NPU must perform packet processing in a very specialized and optimized manner but also provide a degree of flexibility that enables supporting an increasing number of services [35, 22].

The way the NPU of a router series is implemented has an impact on the implementation of the interconnection cards and constrains the overall characteristics of the series [39, 21], such as performance range, functionality, programmability and configurability.

As noted by Aweya [5], high throughput IP routers are possible only if critical tasks are identified and isolated, and special purpose modules are tailored to perform them. In general, the basic principle in routing hardware design and NPU based systems [22, 25, 24, 23] is to exploit parallelism against the main limiting factor imposed on overall performance: the memory access speed. This implies extensive use of pipelining techniques and distribution of tasks among many processing elements. In addition, a major design objective is keeping overall performance stable and predictable. There are two main alternatives for the hardware implementation of an NPU:

- Architectures based on general purpose processors (such as the NPUs from AMCC®, Intel®, IBM®, Motorola®, Vitesse®, Agere® and other vendors [39]).
- Specific architectures, with better performance but lower flexibility. In this case, the development time is usually too long as compared to the fast evolution of services. These architectures can be implemented as ASIC or high-capacity programmable logic devices, specially FPGA devices. The latter case provides a convenient balance between performance and design flexibility and programmability at the expense of a higher cost per unit and power consumption.

In practice, current flexibility requirements lead to a hybrid approach where high-end routers use general purpose units together with co-processing and acceleration elements (processing engines) implemented as ASIC or on specific FPGA devices. In this context, FPGA devices are not only used for developing prototypes but also for final products. The functions of an NPU are usually splitted into two blocks that can be implemented either on a chip or on separate chips:

- A block that carries out the essential and traditional functions of a network processor (NP), such as error correction, classification, address lookup, fragmentation and reassembly of packets.
- An additional block for functions of a higher level generally refered to as traffic management (TM). This block is being more and more developed as quality of service provisioning functions are deployed. TM includes among other functions measurement procedures, network management policies, congestion control and quality of service schemes, queue management and bandwidth allocation.

In order to manage millions of traffic flows in links with bandwidths of the order of 10 Gb/s current NPUs are usually composed of a number of blocks with different degrees of programmability and generality [22, 35]. This way, different options for executing functions on both software and hardware are available, ranging from specific engines implemented as ASIC to general purpose processors, with operating

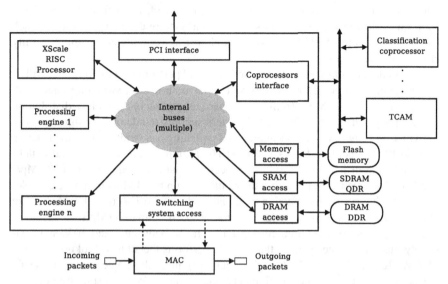

Fig. 6.6 Example NPU architecture (Intel® IXP Series)

systems such as Unix variants and VxWorks that can be programmed in languages
such as C, C++ or Java for functions of the control plane and slow path. Thus,
NPUs are in general a component of development platforms that include tools such
as compilers, debuggers and simulators. Since NPUs employ parallelism and concurrent processes to achieve high performance, the software can be complex and
difficult to analyze and debug. This difficulties are particularly severe in general because of the parallel execution of multiple threads and, more specifically, because
of unexpected interactions among unrelated software components under heavy load
conditions.

Figure 6.6 shows as an example the scheme of an NPU of the IXP (Internet
Exchange Processor) series [42, 22] from Intel®. This series support links of up to
10Gb/s and implements an architecture similar to that of other vendors such as the
Motorola® C-Port, AMCC® nP and Agere® APP.

As in most NPUs, data processing is distributed over a core processor and a set of
specialized processing units. The central processor in this architecture is an Xscale
unit [43], that can run operating systems such as embedded versions of GNU/Linux
and VxWorks. These units are programmed in C and C++, have a heavily constrained memory access model and have specific interfaces for accessing a set of
up to 16 specific processing engines. These engines or processing units [20, 25] are
32 bits RISC processors that work in parallel sharing internal buses. They lack a
full operating system, have a fix number of execution threads and are programmed
in restricted C or microcode but have full access to the memory blocks and other
co-processing and acceleration subsystems.

This way, NPUs are in between flexible specific systems and general purpose processors with a high degree of parallelism and specialization. Specific programming

languages have been proposed such as NCL (Network Classification Language) from Intel. However, there are no standards and mature development tools, which can make designing NPU based systems almost as complex as designing a fully custom ASIC solution.

Let us consider timing constraints and expected performance for high-end NPUs. As an example, series 12000 Cisco routers have an overall switching bandwidth of 1.28 Tb/s or 375 Mp/s (millions of packets per second), with a distribution of 25 Mp/s per line card. Cisco CRS routers support standalone configurations of 320 Gb/s, 640 Gb/s and 1.2 Tb/s aggregated switching bandwidth, with a distribution of up to 40 Mp/s per line. Series T320 and T640 Juniper routers attain a total switching bandwidth of 320 and 640 Gb/s, respectively, which is equivalent to 385 Mp/s and 770 Mp/s, respectively, with a distribution of up to 24 Mp/s in both cases. Series T1600 Juniper routers attain a total bandwidth of 1.6 Tb/s (or 1920 Mp/s) with 100 Gb/s per line card and a packet processing speed of up to 60 Mp/s.

As for power consumption, current NPUs range between 7 and 30 W approximately. Just one of the systems that usually performs the address lookup function, TCAM (Ternary Content-Addressable Memory), normally implemented as ASIC, reaches a mean power consumption between 10 and 15 W. For estimating the impact on the overall consumption of a router, one has to consider that in high-end routers NPUs may be replicated up to hundreds of times. For the case of the Cisco 12816 [19] (that incorporates around 200 ports) overall consumption is 4.7 kW, whereas for the Juniper M160 (that incorporates up to 128 ports), T320, T640 and T1600 overall consumption is 2.6 kW, 2.8 kW, 6.3 kW and 8.3 kW.

6.2.2 Expected Evolution

Current router architectures are ongoing an evolution process that is expected to lead to what is known as the fourth generation of routers. The following changes are expected for the near future to cope with current challenges:

- Links based on opto-electronic technology are increasing available bandwidths at a pace higher than that of the processing speed of elements required to forward, process and store packets.
- Requirements derived from new services such as IPv6, MPLS, Voice over IP and improved security imply modifications in the packet processing path and a dramatical complexity increase of the control plane [39]. This current trend is expected to eventually lead to a convergence between the slow and fast paths since the fast path cannot be kept as simple and monolithic as today.

Current trends clearly discard the overall monolithic design that has characterized routers where all the software runs on a central control block. Instead, new architectures follow a distributed approach where routers are splitted into subsystems [54], making it possible to implement more complex systems and combining subsystems from different vendors by means of standard interfaces [44]. These changes are however not only subject to technological factors as they imply a change in the business model of vendors.

In particular, more and more control functions are being encapsulated as modules and physically distributed among input/output cards [35]. This way, central switching systems must be more open and flexible as they transition into physically independent systems connected to other modules through optical links.

Two important factors enable the design of new routing architectures with overall performance several orders of magnitude above current units. On the one hand, new highly parallel interconnection architectures developed for the switching system [45] increase the overall nominal bandwidth of routing equipment. On the other hand, the surge of opto-electronic technology and the widespread availability of optical links make possible very high link speeds. However, current router architectures can hardly scale up to the speeds possible with technologies available today. A fourth generation of routers is thus sought.

With the aim of easing the design of these complex systems still in an early planning stage, several international consortiums[1] as well as the Forwarding and Control Element Separation (ForCES) working group of the IETF [44] are carrying out the definition of functional blocks and standard interfaces that fourth generation routers should support [54].

6.2.3 Architectures and Platforms for Research

Specific hardware architectures play a key role in supporting passive measurement systems, platforms for experimental network systems design, and virtualization of networks.

Specific hardware for passive network measurement is required in order to record traffic traces at rates of the order of the Gigabit and above, specially in fast backbone links. To the best of our knowledge, packet-level trace recording and precision timestamping matching modern router speeds is available only with Dag cards from Endace [29] and the family of COMBO cards, in particular the COMBO-Precise Timestamp Module (COMBO-PTM) card [51] developed in the framework of the liberouter project. The latest Dag models (4.xx) can reach sub-microsecond accuracy when synchronized to GPS or CDMA [37]. In addition, reconfigurable hardware platforms for real-time monitoring of high-speed networks through flow level measurement have been developed recently [78].

The platform that will be described in section 6.5 has been developed with the aim of enabling the development of fuzzy inference systems applied to networking problems in high-performance equipment.

There exist a number of software platforms with specific hardware support for general education and research projects, such as the Click modular router (http://read.cs.ucla.edu/click/) and the Open Network Laboratory (http://onl.arl.wustl.edu) . These platforms allow for flexible

---

[1] Such as the Multiservice Switching Forum (http://www.msforum.org).

experimentation with a number of aspects dealt with in this monograph, such as queue management schemes, QoS provisioning, network flow measurement, as well as a many other aspects related to traffic management, routing and protocols.

More closely related to the platform described in this monograph, the NetF-PGA project (http://netfpga.org) makes it possible to build prototypes of hardware-accelerated high-speed networking systems. It can be used to build Gigabit Ethernet switches and IP routers that use hardware acceleration for packet forwarding tasks. Originally introduced in 2001, it was intended as an education platform for network systems design. Current versions, introduced in January 2007, rely on a Xilinx® Virtex II-Pro 50 FPGA in order to provide full Gigabit-Ethernet rates for 4 physical interfaces. In a similar way to the platform that is described in this monograph, a PC is used as host to the NetFPGA system through standard PCI interfaces. The user can implement specific packet processing logic in the FPGA and define software networking applications in the host PC.

On a final note, network virtualization techniques allow for the simultaneous operation of multiple and independent logical networks or overlays on a single physical platform. The emergence of these techniques in routing hardware and networking equipment in general opens new ways for experimenting. Recent high performance routing architectures from major vendors have supported virtualization techniques for some years. It is expected that network virtualization will become one of the major paradigms of the future Internet. The deployment of this relatively new technologies in academic networks will enable flexible research in areas where innovation is currently encumbered by the lack of possibilities for disruptive experimentation with networking hardware.

6.3   Inference Rate of Software Implementations

Here it is performed an approximate estimation of the inference rate that software implementations of fuzzy inference systems can attain. This analysis focuses on the maximum inference rate attainable by software running on a dedicated core of a multicore general purpose CPU. This study is by no means intended to be a systematic analysis of the inference rate that can be achieved by software implementations of fuzzy systems, not even with the software synthesis tools included in the Xfuzzy environment. A more thorough analysis of the speed attainable using software implementations generated with the version 3.0 of the Xfuzzy environment for different inference systems can be found in [60].

Table 6.1 shows the maximum inference rate for the fuzzy inference systems described in chapter 6. The overall characteristics of these systems are shown in table 6.5.2 (page 288). In particular, the fuzzy mean defuzzification method was used. Results are shown for C implementations on two system configurations: system A and system B. The C implementations of these systems were generated using the xfsw tool in the Xfuzzy 3.2 environment. The same implementation options as for the hardware systems described in chapter 6 were set.

Table 6.1 Maximum inference rate of software implementations (MFLIPS, C implementation)

Fuzzy System	System A	System B
FAQMBestEffort	1.21	1.70
TCPSS	1.79	2.34
DSSelect	5.47	6.76
AQMDSAF	9.85	10.09
RxBufferSize	2.36	3.41
RTperf	2.06	2.90

System A is based on an Intel© Xeon™ CPU with 4 cores at 3.20 GHz with 2 MB of L1 cache per core, whereas system B is based on a AMD© Opteron™ Processor 248 at 2.2 Ghz with 1 MB of L1 cache. In both cases, the GNU GCC compiler was used on GNU/Linux operating systems. For system A, the version 4.1.2 20071124 (Red Hat 4.1.2-42) of GCC was employed, whereas GCC 4.1.0 (SUSE Linux) was used on system B. The whole range of compiler optimization options was explored. 10 repetitions of 10 million consecutive inferences were performed for each optimization case while there was no significant competing load. The best case for each system is shown in the table. The worst case for the best optimization options set was consistently within a band of 0-6% below the best performance. Compiler optimization options were found to boost performance by up to 300%.

C implementations were found to be consistently faster than C++ and Java implementations. This can be due to several factors, such as the compilers used and the specific implementation scheme that the Xfuzzy software synthesis tools apply for each language.

As conclusion, the inference rate attainable by software implementations running on current general purpose processors is at least one order of magnitude below that of the hardware implementations that will be described later on in this chapter. It should be noted though that it is possible to generate further optimized software implementations that can perform up to approximately two times faster. For instance, using the code optimization techniques implemented in the version 2.1 of xfc, a C implementation of AQMDSAF can achieve an inference rate of 19.67 MFLIPS on system B. These inference rates are however possible with software implementations only at the cost of a fully dedicated core of current high performance general purpose processors.

6.4 Hardware Implementation of Fuzzy Inference Systems

A number of approaches have been proposed to date for implementing fuzzy inference systems on hardware. Different hardware architectures and technologies have been proposed and applied to implementing fuzzy inference systems in an efficient manner [7, 46], optimizing both performance and consumption of resources. These developments have been instrumental to the spread of fuzzy logic based controllers in many industrial fields and a number of innovative applications in consumer electronics and signal processing.

In this section we outline several alternatives and argue which option appears to be the most plausible from our viewpoint for the kind of networking systems this monograph deals with. For a more in depth discussion of the different strategies that can be followed for the hardware realization of fuzzy inference systems we refer refer the interested reader to the chapter 6 of [7].

As discussed above in this chapter, networking applications may require very short inference periods, of the order of the microsecond and lower, and the implementation of tens or even hundreds of parallel fuzzy inference systems for handling packet queues associated to different network interfaces. This can be achieved to a limited extent by software implementations running on current high-performance general purpose CPUs (see previous section). The implied cost is however extremely high and resources are used very inefficiently. Optimization of speed, area and consumption for generic and complex fuzzy systems can only be achieved through dedicated hardware realizations. Both analog and digital approaches have been proposed.

A number of digital approaches have been proposed for different applications, including the use of programmable logic controlers (PLC) and programmable automata in industrial automation, the implementation as optimized code or microcode running on microcontrollers for the field of embedded systems, the so-called hardware expansion of general purpose processors (by extending a generic ALU or adding specific ALUs, fuzzy functional blocks or fuzzy coprocessors), and implementations on digital signal processors (DSP) [9].

Analog implementations can achieve high speed as well as low area and power consumption when implementing multivalued systems such as fuzzy inference systems. Thus, analog VLSI implementation of fuzzy logic controlers in CMOS technologies can fulfill the constraints of fuzzy inference systems in the networking field [28]. We note however that the interest in analog circuits for implementing fuzzy systems is mainly motivated by two factors: the affinity between most fuzzy algorithms and analog circuits as well as to avoid the use of A/D and D/A converters to interface sensors and actuators. Mixed-signal analog computation circuits have been used for the implementation of artificial neural networks and fuzzy systems as well [8, 28].

However, the field of traffic analysis and control in networking does not require analog interfacing. In addition, no analog circuitry is used to implement the processing units in routing equipment. In fact, the limited flexibility and extensibility of analog design schemes as compared to digital ones encumbers the use of analog circuitry in networking equipment. These facts render analog and mixed-signal implementations unpractical for the problems addressed here.

On the other hand, digital implementations of fuzzy inference systems are possible using both sequential and parallel architectures. In particular, in this work we leverage on a specific architecture for implementing fuzzy inference systems

[68, 49, 9, 6]. This architecture has three main characteristics: it uses an active rule-driven inference mechanism, some restrictions are defined on the form of membership functions, and simplified defuzzification methods are used. This architecture can achieve inference rates of the order of the MFLIPS and above, and is suitable for both ASIC and gate arrays-based implementation techniques.

The fuzzy inference systems for traffic control and analysis presented in chapter 5 were designed with these constraints in mind, i.e., only triangular, trapezoidal and singleton membership functions are used, the degree of overlapping between membership functions is limited to 2 and the fuzzy mean method is used for defuzzification. Thus, the fuzzy systems discussed in chapter 5 can be implemented on hardware using the aforementioned architecture without modifications.

6.5 Development Platform for Fuzzy Inference Systems with Applications to Networking

As stated above, diverse research results show that fuzzy systems can help solve current problems in Internet traffic control. Soft computing techniques, and fuzzy systems in particular, are gaining momentum as tools for network traffic modeling, analysis and control. Fuzzy systems find applications in a number of areas such as traffic control in routers [26, 80], admission control [52], support for differentiated services within the DiffServ architecture [77], policy and quality of service evaluation [66], real-time traffic measurement, analysis and monitoring [58], power saving for wireless networks, as well as end-to-end traffic control [56] and end-to-end control for wireless networks [63].

However, while many industrial applications of fuzzy systems in a variety of fields have been reported, fuzzy systems for traffic control have not yet found their way into real-world applications. In particular, despite the good performance of the aforementioned fuzzy logic based mechanisms for traffic analysis and control, there is a lack of architectures and design procedures for implementing them in a systematic manner yet addressing current challenges in high-performance networking systems. As a consequence, although significant results in diverse applications of fuzzy systems in communications and networking have been reported since more than a decade [33], the deployment of these systems in the real world is still a challenge.

In the current Internet, link speeds and thus packet processing rates requirements imposed on routers are quickly increasing. The pace at which the speed of memory elements as well as other components of the router processing units increases is significantly slower. That is, the rates at which two key technological factors evolve have been increasingly diverging, and it is expected that this trend will continue. As a consequence, many data storage and processing elements (or processing engines) in current routers are implemented by means of specialized processing engines using specific hardware architectures [23, 24, 25, 38, 22].

Regarding traffic measurement, analysis and control, specific hardware architectures are sought in order to cope with the increasing packet processing rates. Specific hardware architectures have been proposed for measurement [29], analysis of network flows [78] and implementing a number of common processing engines [23, 24, 25]. In particular, when implementing active queue management schemes, the packet queue control rate must be very close to the maximum packet processing speed attainable by a router. This is a strong requirement that is further aggravated by two coincidental factors:

- Traffic in packet switched networks is inherently bursty. Bursts of several packets at the highest link speed are fairly common in a broad spectrum of network scenarios [65].
- Queue lengths are commonly small, around a few tens of packets [2].

Thus, controlers must have a fine packet processing granularity in order to cope with packet bursts and properly control small packet queues.

Let us now consider the implementation of fuzzy inference systems applied to traffic analysis and control in routers. Though a fully software solution would be obviously more flexible than a hardware solution, it is very straightforward to show that software implementations cannot currently attain inference rates of the order of MFLIPS, tens of MFLIPS and higher even running on high performance general purpose CPUs. Also, considering the trends in several technological factors outlined above, this is unlikely to change in the foreseeable future. In fact, the required inference speeds are expected to keep on growing at a faster pace than that of software based processing units.

In addition, even though software implementations were fast enough, using a dedicated general purpose CPU for implementing active queue management in routers does not seem to be a feasible option. In current routing architectures functional blocks for traffic analysis and control need to be replicated in every input port for each virtual output queue (VOQ), i.e., they need to be replicated in every input port as many times as output ports are enabled (see section 6.2 for an explanation of VOQs). As a result, the number of VOQs and associated queue analysis and control blocks can be around a few tens or hundreds.

This fact imposes hard consumption and cost constraints on the implementations of traffic analysis and control mechanisms. It is very unlikely that vendors will afford the inclusion of tens or hundreds of high performance CPUs (with the implied cost and consumption) for each unit.

Rather than using software implementations running on general purpose hardware (that would imply higher cost and insufficient performance in most cases), in current routing architectures computationally intensive and critical processes are run on hardware processing engines that use specialized architectures. This is the case for instance of TCAM memories for storing and querying routing information bases. Though in the past pure-ASIC implementations have been used, currently many of these engines are often implemented on one or several FPGA devices per interface card for better programmability and design flexibility. Specific tasks are implemented by specialized subsystems whereas general purpose processing units realize various coordination functions as well as high level management tasks.

Thus, efficient hardware implementations of fuzzy inference systems are required to fulfill common operational requirements in traffic analysis and control mechanisms to be implemented in routers. Although fuzzy inference systems are usually computationally intensive, its performance can be boosted by means of hardware implementations based on optimized architectures that exploit the inherent parallelism of fuzzy inference or otherwise simplify the inference process. As detailed in the previous section, the hardware implementation of fuzzy systems is a well established field of microelectronics [9], thus making them a feasible solution for processing massive traffic volumes in real-time.

Here we introduce an open FPGA-based platform for the development of modular fuzzy components of complex systems [57, 55]. The platform has been applied to networking and communications systems. In particular, it has been successfully employed in order to develop intelligent traffic analyzers and regulators that can achieve real-time operation within current high-performance Internet routers. The platform has been developed with a twofold objective in mind:

- Enabling the automated and efficient (in terms of performance and development effort) implementation of a number of fuzzy systems proposed throughout the last years.
- Fostering the research on fuzzy logic based solutions to Internet traffic analysis and control. This is a consequence of the availability of a platform for validating hardware prototypes using inexpensive equipment, which eases testing in experimental facilities in support of disruptive experiments through network virtualization. The widespread availability of such facilities, expected for the near future is further discussed below.

Throughout more than a decade, strategies and methodologies for developing fuzzy logic based controlers have been proposed and applied. To date, most work on this topic has been focused on industrial applications [12] and, more recently, on areas such as signal processing, image processing and switching power control, among others.

Let us consider the task of evaluating a traffic analysis and control system whose performance depends on the nature of traffic. Ideally, one should be able to deploy the system and study its performance in the real world. Of course, this is most often intractable, specially in the case of active queue management schemes. Therefore, the availability of flexible means for evaluating such systems with different traffic patterns (be it through simulation, emulation or implementation) is a key aspect of design.

Considering the specific requirements as well as the high cost and complexity of high performance routers deployed on the Internet today, we have defined a flexible development platform for prototyping network traffic analysis and control systems. The platform is defined so that fuzzy systems are integrated as independent modules into complex networking systems.

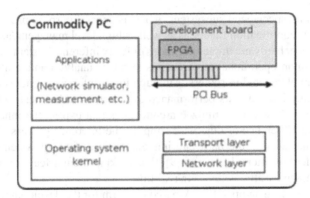

Fig. 6.7 Prototyping platform scheme

Additional requirements taken into consideration are: seamless integration into current router architectures [48], flexibility, and performance scalability up to higher requirements in current and foreseeable network technologies. The platform provides a complete set of tools and an environment for easing the development of fuzzy systems for networking prototypes as well as performing its validation.

The architecture of the platform, outlined in figure 6.7, is based on a commodity PC equipped with an FPGA development board with PCI interface, thus making a flexible and cheap solution with no specific hardware requirements yet able to emulate the behavior of complex and expensive routing equipment. This makes it possible to perform experimental validation by means of prototypes using affordable hardware.

Obviously, these prototypes will suffer from performance limitations and will probably not attain for real traffic a total throughput close to or beyond the Gb/s. This is however a consequence of the limitations in total switching bandwidth currently attainable by a common PC architecture and does not exclude the validation of fuzzy inference modules at higher rates.

When implementing fuzzy systems, two main function blocks are distinguished: those directly related to fuzzy inference and those that can be classified as auxiliary functions, such as initialization, timing, pre- and post-processing, etc. [12]. For the implementation of prototypes of fuzzy systems for analysis and control of Internet traffic the following model is used:

- As fuzzy inference modules (FIM) are the potential system bottlenecks, they are implemented on FPGA devices and described by means of VHDL according to an specific processing architecture [9] tailored for efficient and fast fuzzy inference. The methodology and tools employed for the development of the FIM is described in the next section.
- In the basic configuration of the platform, all auxiliary functions are implemented as software. Software can run on the PC operating system as well as on optional components implemented on the FPGA of the development board.

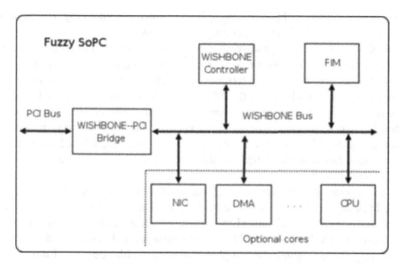

Fig. 6.8 Fuzzy SoPC as a PCI device

A flexible and open architecture for implementing fuzzy systems on the FPGA has been defined. Within this architecture, depicted in figure 6.8, FIM modules are integrated as subsystems of a potentially complex and reconfigurable fuzzy logic based digital system.

Interconnection between the host PC and the fuzzy inference module (FIM) is done through a standard PCI/PCI-E bus. The internal bus of the fuzzy digital system conforms to the WISHBONE logic bus [36] public domain standard. WISHBONE is a SoC interconnection architecture for portable IP cores, that connects a variable number of components.

All top level interfaces are designed to be WISHBONE compatible. Both the WISHBONE bus controler and the PCI-WISHBONE bridge [27] IP cores have been developed under free distribution licenses by the OpenCores [64] organization as well as other entities. An alternative option is a direct mapping between On-Chip Peripheral Bus (OPB) [40] and WISHBONE signals, as most of them map one-to-one. While the WISHBONE-PCI bridge (with PCI 2.2 support) and WISHBONE controler (Conbus [16]) are implemented in Verilog as provided by OpenCores, the FIM module is implemented using VHDL.

WISHBONE systems can easily interface with other SoC bus standards, including OPB through the WISHBONE-OPB bridge [71]. In its basic configuration, the system comprises three cores: a fuzzy inference module (FIM) (as a slave WISH-BONE device), the WISHBONE controler and the PCI-WISHBONE bridge (as a master WISHBONE device). The WISHBONE controler IP module employed supports up to 8 master and 8 slave devices.

This way, software tasks can be defined using common programming languages and can be run on the generic purpose processing units of the PC as well as on specific processing units implemented on the FPGA. For instance, a fuzzy logic based traffic analysis application can be implemented incorporating an OpenRisc Core for which the GNU/Linux operating system is available.

The PCI interface of the prototypes eases integration with routing architectures by major vendors [48, 34]. Within routing architectures currently deployed in the Internet [39, 48], fuzzy traffic analyzers and controlers could be seamlessly integrated as processing engines whether at the NPU and/or the output/input cards depending on the quality of service architecture implemented on the router. In addition, when the development board employed includes a network interface card, as is the case for the board employed for our implementation, a whole fuzzy logic based traffic analysis application can be implemented as a standalone system-on-a-programmable-chip (SoPC) on the FPGA.

As we will describe in section 6.5.2, the configuration presented so far allows for the implementation of fuzzy processing units that can be generally applied to network traffic analysis and control. Additional IP cores (such as network interface card control, DMA devices and CPUs) can be incorporated in order to develop extended fuzzy processing units or complementing them with extensions for other soft computing techniques.

Within the prototyping platform presented, a fully automated design flow has been employed. The fuzzy systems design flow, described in the next section, covers from an initial high level fuzzy system description to an FPGA-based implementation of FIMs.

6.5.1   Development Methodology and Design Flow

A methodology and design flow tailored for the development of fuzzy inference systems applied to Internet traffic analysis and control have been defined. The design flow we have defined and applied for designing fuzzy inference modules is depicted in figure 6.9. The whole development process is covered, from initial specification to final implementation whether as software or digital hardware.

The design flow spans from initial specification in a high-level language to an FPGA implementation of FIMs by means of the tools included in the Xfuzzy development environment as well as tools in the Xilinx® ISE environment [76]. We leverage on the Xfuzzy [62, 61] CAD suite of tools and a methodology [12] for the development of fuzzy controllers to define a methodology and design flow tailored for the development of fuzzy systems applied to Internet traffic analysis and control. The Xfuzzy environment eases the specification, verification and synthesis of fuzzy inference systems. The whole set of tools included in the environment are based on a common high level specification language: XFL3 [59].

The first development stage (description) is performed using the XFL language (or alternatively, using visual interfaces that rely on the XFL language), which can later be turned into C and VHDL code among other implementation options. The tool chain includes:

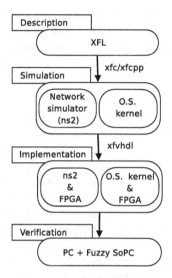

Fig. 6.9 Design flow of fuzzy systems for networking

- The xfc and xfcpp tools (included in Xfuzzy), which turns an XFL specification into C and C++ code that can be employed in both user and kernel space.
- The xfvhdl tool (included in Xfuzzy), which turns an XFL specification into synthesizable VHDL code generated for a specific processing architecture for the efficient implementation of fuzzy systems with a good cost-performance ratio and an extremely short development cycle [61, 6]. xfvhdl applies an active-rule driven architecture for fuzzy inference, using simplified defuzzification methods and parallelization in order to provide high inference rates. The output of xfvhdl can be fed to a number of synthesis tools, such as those from Xilinx® and Synopsys. As it will be show in section 6.5.2, this architecture can provide efficient implementations of fuzzy systems even with a high number of variables, linguistic terms and rules.
- ns-2 [41], an open network simulator widely spread within the Internet research community.
- Operating system kernel (currently GNU/Linux and FreeBSD).

The development stages after specification have been tailored for Internet traffic controler development as follows.

- For network simulation, we have used ns-2. ns-2 is an object oriented discrete event driven simulator with support for a vast variety of transport protocols, queueing systems, routing schemes and access media, thus enabling us to evaluate the performance of traffic controlers under complex and realistic simulated scenarios. Fuzzy controlers are integrated into ns-2 as components implemented in C. This makes it easy to evaluate the influence of various factors, such as the precision of fuzzy modules, in a convenient manner.

- Verification is considered for both software and hardware implementations of fuzzy controlers. Software verification is performed over a controler implementation within the kernel of the general purpose operating system running on the PC. Three different yet complementary approaches can be followed for verifying hardware prototypes of FIMs:

  - Verification by means of network simulators. To this end, code and drivers to access the FIM in the FPGA development card from ns-2 has been developed.
  - Verification through emulated scenarios where a router is emulated by means of the prototyping platform. Validation in real or emulated scenarios is also possible with our prototyping architecture by using the prototyping PC as a router. This accomplished by replacing queue control functionality in the operating system network layer with functionality provided by the FIM in the development board. To this end, kernel drivers have been developed to make it possible to access the fuzzy controler in the FPGA development card from networking modules in the operating system kernel. Drivers have been developed for FreeBSD 6.x and Linux 2.6.x kernels. Alternatively, emulated scenarios can be constructed by means of software packages for network emulation, as was illustrated in chapter 5, where the NIST Net software router was used to emulate a dumbbell network and test different active queue management schemes.
  - Verification in the Internet. This case is hardly feasible in practice. In fact, disruptive experiments are currently not possible in real networks. Besides the lack of control on the network conditions (derived from the current lack of measurement infrastructures for intra- and inter-domain paths), most often there is no possibility to inject significant traffic loads or implement novel mechanisms in real networks.

- We have defined as general hardware-software partition the implementation on hardware of the FIM module and its interfacing logic whereas all other tasks are implemented as software.

The fact that disruptive experiments are not currently possible in real networks has hampered research on novel protocols and architectures during the last years. Although a number of projects and infrastructures concerning virtual networks and novel network architectures, mostly based on overlays, have been running for the last years, these still face the same constraints. Lately, a great deal of interest in infrastructures for network virtualization has raised within researchers, institutions and agencies. It has been recognized a need for supporting clean slate design for the Internet as well as realistic experiments in network science and engineering. A number of related initiatives for developing experimental facilities in support of disruptive experiments through network virtualization are in planning stage or early stages of development at the time of this writing, such as the Global Environment for Network Innovations (GENI, http://www.geni.net), the Federated E-infrastructure Dedicated to European Researchers Innovating in Computing network Architectures (FEDERICA, http://www.fp7-federica.eu) and "Plataforma de prueba de servicios de comunicationes" (PASITO, https://proyectos.rediris.es/pasito/). However, several years of development are required to address the complexity of such infrastructures. We entertain the hope that these initiatives will shortly pave the way for testing in real scenarios techniques such as the hardware implementations of AQM schemes proposed in this monograph.

In addition, the implementation of novel hardware components and experimental deployment on high-end equipment poses major practical problems. Deployment on high-end (around and above 1 million euro cost per unit) routing equipment requires the adoption of a new technology by vendors of routing hardware (a market with high inertia), which is a long term objective of our research. Nonetheless, by means of our prototype architecture, validation can be performed the same way as verification through emulated scenarios as described above.

By following a well defined development methodology, we provide a more efficient and formal approach that those currently used for the development of Internet routers from major vendors [48, 39].

6.5.2   Application to Internet Traffic Analysis and Control

As shown in figure 6.10, currently deployed schemes for traffic control in the Internet, as well as most proposed alternatives, fall into one of the two following approaches [74]:

- Distributed control, with functionality distributed among the end nodes in the network and implemented by means of end-to-end transport protocols. Transmitter and receiver end nodes of packet flows cooperate so as to perform flow and congestion control as well as fair distribution of network resources.
- Queue controlers in intermediate nodes or routers. These mechanisms may discriminate packet flows and enforce resource distribution and reservation in some cases.

Thus, regulation of packet flows from sender to receivers can involve all the network nodes in the end-to-end path and is performed on both an end-to-end and a per-hop basis. Such a scheme leads to a system that comprises multiple feedback loops with complex interactions.

Both aforementioned approaches can be redefined in terms of fuzzy systems. This approach does not only provide a deeply backgrounded engineering approach but also a modeling and analysis framework for Internet traffic, which the current Internet research community lacks [31].

This section provides results, in terms of inference rate, occupation and power consumption, for a set of example applications of the described platform to the area of Internet traffic analysis and control. A summary of implementation results is presented in what follows.

We will summarize the results of the microelectronic implementation of a set of fuzzy systems on FPGAs. The focus is on the implementation results for FIM

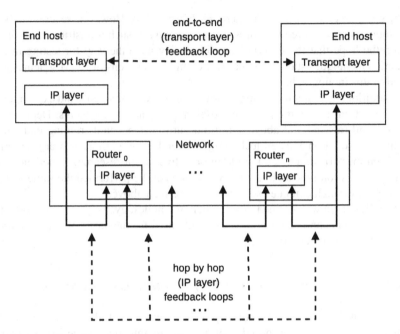

Fig. 6.10 Feedback loops in Internet traffic control systems

modules as they are the key components with higher operational requirements. The systems we will deal with in this section have been implemented on two different Xilinx® FPGA devices, part of two development boards. Both are shown in figure 6.11. The first device is a a Xilinx® Spartan-3 FPGA, xc3s1500-fg456-5 device (1.5 million equivalent gates) included in an AvNet® ADS-XLX-SP3-EVL1500 development board with standard PCI 2.0 interface. It is a very low cost device for our target application. This device has been selected with the aim of evaluating the feasibility of the approach proposed here with constrained resources. The second device, with performance closer to that of the programmable logic devices used in current routing hardware, is a Xilinx® Virtex-5 FPGA, XC5VLX50T-1FF1136-1C-ES device, included in a Xilinx® LXT FPGA ML505 development board with PCI-E interface.

The tool xfvhdl was used to generate VHDL descriptions from XFL specifications as described in section 6.5. xfvhdl provides several FIM implementation options. In particular, we set ROM based storage for the rule base and membership functions. An scheme of the design flow for generating hardware implementations of fuzzy inference systems is shown in figure 6.12. In the synthesis stage, the xfvhdl tool generates a VHDL description from an XFL specification. xfvhdl uses a cell library that contains parameterized VHDL descriptions for the basic building blocks of the specific architecture followed. Two kinds of blocks are included in this cell library: data path blocks, which implement the fuzzy inference, and control blocks, which control memory read and write operations and the operation

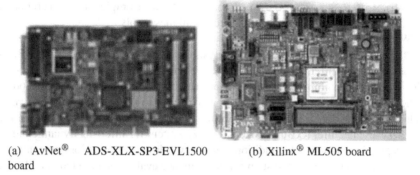

(a) AvNet® ADS-XLX-SP3-EVL1500 board

(b) Xilinx® ML505 board

Fig. 6.11 FPGA development boards used

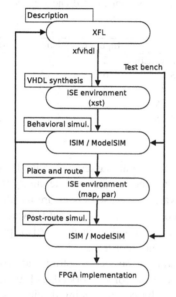

Fig. 6.12 Design flow for hardware fuzzy inference systems

scheduling control signals. These cells are defined in a way compatible with the VHDL restrictions of synthesis tools by major vendors.

The validation stage of the VHDL description generated by xfvhdl, whether behavioral or post-route is performed by means of simulation tools. Several options are available from different vendors for this stage as well as for synthesis, place and routing [75]. In our case, we have used simulators such as ISim from Xilinx and ModelSim from Mentor Graphics, as well as the tools included in the Xilinx® ISE environment [75, 76]. A performance evaluation of the FIM architecture in terms of inference speed, area and power consumption was conducted. Synthesis as well as place and routing were performed by means of the tools included in the Xilinx® ISE environment [76], namely xst and par. ISE 9.2i (more specifically, updated to

9.2.04i), xst J.40, map J.40 and par J.40 were employed for the reported prototypes. Default options were selected for all these tools.

We will focus on the fuzzy systems listed in table 6.2, which implement different intelligent Internet traffic control and analysis mechanisms. In the table, FAQMBest-Effort is a fuzzy inference system for active queue management, as described in chapter 5. TCPSS implements a fuzzy inference system for the slow start update policy in TCP-like end-to-end congestion control, also analyzed in chapter 5. DS-Select and AQMDSAF are traffic controlers for QoS enabled networks within the DiffServ architecture. RxBufferSize is a fuzzy system for inferring dynamic buffer size depending on network conditions (one-way delay and packet loss percentage). RTperf is a fuzzy inference system for performance evaluation targeted at real-time network applications and services.

Table 6.2 Fuzzy inference systems implemented and their complexity in terms of inputs, linguistic terms and rules

System	Inputs	Linguistic terms	Rules
FAQMBestEffort	2	7,7,7	37
TCPSS	6	5,5,5,5,5,5,5	24
DSSelect	2	5,5,2	17
AQMDSAF	2	3,3,4	7
RxBufferSize	2	5,5,5	25
RTperf	4	5,5,5,5,5	27

As outlined in chapter 1, several classes of service have been specified to date within the standard differentiated services architecture [10], such as expedited forwarding and assured forwarding. These classes are usually referred to as per hop behavior (PHB) groups. For instance, the assured forwarding (AF) PHB group defines several drop precedences for data packets. Also, packets within data flows belonging to an AF class are not reordered.

When implemented, quality of service constraints are commonly enforced at the network edges. Routing hardware for network edges [39] usually integrates support for quality of service (QoS) based on queuing disciplines, using separated packet queues for different classes of service. With regards to active queue management, these systems comprise components belonging to two main categories: queuing disciplines (or schedulers) that select among eligible packets from a set of queues, and traffic regulators (or policers).

Common queuing disciplines include weighted priority based scheduling and weighted round robin (WRR), among many others. Common regulators include first in-first out (FIFO), token bucket filter (TBF) and RED, among others. This way, a wide range of service disciplines can be implemented by selecting a set of disciplines and regulators.

Fuzzy inference systems can be designed for implementing both queuing disciplines and schedulers. For classification purposes, DSSelect has been defined as a fuzzy classifier for class of service enabled networks. DSSelect balances priorities between two different classes of service within the DiffServ architecture, outlined in chapter 1. The inference system has two inputs, defined as the number of packets in two separated packet queues: a queue for expedited non-assured (EF) flows and a queue for non-expedited assured (AF) flows, respectively. The rule base of the system balances priorities depending on current queue lengths following the approach proposed in [79]. The output of the system selects the next packet to be forwarded from one of the two queues.

Table 6.3 show the DSSelect rulebase. For the inputs, q_{EF} and q_{AF}, representing the length of the queues of EF packets and AF packets respectively, 5 linguistic terms are defined: ZO, S, M, B, VB, meaning "zero," "small," "medium," "big," and "very big," respectively. Two linguistic terms for the output, C, are defined, one for each class of service: "EF" and "AF". The membership functions for these terms are triangular and uniformly distributed in the input space. The "EF" output value indicates that a packet from the EF queue is selected for forwarding, while the "AF" output value indicates that the next packet to be forwarded should be selected from the AF queue.

The rationale proposed in [79] is applied to define the rule base. When the length of q_{EF} is large and the length of q_{AF} is small, EF packets are selected in order to avoid high delays. In the opposite case, AF packets are selected. When both queues have similar lengths, two possibilities arise. If both queues are short, priority is given to real-time traffic and thus EF packets are selected. If both queues are large, AF packets are forwarded as the drop probabilities increase and thus the drop priority of AF packets becomes more important than the delay priority of EF packets.

Table 6.3 DSSelect Rule Base

C		q_{EF}				
		Z	S	M	B	VB
q_{AF}	Z	EF	EF	EF	EF	EF
	S	AF	EF	EF	EF	EF
	M	AF	AF	EF	EF	EF
	B	AF	AF	AF	EF	EF
	VB	AF	AF	AF	AF	AF

In an analogous manner to the FAQMBestEffort system described in the revious chapter, we have defined a controler for the assured forwarding PHB within the differentiated services architecture. This system, AQMDSAF, has the same two inputs and one output as FAQMBestEffort. The rule base of AQMDSAF is also similar to that of FAQMBestEffort. However, the number of linguistic labels is lower and the rule base is defined so that a higher guarantee of delivery is provided, i.e., higher forwarding probabilities are considered. Both DSSelect and AQMDSAF were defined as initially proposed in [79, 80]. These systems were evaluated in the same scenario used in chapter 5 in order to evaluate FAQMBestEffort. Results confirmed satisfactory performance and the ability to solve open problems in the differentiated services architecture [80].

RTperf is a fuzzy inference system for overall network performance evaluation targeted at real-time network applications and services. Many approaches to network performance evaluation have been proposed to date [13, 67]. Besides their direct application to analysis, monitoring, traffic engineering and capacity planning, among many other applications, the results of a performance evaluation system can be used as additional inputs to advanced traffic control systems, both centralized and end-to-end.

The RTperf fuzzy inference system provides a measure of the adequacy of current network conditions for time constrained traffic or real-time network applications. Based on four scalar numerical measures of current network performance (one-way delay, round-trip delay, packet loss percentage and inter-packet delay variation), RTperf provides as output a fuzzy degree of certainty about the adequacy of current conditions for this class of applications. RTperf has been defined on the basis of prior accepted linguistic knowledge about network performance. Figure 6.13 shows a few sample rules (in XFL3 format) from the rulebase of the RTperf inference system.

```
rulebase RTperf (Towd owd, Trtt rtt , Tjitter jitter , Tloss loss : Tperf perf)
{
    if ( owd == ZO & rtt == ZO & jitter == ZO & loss == ZO ) -> perf = OPTIMUM;
    if ( owd == S & rtt == S  & jitter == S  & loss == S )   -> perf = GOOD;
    if ( owd == M & rtt == ZO & jitter == ZO & loss == ZO )  -> perf = POOR;

                              .
                              .

    if (owd == VB)       -> perf = BAD;
    if (rtt == VB)       -> perf = BAD;
    if (jitter == VB)    -> perf = BAD;
    if (loss  == VB)     -> perf = BAD;
}
```

Fig. 6.13 Sample rules (in XFL language) from the rulebase of the RTperf fuzzy inference system

Figure 6.14 shows a summary of implementation results for the above listed systems. A precision of 8 bits is used for inputs, outputs and membership functions. An 8 bits configuration was found to satisfy overall precision requirements with no practical difference as compared to a 16 bits configuration. These systems perform dynamic adjustment of reception buffers, active queue management in best-effort and differentiated services schemes, and analysis of the network performance from a real-time application viewpoint.

Tables 6.4 and 6.5 show post-synthesis and post-implementation area and estimations. Tables 6.5 and 6.7 show post-synthesis and post-implementation timing estimations, respectively. The equivalent gate count shown excludes the equivalent gate count for the input-output blocks, i.e., only the equivalent gates for the fuzzy inference system design are accounted. The timing parameters considered for post-synthesis are defined as follows:

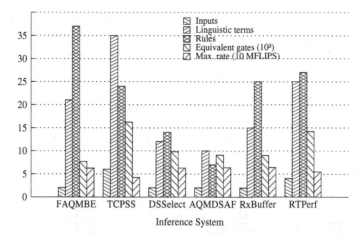

Fig. 6.14 FPGA implementation results summary for the Spartan-3 device. Complexity of systems is given in terms of the number of inputs, linguistic terms and rules. Complexity of the implementations is given in terms of equivalent gates and maximum inference rate.

- T_{min}: minimum period.
- v_{max}: maximum frequency.
- $t_{input\,clk}$: minimum input arrival time before clock.
- $t_{output\,clk}$: maximum output required time after clock.

The timing parameters included in the place and route reports below are defined as follows:

- cd_{avg}: average connection delay.
- pd_{max}: average pin delay.
- cd_{avg}^{10}: average connection delay on the 10 worst nets.
- $t_{max}^{setup-clk}$: maximum setup time to the edges of the clock signal in the inputs. Hold times in inputs are in general significantly lower (around 60 or 70%) than setup times.
- $t_{,max}^{clk-pad}$: maximum delay after clock edges to output pads.

Table 6.4 Post-synthesis area estimations (Spartan-3). Inputs: 8 bits, outputs: 8 bits, MFs: 8 bits.

System	Area (total/%)			
	Slices	Slice Flip Flops	LUTs	Bonded IOBs
FAQMBestEffort	187 (1%)	147 (0%)	339 (1%)	28 (8%)
TCPSS	847 (6%)	154 (0%)	1550 (5%)	60 (18%)
DSSelect	320 (2%)	144 (0%)	586 (2%)	28 (8%)
AQMDSAF	213 (1%)	145 (0%)	387 (1%)	27 (8%)
RxBufferSize	267 (2%)	148 (0%)	488 (1%)	28 (8%)
RTperf	542 (4%)	152 (0%)	990 (3%)	44 (13%)

Table 6.5 Post-synthesis timing estimations (Spartan-3). Inputs: 8 bits, outputs: 8 bits, MFs: 8 bits.

System	$T_{min}\,(ns)$	$v_{max}\,(Mhz)$	$t_{input\,clk}\,(ns)$	$t_{output\,clk}\,(ns)$
FAQMBestEffort	16.048	62.312	18.174	6.216
TCPSS	23.742	42.119	30.368	6.212
DSSelect	15.918	62.822	19.784	6.216
AQMDSAF	15.665	63.838	19.353	6.216
RxBufferSize	15.648	63.908	20.007	6.216
RTperf	18.347	54.503	25.824	6.216

Table 6.6 Post-implementation area results (Spartan-3). Inputs: 8 bits, Outputs: 8 bits, MFs: 8 bits.

System		Area (total/%)			
	E.G.	Slices	Slice Flip-Flops	LUTs	Bonded IOBs
FAQMBestEffort	7676	202 (1%)	139 (1%)	314 (1%)	28 (8%)
TCPSS	16228	813 (6%)	146 (1%)	1558 (1%)	60 (18%)
DSSelect	9365	360 (2%)	136 (1%)	588 (2%)	28 (8%)
AQMDSAF	8014	226 (1%)	137 (1%)	390 (1%)	28 (8%)
RxBufferSize	8800	278 (2%)	140 (1%)	467 (1%)	28 (8%)
RTperf	12111	577 (4%)	144 (1%)	970 (3%)	44 (13%)

Table 6.7 Post-implementation timing results (Spartan-3). Inputs: 8 bits, Outputs: 8 bits, MFs: 8 bits. Times are expressed in ns.

System	cd_{avg}	pd_{max}	cd_{avg}^{10}	$t_{max}^{setup-clk}$	$t_{max}^{clk-pad}$
FAQMBestEffort	1.412	6.043	4.428	20.888	6.421
TCPSS	1.928	6.317	5.605	35.560	6.473
DSSelect	1.769	6.149	4.903	22.378	6.415
AQMDSAF	1.560	8.663	5.330	23.898	6.444
RxBufferSize	1.708	6.545	5.209	24.269	6.421
RTperf	1.846	6.444	5.811	29.513	6.421

Figure 6.15 shows a summary of implementation results for the above listed systems using the Virtex-5 device. As before, a precision of 8 bits is used for inputs, outputs and membership functions.

Tables 6.8 and 6.9 show post-synthesis and post-implementation area and estimations. Tables 6.9 and 6.11 show post-synthesis and post-implementation timing estimations, respectively.

Table 6.8 Post-synthesis area estimations (Virtex-5). Inputs: 8 bits, outputs: 8 bits, MFs: 8 bits.

System	Area (total/%)		
	Slice Registers	Slice LUTs	Bonded IOBs
FAQMBestEffort	120 (0%)	263 (0%)	28 (5%)
TCPSS	130 (0%)	879 (3%)	60 (12%)
DSSelect	117 (0%)	282 (0%)	28 (5%)
AQMDSAF	118 (0%)	257 (0%)	28 (5%)
RxBufferSize	120 (2%)	271 (0%)	28 (5%)
RTperf	126 (0%)	497 (1%)	44 (9%)

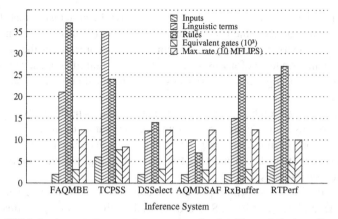

Fig. 6.15 FPGA implementation results summary for the Virtex-5 device. Complexity of systems is given in terms of the number of inputs, linguistic terms and rules. Complexity of the implementations is given in terms of equivalent gates and maximum inference rate.

Table 6.9 Post-synthesis timing estimations (Virtex-5). Inputs: 8 bits, outputs: 8 bits, MFs: 8 bits.

System	$T_{min}(ns)$	$v_{max}(Mhz)$	$t_{input\,clk}(ns)$	$t_{output\,clk}(ns)$
FAQMBestEffort	8.126	123.062	10.333	3.391
TCPSS	12.012	83.250	15.643	3.393
DSSelect	8.170	122.399	10.634	3.391
AQMDSAF	8.136	122.911	10.631	3.391
RxBufferSize	8.126	123.062	10.442	3.391
RTperf	10.031	99.691	13.811	3.392

Table 6.10 Post-implementation area results (Virtex-5). Inputs: 8 bits, Outputs: 8 bits, MFs: 8 bits.

System	Area (total/%)			
	E.G.	Slice Registers	Slice LUTs	Bonded IOBs
FAQMBestEffort	3077	120 (1%)	264 (1%)	28 (5%)
TCPSS	7721	130 (1%)	896 (1%)	60 (12%)
DSSelect	3229	117 (1%)	297 (1%)	28 (5%)
AQMDSAF	3031	118 (1%)	264 (1%)	28 (8%)
RxBufferSize	3165	120 (1%)	309 (1%)	28 (5%)
RTperf	4764	126 (1%)	501 (3%)	44 (13%)

Table 6.11 Post-implementation timing results (Virtex-5). Inputs: 8 bits, Outputs: 8 bits, MFs: 8 bits. Times are expressed in ns.

System	cd_{avg}	pd_{max}	cd_{avg}^{10}	$t_{max}^{setup-clk}$	$t_{max}^{clk-pad}$
FAQMBestEffort	1.066	2.869	2.441	9.353	9.930
TCPSS	1.167	2.642	2.446	16.054	8.875
DSSelect	1.356	3.149	2.839	10.013	8.972
AQMDSAF	1.135	2.827	2.489	9.378	9.401
RxBufferSize	1.206	3.010	2.659	11.417	8.861
RTperf	1.166	2.702	2.395	13.849	8.895

From the results shown in the tables we can draw the conclusion that the speed variability with regards to the fuzzy system complexity (measured as number of inputs, rules, membership functions and accuracy) that characterizes the architecture and implementation technology used are within the required bounds. This makes it possible to achieve high inference rates even for the most complex systems developed here.

As detailed in section 6.2, in current router architectures traffic analysis and regulation subsystems are integrated into output interface cards together with the virtual output queue processing logic. In general, these subsystems can be thought of as queue schedulers that run, at most, at frequencies around the maximum per interface packet processing speed.

As for inference rate, prototypes implemented on a Xilinx® Spartan-3 FPGA could achieve above 60 MFLIPS. As discussed at the beginning of this section, packet queue controlers must have a fine packet processing granularity in order to cope with packet bursts and properly control small packet queues. The finest processing granularity is only possible if the maximum packet processing rate is equal to or greater than the maximum number of packets per second accepted by packet queues. Routers from the Cisco 12000, Cisco CRS series as well as Juniper M and T series process up to 25 Mp/s, 40 Mp/s, 24 Mp/s and 60 Mp/s, respectively, per interface output queue [19, 17, 32]. Thus, even a prototype implementation using a low cost FPGA can provide the required inference speed for the finest packet processing granularity in current high performance router families.

The tool Xilinx® XPower Analyzer 10.1.03, version K.39 was used in order to study the power consumption of the implementations described above. The tool uses so-called production characterization data for both the Spartan-3 and the Virtex-5 devices. That is, enough production silicon of these devices have been characterized to provide full power correlation over numerous production lots. Also, characterization data for all blocks in the device fabric are included. The power analysis was performed for an ambient temperature of 25°C and voltage sources set to $V_{CCINT} = 1.2V$, $V_{CCAUX} = 2.5V$, $V_{CCO25} = 2.5V$, for the case of the Spartan-3 device, and $V_{CCINT} = 1V$, $V_{CCAUX} = 2.5V$, $V_{CCO25} = 2.5V$, for the case of the Virtex-5 device.

Quiescent power consumption, around 150 and $443\,mW$, for the Spartan-3 and Virtex-5 devices, respectively, are negligible for current high performance routers as it is more than three orders of magnitude below the overall consumption of an output interface card.

Regarding dynamic power of the fuzzy inference modules, two cases, shown in figures 6.16(a) and 6.16(b), respectively, are considered. In both cases, dynamic power is analyzed for frequencies ranging from 16 Mhz through 150 Mhz. Figure 6.16(a) shows the power consumption results when input stimuli are generated using the standard test bench generated by the xfvhdl tool. Figure 6.16(b) shows the results for a pessimistic power analysis where the toggle rates for inputs and outputs as well as flip-flops are set to 100% of the clock frequency.

In the xfvhdl test bench simulation, the values of the inputs are progressively increased starting from all inputs set to 0 and ending when all input signals are set

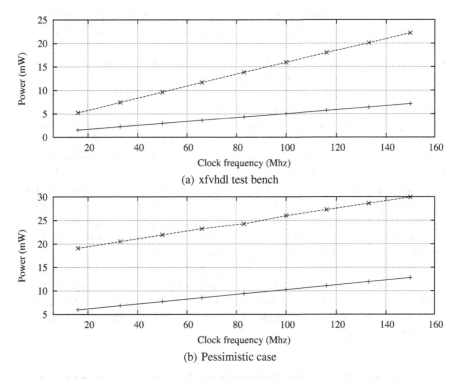

Fig. 6.16 Power analysis of the FAQMBestEffort system for different frequencies. Two stimuli are used: xfvhdl test bench and a pessimistic case. Spartan-3 device: +, continuous line; Virtex-5 device: ×, dashed line.

to their maximum value. This way, all the possible combinations of values for the inputs are explored in $2^{i \cdot n}$ cycles, where i is the number of inputs to the system and n the bit precision of the inputs, assuming the same precision is used for every input.

As expected [1], dynamic power consumption depends linearly on the system clock frequency and the number of toggling nodes. Besides, it can be observed that the power consumption of the fuzzy inference modules is negligible as compared to the overall power requirements of the systems they are aimed to be integrated in. Junction temperature ranges from 29.6°C through 29.8°C for the Spartan-3 device, and from 27.8°C through 28.0°C for the Virtex-5 device.

A number of hardware implementations of fuzzy systems for tasks belonging to the physical and link layers of communications systems (such as signal filtering) have been reported in the literature. Some of them are based on FPGAs. However, we are not aware of proposals of FPGA based implementations of fuzzy systems applied to network traffic control and network layer tasks in general.

The most closely related work we are aware of [3] reports an VLSI implementation that attains an inference rate of 3.3 MFLIPS for a 60 Mhz clock, which would not fulfill current requirements. However, this work dates back more than 10 years. Also, we note though there are major differences between our proposal and the aforementioned work. Thus, a direct comparison cannot be done. In the latter case, the target application is traffic control for ATM networks. Additionally, it is based on a substantially different architecture (using the concept of fuzzy processor) and the system is implemented as an ASIC. The fuzzy processor employed requires 18 cycles for a fuzzy inference. We note though that in a more general context, digital realizations achieving inference rates of the order of the MFLIPS were reported during the 1990s [46]. In particular, the authors of [4] report an inference rate of the order of the MFLIPS for an architecture able to implement complex fuzzy systems. More recently, the potential to achieve inference rates of the order of 50 MFLIPS with digital realizations and FPGA based systems in particular has been widely reported in the literature [53, 14].

In addition, our solution provides a development methodology and a tool chain that fulfill an important gap in current custom, unscalable and inefficient design schemes [39]. The complexity introduced into a routing system is negligible as compared to the complexity increment that is taking place at present and is expected to take place in foreseeable future high performance routers.

In fact, an FPGA approach to the implementation of router components is in line with the current trend towards FPGA based development router design of major vendors [48, 34]. In particular, providing a PCI/PCI-E compliant interface eases integration of fuzzy inference modules as processing units within current network processing architectures.

6.6 Computational Intelligence Based Processing Subsystems in Routing Architectures

There are a number of implementation options for incorporating computational intelligence systems and fuzzy inference systems in particular as components of Internet routers. In the context of the development platform described above, here we discuss these options with the focus on controls systems of the kind described in chapter 5 .

As a first general option, the software implementations (in C, C++ or Java) that can be generated with the tools included in the Xfuzzy environment can be used in those routers with processing units supporting these programming languages. This is the case of the units currently used by most vendors, such as the Xscale systems of the Intel® NPUs [43].

Nonetheless, software implementations running on high level processing units can be practical for some high level control functions, functions of the slow path, or functions with a low execution rate. However, they cannot attain the high execution rates (above millions of inference per second for current technologies) needed for low level processing of high bandwidth network links.

The aforementioned architecture for hardware implementation of fuzzy inference modules [11] is a simplified architecture where the degree of overlapping is limited and simplified defuzzification methods are used. This fact has been shown to solve the general challenge currently posed in the design of packet processing components for high performance routers: developing sufficiently fast implementation schemes. In this aspect, the FIM implementation architecture used here fits in general in current and future generation routing architectures.

This way, it is possible to integrate FIMs in Internet routers as additional processing units in current NPUs. In addition, using a PCI/PCI-E interface for the whole fuzzy inference system eases its incorporation into current and next generation NPUs that are in many cases based in the PCI standard, such as the general architecture of the Intel® NPUs [43].

A related issue we take into consideration here is the physical implementation and distribution of hardware FIMs. Current routers usually include hardware modules for implementing queue management schemes in the circuitry of each connection port in input/output cards which in most cases utilize specific high performance FPGA devices as implementation option.

The fuzzy systems described in this monograph can be implemented as follows:

- Performance measurement and analysis such as RTperf can be implemented following two alternatives:

  - Direct processing of traffic. In this case FIM must be integrated into input cards.
  - High level processing and analysis. In this case FIMs can be implemented as processing units physically located in the NPU.

- Queuing control systems, including FAQMBestEffort, DSSelect, AQMDSAF and AQMDSBE, can be integrated into two components depending on the level of operation.

  - For processing packet queues, a processing module has to be integrated for each and every output queue, i.e., into the enabled output cards. However, in those systems implementing virtual output queuing these modules would be physically located in the input cards.
  - For high level processing within the DiffServ architecture, fuzzy systems must be implemented as control modules of the NPU.

Following this scheme it is possible to define processing units for other applications of computational intelligence techniques such as traffic analysis and packet classification.

By comparing the performance of the FPGA-based prototypes described in section 6.5.2 against operational requirements of current high-end routers the following conclusions can be drawn:

- Power consumption is negligible in the case of the prototypes described, being more than three orders of magnitude below the overall consumption of the system.
- Regarding the inference rate, the prototypes developed on FPGAs belonging to the Spartan-3 family can attain inference rates around 50 MFLIPS. Routers of the

Cisco 12000 and CRS series as well as Juniper M and T series can attain processing rates of around 25 Mp/s and up to 60 Mp/s [18, 69]. Under these conditions, it is clear that even prototypes implemented on medium-low cost FPGA devices can attain the inference rates currently required.

- By incorporating fuzzy inference processing units the complexity increase of the overall routing system is small as compared to the increase that is currently taking place as well as the anticipated complexity growth derived from further developments in routing architectures for high performance [54]. In addition, the availability of specific languages, methodologies and CAD tools for fuzzy systems is an significant advantage over common custom design procedures, highly inefficient and lacking scalability, currently applied to the development of router components [54].

This considerations lead us to remark that the described hardware prototypes of fuzzy inference systems are also suitable for medium performance routers and communications equipment, such as the Catalyst series from Cisco [30] where incorporating intelligent systems for active queue management and class based queuing (CBQ) is specially relevant.

We remark finally that developing computational intelligence based mechanisms for high level decision making in routers is becoming more and more an attractive venue for research and development, with increasing potential to be eventually implemented in commercial products. In fact, considering the evolution that is taking place in router design as well as research and vendors plans for the future expert systems for supporting traffic engineering will become an important area of research. Thus, a possible approach to the design of computational intelligence techniques applied to networking is to differentiate layers for intelligent measuring, analysis and processing of network traffic.

6.7 Conclusions

Here we have described the major issues to be considered when implementing fuzzy systems for network traffic analysis, modeling and control in routers. Current technological constraints and trends as well as architectural factors have been addressed. A platform that satisfies these constraints has been proposed for the development of FPGA-based implementations of fuzzy systems for traffic analysis and control.

We have first described the architectural aspects constraining the implementation of fuzzy methods. With this in mind, we have defined a PCI/PCI-E compatible modular architecture for the integration of fuzzy subsystems into current router architectures. We have then described a generic and open platform that eases the development of fuzzy systems and its implementation as SoPC on FPGAs. The platform integrates both open tools and open IP cores. In addition, a systematic development methodology and design flow is followed. The Xfuzzy environment automates development from initial high-level specifications to synthesizable VHDL. Fuzzy inference modules are integrated into a SoPC architecture made of open IP cores that is suitable for developing fuzzy systems applied to networking among many other possible areas. Those SoPC developed using the outlined architecture can be integrated in current router architectures as processing units.

Then, we have looked at the inference speed and resource consumption aspects, with the focus on inference speed. The fuzzy inference systems analyzed have been shown to satisfy operational requirements of current and future high performance routing hardware in terms of both inference speed and resource consumption. Even prototype implementations using low cost FPGAs can provide the required inference speed in current high performance routers with a very low cost in terms of area and consumption. In addition, the prototypes have been designed for easy integration with routing architectures currently deployed in the Internet.

The open development platform presented paves the way for further development of efficient intelligent traffic controlers. We also entertain the hope that the availability of such a platform will foster the development of fuzzy systems for a number of areas where intelligent analysis systems are sought, such as packet and flow identification, classification and filtering, among many others. The natural continuation of the work presented in this chapter is the application of the platform in experimental network infrastructures for clean-slate design of network protocols and mechanisms, which are expected to be generally available in the near future.

References

[1] Abusaidi, P., Klein, M., Philofsky, B.: Virtex-5 FPGA system power design considerations. Tech. Rep. WP285 (v1.0), Xilinx Inc. (2008),
http://www.xilinx.com/support/documentation/
white_papers/wp285.pdf

[2] Appenzeller, G., Keslassy, I., McKeown, N.: Sizing Router Buffers. In: ACM Special Interest Group on Data Communications (SIGCOMM) Conference, Portland, OR, USA, pp. 281–292 (2004)

[3] Ascia, G., Catania, V., Ficili, G., Palazzo, S., Panno, D.: A VLSI Fuzzy Expert System for Real-Time Traffic Control in ATM Networks. IEEE Transactions in Fuzzy Systems 5(1), 20–31 (1997)

[4] Ascia, G., Catania, V., Russo, M.: VLSI hardware architecture for complex fuzzy systems. IEEE Transactions in Fuzzy Systems 7(5), 553–570 (1999)

[5] Aweya, J.: IP router architectures: an overview. International Journal of Communication Systems 14(5), 447–475 (2001)

[6] Barriga, A., Sánchez-Solano, S., Brox, P., Cabrera, A., Baturone, I.: Modelling and implementation of fuzzy systems based on VHDL. International Journal of Approximate Reasoning 41(2), 164–178 (2006)

[7] Baturone, I., Sanchez-Solano, S.: Microelectronic Design of Universal Fuzzy Controllers. In: X Congreso Español sobre Tecnologías y Lógica Fuzzy (ESTYLF 2000), pp. 247–252 (2000)

[8] Baturone, I., Sánchez-Solano, S., Barriga, A., Huertas, J.L.: Implementation of CMOS Fuzzy Controllers as Mixed-Signal Integrated Circuits. IEEE Transactions on Fuzzy Systems 5(1), 1–19 (1998)

[9] Baturone, I., Barriga, A., Sanchez-Solano, S., Jimenez, C.J., Lopez, D.R.: Microelectronic Design of Fuzzy Logic-Based Systems. CRC Press, Boca Raton (2000) ISBN: 0-8493-0091-6

[10] Blake, S., Black, D.L., Carlson, M.A., Davies, E., Wang, Z., Weiss, W.: An Architecture for Differentiated Services. RFC 2475, Internet Engineering Task Force, Network Working Group, category: Informational (1998)

[11] Cabrera, A., Sánchez-Solano, S., Senhadji, R., Barriga, A., Jimenez, C.J.: Hardware/Software Codesign Methodology for Fuzzy Controller Implementation. In: 11th IEEE International Conference on Fuzzy Systems (FUZZ-IEEE 2002), pp. 464–469 (2002)

[12] Cabrera, A., Sánchez-Solano, S., Brox, P., Barriga, A., Senhadji, R.: Hardware/ Software Codesign of Configurable Fuzzy Control Systems. Applied Soft Computing 4(3), 271–285 (2004)

[13] Calyam, P., Krymskiy, D., Sridharan, M., Schopis, P.: Active and Passive Measurements on Campus, Regional and National Network Backbone Paths. In: 14th IEEE International Conference on Computer Communications and Networks (ICCCN 2005), San Diego, California, USA, pp. 537–542 (2005)

[14] Cao, Q., Lim, M.H., Li, J.H., Ong, Y.S., Ng, W.L.: A context switchable fuzzy inference chip. IEEE Transactions on Fuzzy Systems 14(4), 552–567 (2006)

[15] Chao, H.J., Liu, B.: High Performance Switches and Routers. Wiley-IEEE Press (2007) ISBN: 978-0470053676

[16] Chi J, et al.: WISHBONE Conbus IP Core (2008),
http://www.opencores.org/projects.cgi/web/wb_conbus/

[17] CISCO CRS-1, Cisco CRS-1 Distributed Route Processor Datasheet. Cisco Systems, Inc. (2008),
http://www.cisco.com/en/US/prod/collateral/routers/ps5763/product_data_sheet0900aecd80501c66.html

[18] Cisco Systems, Inc., The Evolution of High-End Router Architectures. Tech. rep., Basic Scalability and Performance Considerations for Evaluating Large-Scale Router Designs (2001),
http://www.cisco.com/en/US/products/hw/routers/ps167/

[19] Cisco Systems, Inc., Portable product sheet: Router switching performance in packets per second (pps). Tech. rep., Cisco Systems (2006),
http://www.cisco.com/web/partners/downloads/765/tools/quickreference/routerperformance.pdf

[20] Comer, D.E.: Network Systems Design Using Network Processors. Prentice-Hall, Upper Saddle River (2003) ISBN: 0-13-141792-4

[21] Comer, D.E.: Network Processors: Programmable Technology for Building Network Systems. The Internet Protocol Journal 7(4), 3–12 (2004)

[22] Comer, D.E.: Network Systems Design Using Network Processors: Intel 2XXX Version. Pearson Prentice Hall, Upper Saddle River (2005) ISBN: 9780131872868

[23] Crowley, P., Franklin, M.A., Hadimioglu, H., Onufryk, P.Z. (eds.): Network Processor Design: Issues and Practices, Computer Architecture and Design, vol. 1. Morgan Kaufmann Publishers, San Francisco (2002) ISBN: 978-1558608757

[24] Crowley, P., Franklin, M.A., Hadimioglu, H., Onufryk, P.Z. (eds.): Network Processor Design: Issues and Practices, Computer Architecture and Design, vol. 2. Morgan Kaufmann Publishers, San Francisco (2003) ISBN: 978-0121981570

[25] Crowley, P., Franklin, M.A., Hadimioglu, H., Onufryk, P.Z. (eds.): Network Processor Design: Issues and Practices, Computer Architecture and Design, vol. 3. Morgan Kaufmann Publishers, San Francisco (2005) ISBN: 978-0120884766

[26] Di Fatta, G., Hoffmann, F., Re, G.L., Urso, A.: A Genetic Algorithm for the Design of a Fuzzy Controller for Active Queue Management. IEEE Transactions on Systems, Man and Cybernetics, Part C: Applications and Reviews 33(3), 313–334 (2003)

[27] Dolenc, M., Markovic, T.: PCI IP Core Specification. Tech. Rep. Rev. 1.2, Open-Cores.Org Free Open Source IP Cores and Chip Design (2004)

[28] Dualibe, C., Verleysen, M., Jespers, P.G.: Design of Analog Fuzzy Logic Controllers in CMOS Technologies. Kluwer Academic Publishers, Dordrecht (2003) ISBN: 1-4020-7359-3

[29] Endace Limited, DAG Network Monitoring Cards (2008),
http://www.endace.com/our-products/
dag-network-monitoring-cards/

[30] Flannagan, M., Froom, R., Turek, K.: Cisco Catalyst Qos. Quality of Service in Campus Networks. Cisco Systems, Inc. (2003) ISBN: 1-58705-120-6

[31] Floyd, S., Kohler, E.: Internet Research Needs Better Models. ACM SIGCOMM Computer Communication Review 33(1), 29–34 (2003)

[32] Gardner, E., et al.: T1600© Internet Routing Node Hardware Guide, 2nd edn. Juniper Networks, Inc. (2008),
http://www.juniper.net/techpubs/hardware/t-series.html

[33] Ghosh, S., Razouqi, Q., Schumacher, H.J., Celmins, A.: A Survey of Recent Advances in Fuzzy Logic in Telecommunications Networks and New Challenges. IEEE Transactions on Fuzzy Systems 6(39), 443–447 (1998)

[34] Goralski, W.J.: Juniper and Cisco Routing. Policy and Protocols for Multivendor IP Networks. Wiley Publishing Inc., Indianapolis (2002) ISBN: 0-471-21592-9

[35] Grosse, E., Lakshman, Y.N.: Network Processors Applied to IPv4/IPv6 Transition. IEEE Network 17(4), 35–39 (2003)

[36] Herveille, R., et al.: WISHBONE System-on-Chip (SoC) Interconnection Arhitecture for Portable IP Cores. Tech. Rep. Revision B.3, OpenCores Organization (2002)

[37] Heyde, A.A.: Investigating the performance of Endace DAG monitoring hardware and Intel NICs in the context of Lawful Interception. Tech. Rep. 080222A, Centre for Advanced Internet Architectures (CAIA), Swinburne University of Technology (2008),
http://caia.swin.edu.au/reports/

[38] Hidell, M.: Decentralized Modular Router Architectures. PhD thesis, KTH-Royal Institute of Technology (2006)

[39] Hidell, M., Sjödin, P., Hagsand, O.: Control and Forwarding Plane Interaction in Distributed Routers. Tech. Rep. TRITA-S3-LCN-0501, Laboratory for Communication Networks, Department of Signals, Sensors, and Systems. KTH Royal Institute of Technology, Stockholm, Sweden (2005)

[40] IBM:SA-14-2528-02 On-Chip Peripheral Bus: Architecture Specifications. Version 2.1. International Business Machines Corporation (2001)

[41] Information Sciences Institute University of Southern California, Viterbi School of Engineering, The Network Simulator – ns-2 (2008),
http://www.isi.edu/nsnam/ns/

[42] Intel Corporation, Intel IXP12XX Product Line of Network Processors. Tech. rep., Intel Corporation (2008a),
http://www.intel.com/design/network/products/
npfamily/ixp1200.htm

[43] Intel Corporation, Intel XScale Technology (2008b),
     http://www.intel.com/design/intelxscale/
[44] Internet Engineering Task Force, Transport Area Forwarding and Control Element Sep-
     aration (ForCES) Working Group. Internet Engineering Task Force, Transport Area
     (2008),
     http://www.ietf.org/html.charters/forces-charter.html
[45] Iyer, S., McKeown, N.: Analysis of the Parallel Packet Switch Architecture. IEEE/ACM
     Transactions on Networking 11(2) (2003)
[46] Kandel, A., Langholz, G.: Fuzzy Hardware, Architectures and Applications. Kluwer
     Academic Publishers, Norwell (1997)
[47] Keshav, S., Sharma, R.: Issues and trends in router design. IEEE Communications Mag-
     azine 36(5), 144–151 (1998)
[48] Kloth, A.K.: Advanced Router Architectures. CRC Press, Boca Raton (2005) ISBN:
     0849335507
[49] Lago, E., Jiménez, C., Lopez, D., Sánchez-Solano, S., Barriga, A.: XFVHDL: A Tool
     for the Synthesis of Fuzzy Logic Controllers. In: Design Automation and Test in Europe
     (DATE 2008), Paris, France, pp. 102–107 (1998)
[50] Lekkas, P.A.: Network Processors: Architectures, Protocols and Platforms, 1st edn.
     McGraw-Hill Professional, New York (2003) ISBN: 978-0071409865
[51] Lhotka, L., Novotný, J., et al.: Description of COMBO cards (2008),
     http://www.liberouter.org/hardware.php
[52] Liang, Q., Karnik, N., Mendel, J.M.: Connection Admission Control in ATM Networks
     Using Survey-Based Type-2 Fuzzy Logic Systems. IEEE Transactions on Systems, Man
     and Cybernetics Part C: Applications and Reviews 30(3), 329–339 (2000)
[53] Manzoul, M.A., Jayabharathi, D.: FPGA for fuzzy controllers. IEEE Transactions on
     Systems, Man and Cybernetics 15(1), 213–216 (1995)
[54] McKeown, N.: Growth in Router Capacity. In: IPAM Workshop on Large-Scale Com-
     munication Networks, Lake Arrowhead, CA, USA (2003),
     http://tiny-tera.stanford.edu/~nickm/talks/index.html
[55] Montesino-Pouzols, F., Barriga, A., Lopez, D.R., Sánchez-Solano, S.: FPGA Based Im-
     plementation of Fuzzy Controllers for Internet Traffic. In: XII IBERCHIP Workshop,
     San José, Costa Rica, pp. 34–41 (2006)
[56] Montesino-Pouzols, F., Lopez, D.R., Barriga, A., Sánchez-Solano, S.: Fuzzy End-to-
     End Rate Control for Internet Transport Protocols. In: 15th IEEE International Confer-
     ence on Fuzzy Systems (FUZZ-IEEE 2006), Vancouver, Canada, pp. 1347–1354 (2006)
[57] Montesino-Pouzols, F., Barriga, A., Lopez, D.R., Sánchez-Solano, S.: Open FPGA-
     Based Development Platform for Fuzzy Systems with Applications to Communica-
     tions. In: XXII Conference on Design of Circuits and Integrated Systems (DCIS 2007),
     Seville, Spain, pp. 323–328 (2007)
[58] Montesino-Pouzols, F., Barriga, A., Lopez, D.R., Sánchez-Solano, S.: Linguistic Sum-
     marization of Network Traffic Flows. In: 17th IEEE International Conference on Fuzzy
     Systems (FUZZ-IEEE 2008), IEEE World Congress on Computational Intelligence,
     Hong Kong, China, pp. 619–624 (2008)
[59] Moreno-Velo, F., Sánchez-Solano, S., Barriga, A., Baturone, I., López, D.: XFL3:
     a New Fuzzy System Specification Language. In: 5th WSEAS/IEEE Multiconfer-
     ence on Circuits, Systems, Communications and Computers (CSCC 2001), Rethymon,
     pp. 361–366 (2001)

[60] Moreno-Velo, F.J.: Un entorno de desarrollo para sistemas de inferencia complejos basados en lógica difusa. PhD thesis, University of Seville (2003)

[61] Moreno-Velo, F.J., Baturone, I., Sánchez-Solano, S., Barriga, A.: Rapid Design of Fuzzy Systems With Xfuzzy. In: 12th IEEE International Conference on Fuzzy Systems (FUZZ-IEEE 2003), St. Louis, MO, USA, pp. 342–347 (2003)

[62] Moreno-Velo, F.J., Baturone, I., Barriga, A., Sánchez-Solano, S.: Automatic Tuning of Complex Fuzzy Systems with Xfuzzy. Fuzzy Sets and Systems 158(18), 2026–2038 (2007)

[63] de Oliveira, R., Braun, T.: A Delay-based Approach Using Fuzzy Logic to Improve TCP Error Detection in Ad Hoc Networks. In: IEEE Wireless Communications and Networking Conference, Atlanta, USA (2004)

[64] OpenCores Organization, OpenCores.Org: Free Open Source IP Cores and Chip Design (2007), http://www.opencores.org

[65] Park, K., Willinger, W. (eds.): Self-Similar Network Traffic and Performance Evaluation. Wiley Interscience, New York (2000) ISBN: 0-471-31974-0

[66] Resende, R.A., Nassif, N.A., de Siquira, M.A., da Silva, A.E., Lima-Marques, M.: Quality of Service Control in IP Networks Using Fuzzy Logic for Policy Condition Evaluation. In: IEEE International Conference on Fuzzy Systems (FUZZ-IEEE 2005), Reno, USA, pp. 448–453 (2005)

[67] Rolls, D., Michailidis, G., Hernández-Campos, F.: Queueing Analysis of Network Traffic: Methodology and Visualization Tools. Computer Networks 48(3), 447–473 (2005)

[68] Sánchez-Solano, S., Barriga, A., Jiménez, C.J., Huertas, J.L.: Design and application of digital fuzzy controllers. In: Sixth IEEE International Conference on Fuzzy Systems (FUZZ-IEEE), vol. 2, pp. 869–874 (1997)

[69] Semeria, C.: T-series routing platforms: System and packet forwarding architecture. Tech. Rep. 200027-001, Juniper Networks, Inc (2002),
http://www.arl.wustl.edu/jst/cse/577/
readings/juniperTseries.pdf

[70] Stringfield, N., White, R., McKee, S.: Cisco Express Forwarding, 1st edn. Networking Technology. Cisco Press (2007) ISBN: 978-1-58705-236-1

[71] Usselmann, R., et al.: WISHBONE/OPB & OPB/WISHBONE Interface Wrapper: Overview (2004),
http://www.opencores.org/projects.cgi/web/opb_wb_wrapper

[72] Varghese, G.: Network Algorithmics: An Interdisciplinary Approach to Designing Fast Networked Devices. Morgan Kaufmann, San Francisco (2004) ISBN: 978-0120884773

[73] Waldwagel, M., Varghese, G., Turner, J., Plattner, B.: Scalable High Speed Prefix Matching. ACM Transactions on Computer Systems 19(4), 440–482 (2001)

[74] Wang, J., Wei, D.X., Low, S.H.: Modelling and Stability of FAST TCP. In: 24th Annual Joint Conference of the IEEE Computer and Communications Societies (INFOCOM), Miami, FL, USA, pp. 938–948 (2005)

[75] Xililnx Synthesis, Synthesis and Simulation Design Guide 9.2i. Xilinx© (2007),
http://toolbox.xilinx.com/docsan/xilinx92/
books/docs/sim/sim.pdf

[76] Xilinx ISE, Xilinx ISE 9.2i Software Manuals and Help - PDF Collection. Xilinx© (2007), http://www.xilinx.com/support/sw_manuals/
xilinx92/index.htm

[77] Yaghmaee, M.H.: Design and Performance Evaluation of a Fuzzy Based Traffic Conditioner for Differentiated Services. Computer Networks 47(6), 847–869 (2005)

[78] Yusuf, S., Luk, W., Sloman, M., Dulay, N., Lupu, E.C., Brown, G.: Reconfigurable Architecture for Network Flow Analysis. IEEE Transactions on Very Large Scale Integration (VLSI) Systems 16(2), 57–65 (2008)

[79] Zhang, R., Phillis, Y.A., Ma, J.: A Fuzzy Approach to the Balance of Drop and Delay Priorities in Differentiated Services Networks. IEEE Transactions on Fuzzy Systems 11(6), 840–846 (2003)

[80] Zhang, R., Phillis, Y.A., Kouikoglou, V.: Fuzzy Systems for Queuing Control. Springer, London (2005) ISBN: 978-1-85233-824-4

Index

Symbols

α-cut   166

A

Absolute percent error   *see* APE
ACK   157
Active measurement   4–5
Active Queue Management   *see* AQM
Additive Increase Multiplicative Decrease
   see AIMD
AIC   90
AIMD   204, 215
Akaike information criteria
   see AIC
ALBP   5
ALU   276
AMPATH   111, 168–170, 173–174, 183,
   195
Analog VLSI   276
Anonymization   6, 170
ANOVA   40–41
APE   91
Appropriateness   164
AQM   20, 227, 228
AR   29, 53
ARCH   27
ARIMA   27, 90, 92
ARMA   30, 53, 89
ASIC   277, 278, 296
Association rules   167–168, 183
Asymmetric Link Bandwidth Probing   *see*
   ALBP
Asynchronous Transfer Mode   *see* ATM

ATM   227, 296
Autoregressive models   *see* AR
Autoregressive Moving Average models
   see ARMA

B

Backpropagation   60
Bell Labs   120
Best effort   231–256
BIC   204
Binary Increase Congestion control
   see BIC

C

C   82, 213, 215, 222, 229, 250, 271,
   274–275, 282, 283, 296
C++   271, 275, 283, 296
CAD   213, 229, 282
CBQ   20, 298
CBR   198, 216, 235
CCDF   173
CDF   156
CDMA   120, 273
Class Based Queuing   *see* CBQ
CMOS   276
Code Division Multiple Access   *see*
   CDMA
COMBO network monitoring cards   273
Complementary cumulative distribution
   function   *see* CCDF
Computational intelligence   VI
Computer-aided design   *see* CAD
Confidence   163, 167–168, 183

Constant bit rate   *see* CBR
Cross-validation   38, 67
CUBIC   204
Cummulative distribution function
    see CDF

D

DAG   273
Dag network monitoring cards   273
Day in the life of the Internet   *see* DITL
Delay-bandwidth product   203
Delta test   55, 65, 92
Demilitarized zone   *see* DMZ
Denial-of-service   *see* DoS, *see* DoS
DiffServ   23
Digital signal processor   *see* DSP
DITL   7, 116
DMA   282
DMZ   93, 99
DoS   20, 170
Drop-Tail   227
DSP   276
Dumbbell   194

E

ECN   230
El Niño-Southern Oscillation   61
ELM   38, 88, 90, 92
End-to-end
    Congestion control, 16, 19–20
    Measurement, 4
    Metrics, 14
    Predictability, 89
Equinix   111–116, 168–170, 174
ESTSP   61
Ethernet   7, 27, 93, 94, 99, 120, 135
    Gigabit, 125, 274
European Symposium on Time Series
    Prediction   *see* ESTSP
Evolving intelligent systems   228
Explicit congestion notification   *see* ECN
Exponential decay   198
Extreme Learning Machine   *see* ELM

F

Fairness   193, 196, 219
FARIMA   32–33
FAST TCP   204

fBm   31
FCFS   21
FCM   174
fGn   32
FIFO   227
First Come, First Served   *see* FCFS
FLIPS   274, 277, 278, 294–296, 298
flow-lsummary   168–173, 184–185
ForCES   273
Forward-Backward search   58, 75, 80
FPGA   278
Fractional ARIMA   *see* FARIMA
Fractional Brownian motion   *see* fBm
Fractional Gaussian noise   *see* fGn
FRBS   VI
FreeBSD   283
Full-duplex   222
Fuzzy c-means   *see* FCM
Fuzzy Logic Inferences per Second   *see*
    FLIPS
Fuzzy Rule-Based Systems   *see* FRBS

G

GARMA   33
Gate arrays   277
Generalized ARMA   *see* GARMA
Genetic algorithms   228
Global Research and Education Network
    see GREN
GNU/Linux   271, 282, 283
Goodput   233
GPS   273
Gradient descent methods   60
GREN   194–200

H

H-TCP   204
Hard c-means   *see* HCM
HCM   174
HCPLD   270
High-capacity programmable logic devices
    see HCPLD
HighSpeed TCP   204
Histograms   156

I

IAB   226
IETF   214, 273

Induced OWA operators   *see* IOWA
   operators
Input/Output Block   *see* IOB
Intel IXP NPU   271
IOB   292
IOWA operators   36, 90
IP core   281, 282
IP module   *see* IP core, 281
IP MTU   *see* MTU
ISP   195

J

Java   82, 271, 275, 296
Junction temperature   295

K

K-means   *see* HCM
Kernel methods   37

L

Least Squares Support Vector Machine
   see LS-SVM
Levenberg-Marquardt Method   60–62
Linguistic summary
  Appropriateness, 164
  Confidence, 163
  Degree of truth, 153
  Length, 165
  Preciseness, 164
  Qualifier, 153
  Quantifier, 153
  Summarizer, 153
Link MTU   *see* MTU
Long-memory models   31
Long-range dependencies   3, 7, 9, 29–33
Look-up table   *see* LUT
LS-SVM   36–38, 54
LS-SVMlab   82
LUT   292
Lyapunov exponents   89

M

map   288
Markov chain   226
Markov models   29, 89
Markov-modulated Poisson process   *see*
   MMPP
MCN   192

Mean square error   34–35, 38–40
Micro-congestion   192
Microcode   271
Microcontroller   276
Mission Critical Networks   *see* MCN
MLP   38
MMPP   89
MPLS   23
MSE   55, 139
MTU   157
Multilayer Perceptron   *see* MLP
Multiprotocol Label Switching   *see* MPLS
MxAE   139

N

NAR   31
National Center for Atmospheric Research
   see NCAR
National Research & Education Network
   see NREN
NCAR   125
NCL   272
Nearest neighbor   90
NetFPGA   274
Neural networks   53, 54, 89, 276
  Feed-forward, 38, 89
  Single-hidden-layer, 38
NIST Net   250
Nonparametric autoregressive models   *see*
   NAR
Nonparametric noise estimation   55
Nonparametric residual variance   55
NPU   269–272, 297
NREN   195
ns-2   13, 214, 229, 283
NSFNET   27, 89

O

OC48 traces   100–105
On-Chip Peripheral Bus   *see* OPB
OP-ELM   38, 88
OPB   281
Open Shortest Path First   *see* OSPF
OpenCores   281
OpenRisc   282
Optimally-Pruned Extreme Learning
   Machine   *see* OP-ELM
OSPF   200

Overprovisioning   192
OWA operators   36, 88, 90, 92

P

p/s   *see* pps
Packets per second   *see* pps
par   288
Pareto distribution   198
Passive measurement   6
PCI   280, 282, 296, 297
PCI Express   *see* PCI-E
PCI-E   263, 296, 297
Perl   168
PLC   276
Point process   91
pps   170, 264, 272, 294
Preciseness   164
Processing engine   270–271, 282
Programmable automata   276
PSTN   8–9
Public Switched Telephone Network   *see* PSTN

Q

QoS   17

R

Radial Basis Function   *see* RBF
Random Early Detection   *see* RED
RBF   38, 65, 67
RED   16, 21, 193, 231, 233–250
Request for Comments   *see* RFC
RFC   14, 196
RIB   267
RISC   271
ROM   286
Round-Trip time   *see* RTT
Routing Information base   *see* RIB
RTT   5, 195, 206, 216–219, 225
Runge-Kutta method   78

S

SAPE   91, 92
SCTP   204
Short-memory models   28
SIGCOMM   130
SMAPE   40, 139
SoC   281

SoPC   263, 282, 298–299
Spartan-3   290–292, 294
Support   167–168, 183
SVM   36–38, 54
Swarm optimization   228
Symmetric absolute percent error   *see* SAPE
Symmetric mean absolute percent error   *see* SMAPE
System-on-a-chip   *see* SoC
System-on-a-programmable-chip   *see* SoPC

T

Takagi-Sugeno-Kang   *see* TSK
TCAM   272, 278
TCP Reno   219, 235
TCP SACK   219, 235
TCP Vegas   204
TCP-Friendly Rate Congestion Protocol   *see* TFRC
Teletraffic Theory   8–9
Ternary Content-Addressable Memory   *see* TCAM
TFRC   204
thrulay   5, 250
TSK   61, 202, 208, 227
Type-2 fuzzy logic   227

U

UDP   219, 222
UKERNA   89
United Kingdom Education and Research Network   *see* UERNA89

V

Variable bit rate   *see* VBR
Variable Packet Size   *see* VPS
VBR   35, 89, 227
Verilog   281
VHDL   213, 229, 280–283, 286, 298
Virtex-5   292, 294
Virtualization   274, 284
VLSI   168, 264, 275–277, 295
VOQ   268, 278
VPS   5
VxWorks   271

W

Westwood TCP   204
Westwood+ TCP   204
WIDE   116–120
Widely Integrated Distributed Environment
    see WIDE
WISHBONE   263, 281

X

xfc   215, 250, 275, 283
xfcpp   283
XFL   213, 229, 282, 283, 286
xfsw   274, 283
xftsp   82, 141
Xfuzzy   38, 59, 60, 82, 141, 213, 215,
    222, 229, 250, 274, 282–283, 298

xfvhdl   283, 286
Xilinx
    ISE, 282, 288
    map, 288
    par, 288
    Spartan-3, 286, 290–292, 294
    Virtex II, 274
    Virtex-5, 286, 292, 294
    XPower Analyzer, 294
    xst, 288
xst   288

Z

Zipf's law   10, 198